POLYMERS IN MEDICINE II

Biomedical and Pharmaceutical
Applications

POLYMER SCIENCE AND TECHNOLOGY

Recent volumes in the series:

A Continuation Order Plan is available for this series. A continuation order will bring delivery of each new volume immediately upon publication. Volumes are billed only upon actual shipment. For further information please contact the publisher.

POLYMERS IN MEDICINE II

Biomedical and Pharmaceutical Applications

Edited by

E. Chiellini

and

P. Giusti
University of Pisa
Pisa, Italy

C. Migliaresi

and

L. Nicolais
University of Naples
Naples, Italy

PLENUM PRESS • NEW YORK AND LONDON

Library of Congress Cataloging in Publication Data

International Conference on Polymers in Medicine Biomedical and Pharmaceutical Applications
(2nd: 1985: Capri, Italy)
  Polymers in medicine.

  (Polymer science and technology; v. 34)
  "Proceedings of the Second International Conference on Polymers in Medicine: Biomedical
and Pharmaceutical Applications, held June 3–7, 1985, in Capri, Italy"—T.p. verso.
  Bibliography: p.
  Includes index.
  1. Polymers in medicine—Congresses. 2. Polymeric composites—Congresses. 3. Polymers
and polymerization—Therapeutic use—Congresses. 4. Drugs—Dosage forms—Congresses. I.
Chiellini, Emo. II. Title. III. Series.
  R857.P6I586 1985   610.28   86-25155
  ISBN-13: 978-1-4612-9012-4   e-ISBN-13: 978-1-4613-1809-5
  DOI: 10.1007/ 978-1-4613-1809-5

Proceedings of the Second International Conference on Polymers in
Medicine: Biomedical and Pharmaceutical Applications, held June 3–7, 1985,
in Capri, Italy

© 1986 Plenum Press, New York
Softcover reprint of the hardcover 1st edition 1986
A Division of Plenum Publishing Corporation
233 Spring Street, New York, N.Y. 10013

PREFACE

Polymers and polymer based composites have gained increasingly larger applications in medicine and surgery.

Presently, most biomaterials applications rely on industrial substances that were initially developed by industry for non-medical purposes.

Moreover, polymers have been often used regardless of their peculiar characteristics which can be viceversa and very attractive for some specific applications.

In the past years we have assisted to a significative and faster development of polymer science as well as of medicine and surgery. The assistance of computer aided apparatus, the use of always more advanced instruments, the larger interest of the academic and industrial world, bring continuously new contributions to the research on biomedical and parmaceutical use of polymers.

The need of a forum where these specific researchs can be presented and discussed, and the success of the 1st Conference on Polymers in Medicine , held in Porto Cervo in 1982, have encouraged the Editors to plan a periodical meeting, focused on polymers and composites, to be held every odd year.

This book contains papers selected by an International Scientific Committee among those presented at the 2nd International Conference on Polymers in Medicine, Biomedical and Pharmaceutical Applications, held in Capri, Italy, 3-7 June, 1985.

In addition to contributed papers, several Authors were invited to present the "state of the art" as well as their personal contibution on specific key arguments.

The level of all contributions was high, the participation well qualified, and the meeting interesting and hopefully pleasant.

The papers published in this book have been grouped in two parts: i)Biomedical Polymers, which refers to the synthesis, the characterization and the applications of polymers and/or composites for biomedical prostheses, and ii) Polymer Drugs and Drug Delivery Systems, where pharmaceutical application of polymers are considered.

The Editors express their thanks to the National Research Council of Italy, to the Companies that contributed to the success of the meeting, to the European Society for Biomaterials and to the Society for Plastics Engineers, that sponsored the Conference.

Finally, the Editors wish to thank J.M. Anderson, A. Bantjes, F. Ciardelli, J.R. de Wijn, S. Dumitriu, J. Feijen, P. Ferruti, P.M.

Galletti, E.P. Goldberg, G. Guida, G.W. Hastings, J. Heller, A.S. Hoffman, S.J. Huang, M. Josefowicz, J. Kopecek, A. La Manna, D.J. Lyman, E. Martuscelli, R.M. Ottenbrite, N.A. Peppas, E. Piskin, H. Ringsdorf, J. Sunamoto, P.J. van Mullem, H.G. Willert, D.F. Williams, serving either on the Organizing or Scientific Committee, the partecipants to the Conference, and N. Coseglia, I. Havlicek, I. Hirsch, J. Kenny, M. Montanino, C. Schettini and H. Younes who enthusiastically gave their support to the local organization and secretariat.

Emo Chiellini
Paolo Giusti
Claudio Migliaresi
Luigi Nicolais

CONTENTS

PART I

Ia. BIOMEDICAL POLYMERS: SYNTHESIS and CHARACTERIZATION

PART II

POLYMER DRUGS and DRUG DELIVERY SYSTEMS

INTERFACE EFFECTS AND POLYMERIC BIOMATERIALS

Garth W. Hastings

Bio-Medical Engineering Unit, North Staffordshire
Polytechnic/North Staffordshire Health Authority
c/o The Medical Institute, Hartshill, Stoke on Trent
Staffordshire, England

The extensive clinical experience we now have with carbon fibre epoxy resin composites [1,2,3] reveals that they are generally well tolerated by the body and this has been born out by independent studies conducted by T. Rae using in vitro and in vivo methods. The material used in internal fixation of bone plates is so well accepted that a process of osseo-integration often occurs. The composite bone plates are found to be embedded in bone and in one case impossible to remove. Histology from these human cases reveals the generally good biological performance of the materials.

The origin of this biological acceptability is the critical question to be answered, for it is only by so doing that a unifying principle can be developed to explain and to predict in vivo performance of polymers.

The way in which applications of polymers have changed is illustrated in Table 1.

The exploitation of bulk mechanical properties (strength, compliance, fatigue, wear and friction) depending on the material itself and the form in which it was used, led to surgical progress in areas of joint replacement, vascular and cardiac surgery and related uses such as extracorporeal devices. Many attempts to improve in vivo performance were based on modifications to design e.g. the weave or knit of vascular prostheses and many concepts such as the Gossamer type were evaluated [4] Once the work of Gott [5] showed that surface properties were important this marked the first change in appreciation of the potential role for polymers. Classification in terms of molecular structure assists in developing an understanding of mechanical and physical properties yet the further dimension of chemical activity, [6] first explored via adhesives [7], dental cements [8], began with the era of hydrogels [10] and has continued to be exploited more fully with suture materials and pharmaceutical systems.

1

Following the initial use as engineering materials for replacement or repair of tissues, there has come a realisation that the variation possible in molecular structure can provide for more subtle uses in which there is a co-operative symbiosis between implanted material and living organism. Not only is ease of degradation possible, as with absorbable sutures, but

Table 1.

Bulk Materials	Load bearing	Joint Replacements
		Gap fillers
	Soft tissue	Reconstruction/ replacement
	Cardiac	Valves, vessels
	Joint prosthetics	Cements
Interactive	Cardiac	Heparinised surface
	Sutures	Degradable polymers
	Ophthalmic	Hydrogels
	Extra corporeal	Membranes
Reactive	Pharmaceutical	Drug encapsulation
		Drug carriers
	Blood contact	Surface grafted Copolymers
	Tissue repair	Adhesives
Composites	Permanent	cfrp
	Degradable	Polylactate

mechanical and/or chemical interactions can be adapted to specific requirements. In fact, one of the most rapidly developing areas is that of drug delivery systems in which the multiple reactivity of macromolecular systems is utilised effectively.

CLINICAL EXPERIENCE WITH COMPOSITES

The experience with the clinical use of carbon fibre reinforced epoxies has demonstrated several important features related to the mechanical and chemical/physical nature of the material.

The use of carbon fibre composite plates can now be reported for a series of 40 fractures in 28 patients treated between September 1980 and March 1984. All were followed in a special implant clinic until routine removal of plates. Patient ages ranged between 20 and 80 years, mainly below 30.

Table 2.   Summary of Clinical Series

No. of Patients		28	
Type of Fracture		Transverse/ short oblique	80%
		Butterfly fragment	16%
Site	Radius	Ulnar	
Proximal ⅓	8	3	
Middle	6	7	
Distal ⅓	8	8	

Radiological union for this series was defined as the point at which the fracture line was no longer visible and the medullary canal was reformed. All fractures in the series united with no refractures and 67% united within six months. Three patients aged more than 55 years took more than one year but less than eighteen months. Although not statistically different from another series reviewed using stainless steel plates (AO) note should be taken of the formation of strong close-knit callus which allows for early activity in patients treated with composite plates.

Macroscopically at reoperation five fracture sites showed reaction in the form of opalescent fluid and grey-yellow granulation tissue with one having excavation of bone at a screw site. The others revealed only a thin fibrous capsule round the plate with minimal soft tissue between plate and bone.

Histology showed carbon rods and occasionally granules to be generally present lying inertly in fibrous tissue with occasional giant cells. Six fractures showed reaction varying from a few polymorphs and lymphocytes and plasma cells to one case of frank inflammatory reaction. The extreme reaction was accompanied by S. aureus infection in one case.

Compound fractures, 20% of the series, showed no callus response and all these had received compression on fixation. 70% of the series showed external callus, the majority of these (75%) being uncompressed the rest inadequately so. A prolific callus response in four cases was due, most likely, to soft tissue damage. Generally there was bone growth round the edge of the plate which in one case was completely buried and showed osseo-integration. All patients used their arms functionally from an early stage.

The mechanical interaction of the plates with the healing fracture is shown by the callus response. The early use of the arm and the movement possible with this more flexible plate without the likelihood of fatigue failure has been a factor in callus formation. The results compare well with the review made of a series of one hundred forearm fractures treated by AO techniques in which there was 9% mechanical failure (plate fracture, implant loosening) and 13% biological failure (non-union, cross union, infection, refracture)[11].

The biological response was generally acceptable and the mild fibrous tissue response and varying degrees of osseointegration show the tolerance to the material. There remain the five reactions unrelated to any specific cause. The tissue culture and animal results of Rae[1] did not reveal any adverse reaction, nor did previous work[12,13].

Is this a reaction to epoxy or a reaction to carbon fibres released into the tissue by movement between the stainless steels screws and the countersink? The work on ligament repair shows the reaction to carbon fibre. Epoxy resins have an uneventful career as implanted materials. One explanation therefore is that there is a reaction to carbon fibre but the absence of reaction in the previous studies needs to be related to the clinical findings. The treatment given to a material, e.g. method of cleaning or sterilisation, can considerably modify the biological response. Furthermore, the shape of particles produced in patients is likely to be quite different from those prepared experimentally thus giving an Oppenheimer effect. We do not yet know whether the bone overgrowth is a response to carbon or to epoxy resin.

BIOLOGICAL ACCEPTABILITY

The clinical series showed various stages in the progressive healing of a series of fractures of radius and ulna. The superior fatigue life of these materials shown in laboratory tests and in clinical use [14,15] permits more movement at the fracture site than would be possible with steel plates and this leads to callus formation and early consolidation. This is a mechanical aspect which accords with the views of McKibbin [16] about the sequence of events in fracture healing. The biochemical processes associated with the healing are not understood and McKibbin [17] and Sevitt [18] have differing views on the significance of various stages. Some of our recent observations [19] show an interruption in the sequence of changes from one collagen type to another in the formation of non-unions which may be mechanically mediated.

The osseo-integration seen with some of the forearm series and reported on extensively by Branemark and co-workers for titanium implants appears to be a surface phenomenon. In the case of the metal it is the titanium oxide surface rather than the metal itself in contact with tissue. The composites present a mixture of epoxy resin and carbon fibre at the interface. Carbon fibre is known to produce a scaffold for tissue growth utilised by various groups for repair of tendons, ligaments or articular cartilage [20,21]

It is at present seldom possible to judge a material completely acceptable for all purposes. Such a decision will be closely related to end use and different requirements will predominate. The comparison between polymethyl methacrylate used as the bearing surface in a hip prosthesis and its use as bone cement for fixation of metallic hip prostheses shows failure in the first case and long-term success in the latter. What can be said in this case about its so called "biocompatibility". Only that it is acceptable for a particular end application and that acceptability is a multi facetted subject. Fig. 1.

Biological Acceptability
(Biocompatibility)

Surface effects	Mechanical factors	Chemical
Chemical	Site of use	Properties
Physical	Duration of use	Adhesiveness
Mechanical	Effect on stress	Therapeutic
	distribution	effects
		Solubility
		Degradability

Fig. 1. Biological acceptability is a multifactorial
concept related to intended use.

One of the problems in assessment is that we are always looking at past events. It has been compared with historical attempts to reconstruct the events of a battle several days later[22]. By viewing the collection of rusting weapons and shell holes it is not generally possible to reconstruct inner motivations and the intimate details of the action. But by missing these something essential to the scene has been lost. Furthermore, though we may learn about the numbers involved and the ground lost or won, little may be learned about prevention of future engagements.

It is the final point which bears most on the discussion of polymers as biomaterials, because the aim is to improve in vivo biomaterials performance and to prevent unsought disasters. Both mechanical and surface properties require consideration and will be discussed in turn.

MECHANICAL PROPERTIES

There must always be the necessity for adequacy of mechanical properties though which particular one is important will vary according to the application. For the component of a major joint replacement strength and tribological factors will predominate whereas elastic recovery will be more significant for a tendon or ligament. There is abundant information to relate polymer structure to these properties but there is still need to test them in an environment and manner which adequately simulates the intended use. All laboratory testing presents an investment in time, often considerable and there is still little progress in devising standard tests. The British Standards Institution has produced two interim procedures for testing bone plates and endurance testing of joint prostheses[23,24]. For the evaluation of ultra high molecular weight polyethylene (UHMWPE)[25] tests developed for industrial applications were adopted even though they had no direct clinical relevance.

There are two stages in the testing, one for quality assurance of basic materials, the second for device evaluation. It is the second which presents the problems. In a recent article Frisch[26] reviews the cost dilemma of biomaterials development and he concludes that "most of the current unmet needs in biomaterials and medical devices will probably not be satisfied in the near future". Yet it is vital that the problem be addressed since the mechanical considerations of implants include not just a catalogue of properties but the actual mechanical interaction with the living tissue. The load-growth relationships have been demonstrated for living organisms and when the natural pattern is disturbed by the presence of an implant there are observable sequelae. The effect on fracture healing has been described and there is the longer-term effect yet to be studied when the bone plate remains in situ.

Cell function, which ultimately will determine tissue properties is affected by internal and external factors which are often related to cell membrane behaviour. Ionic environment inside and outside the cell, and the flux of ions and small molecules across the membrane can be modified by changes in cell membrane structure. Chemical, electrical and mechanical factors all play a part in this and the macromolecular domains of the membrane can be modified readily. Mechanical deformation will modify transport phenomena and changes from a normal mechanical environment could feasibly change cell behaviour directly. It is observed that electrical phenomena can be mechanically induced and the resulting generation of charges could also have similar effects. The magnitude may be several millivolts and they are dependent on strain gradients in the tissue and therefore modified by disease, injury or the presence of an implant.

Fibre composite materials are attractive because the mechanical behaviour is anisotropic and therefore closer to that of tissue. Our work has concentrated on crosslinked epoxy polymers for the matrix. Others have developed absorbable matrices[27] or thermoplastics[28]. There are newer polymers which require investigation and which may help to evaluate the nature of mechanical interactions as well as providing improved mechanical properties for a particular device. Carbon fibre appears to be satisfactory for implant purposes and there are different fibre forms that should be examined. There is increasing interest in composites for dental/maxillo facial[29] as well as orthopaedic and cardiovascular uses and the potential has yet to be realised. When the polymer chemists begin to develop matrix polymers to provide specific biological requirements, optimum matrix-fibre interactions and specific surface interactions, this will also become a growth area.

SURFACE PROPERTIES

Surface properties are critical for determining not only short term (immediate) acceptance but in setting the pattern for longer-term behaviour. Our studies on collagen type variation at fracture sites illustrates the way in which small changes in macromolecule structure can affect, in this case, the formation of a biological interface. Conversely, a synthetic interface affects biological performance.

The aim is to establish a unifying principle for what may be seemingly disparate requirements, cell attachment and growth in some situations, "non-stick" properties in another. This takes the biomaterials practitioner away from early generation materials and "bulk" uses, to a consideration of a much more subtle chemistry and hence to methods of control. What can be done is being shown in blood contact materials and controlled drug release systems, in particular targettable macromolecular drugs. The question to be addressed is whether the insights being gained are applicable to developments of functional implants.

Composites were referred to above and the potential for mechanical property variation discussed. The addition of surface treatment has not yet been readily explored even though tissues are themselves surface active composites. Bonfield has developed polyethylene-apatite materials[30] which may be expected to show some surface activity at the inorganic filler sites. Polymethyl methacrylate bone cement has been mixed with Ceravital$_R$ glass ceramic particles and glass fibre and though it is not strictly a composite should combine physicochemical adhesion with the normal mechanical grouting action of bone cement[31]. The long-term resistance to implant loosening is important and may be provided by this system. However, these are not surface modifications or coatings. The work of Hench in developing reactive glasses has shown the importance of reactive coatings on substrate materials[32] and showed that the interfacial bond is usually stronger than the implant or the bone in which it has been placed. The bonding mechanism for these reactive glasses is related to ion exchange, with a surface Ca,-P- rich layer being formed rapidly and the involvement of macromolecules from the tissues in developing an intermediate region. These glasses have been used to coat metallic substrates and should be one of the rapidly developing areas in the future.

Coatings have been generally applied to polymer materials in order to prevent cell adhesion for blood contact purposes and heparin and heparin analogues [33] are being studied as coatings, grafts and copolymers. Where tissue attachment is required this has been achieved by mechanical means e.g. polyester fabric patches on the surface. Natural macromolecules provide interesting examples of cell adhesion for example the use of spiders webs as wound dressings.

CAN A UNIFYING PRINCIPLE BE IDENTIFIED?

To develop the applications of polymers as multi-potential biomaterials a unifying principle is needed. Before examining progress in this direction we should review the reasons why polymers are of interest as biomaterials.

i   Versatility in structure-property combinations.
ii   Absence of metallic ions having potential short or long-term adverse effects.
iii   Stability/degradation is controllable.
iv   Balance between hydrophilic-hydrophobic properties is variable.
v   Chemical modification is used to change properties or to attach active agents.
vi   Can be solubilised.
vii   Mechanical/physical properties closer to tissue norms.
viii The same material can often be produced in differing forms; block, foam, fibre, film.
ix   Matrix materials for composites
v   Membrane properties
xi   Large scale fabrication is relatively easy.

It is the combination of structural properties and chemical activity that gives polymers the advantage over other materials. Ceramics in contrast are more restricted in versatility. The use of polymers as chemically active biomaterials is likely to increase in future.

The basic reasons for which polymers are required include:

i   Tribological reasons.
ii   Protection of substrates.
iii   Fibre properties (sutures etc.)
iv   Hardening in situ (adhesives, grouts etc.)
v   Membranes
vi   Encapsulants for release of other agents.

For these to be effective the controlling factors for inter-face reactions must be identified. Baier [34] has drawn attention to surface energy considerations and has identified the range of surface energy within which tissue acceptability is shown. More recently Lydon and coworkers have shown the importance of equilibrium water content for reconciling apparently anomalous behaviour between hydrophilic or hydrophobic materials and cell adhesion [35]

8

These approaches will help formulate the conditions for surface adhesion but the mechanism will still need definition. Baier points to the complex deposition of protein and other materials from the surrounding tissue environment in a process of "interface conversion". There are therefore two stages, the first is the attachment of macromolecules from the tissue and the second the recognition of the surface by the cells. Cell recognition studies are an area of macromolecular pharmaceuticals receiving attention and attachment of monoclonal antibodies and specific chemical residues is being investigated.

The studies on synthetic polymerisable liposomes should further assist in our understanding of cell surface reactions, including recognition. Precise targetting of polymer-carried drugs will certainly require this.

FUTURE PROSPECTS

There is another aspect. The above has concentrated upon the effects of polymers on tissues from the point of mechanical and surface interactions. Polymers synthesised by micro-organisms have often been of interest[36] and commercial production of bacterial polyhydroxybutyrate has been achieved[37]. This paper has pointed to the consequences of an interruption in in vivo synthesis of collagen and the non-union of fractures. As commercialisation of naturally produced polymers proceeds it is hoped that the mechanisms of in vivo polymerisation will be investigated and may provide for alleviation of collagen-based disorders (e.g. fracture non-union, osteogenesis imperfecta, rheumatoid diseases). This will be a full circle indeed in biomaterials studies.

REFERENCES

1. G. W. Hastings. Composites as Implant Materials. Deutsche Verband fur Material Prufung. In press. (1985).
2. K. Tayton, C. Johnson-Nurse, B. McKibbin, J. S. Bradley, G. W. Hastings. Results of preliminary trials using CFRP plates for human fracture fixation. J. Bone Jt. Surg. 64B, 105-111. (1982).
3. M. S. Ali, T. French, G. W. Hastings, C. Wynn Jones, T. Rae. A preliminary report on the use of carbon composite plates in forearm fractures. Presented to Brit. Orthop. Assoc. Meeting, April (1985).
4. S. A. Wesolowski, C. C. Fries, A. Martinez, J. D. McMahon. Arterial prosthetic materials. Am. NY Acad. Sci., 146, 325 (1968).
5. V. L. Gott, R. L. Daggett, J. D. Whiffen, D. E. Koepke, G. G. Rowe, W. P. Young. A hinged leaflet valve for total replacement of the human aortic valve. J. Thoracic Cardiovasc. Surg., 48, 713 (1964).
6. G. W. Hastings, B. Bloch. Plastics Materials in Surgery. 2nd Edition. C. C. Thomas, Springfield (1972).
7. op cit.
8. G. W. Hastings. Surgical adhesives. Plastics and Rubber Intnl., 9, 38-41 (1983).
9. H. J. Prosser, A. D. Wilson. The compatibility of an adhesive ceramic polymer cement with dental tissue in "Mechanical Properties of Biomaterials", G. W. Hastings, D. F. Williams (Eds). J. Wiley, Chichester, Chapter 31, (1980).

10. D. Wichterle, D. Lim. Hydrophilic gels for biological use. Nature (Lond), 185,117 (1960).
11. E. R. S. Ross. Unpublished results.
12. J. S. Bradley, G. W. Hastings, C. Johnson-Nurse. Carbon fibre reinforced epoxy as a high strength low modulus material for internal fixation plates. Biomaterials, 1, 38-40 (1980).
13. M. S. Ali, T. A. French, G. W. Hastings, T. Rae, C. Wynn Jones Clinical and laboratory biocompatibility of carbon fibre reinforced epoxies used for less rigid fracture fixation in the forearm. Presented to Biointeractions '84 London. Sponsored by Biomaterials. (1984).
14. P. Bell. Fatigue studies of carbon fibre reinforced epoxy bone plates. Project thesis. North Staffordshire Polytechnic.
15. K. Tayton, C. Johnson-Nurse, B. McKibbin, J. S. Bradley, G. W. Hastings. Results of preliminary trials using CFRP plates for human fracture fixation. J. Bone Jt. Surg. 64B, 105-111 (1982).
16. B. McKibbin. Biology of fracture healing in long bones. J. Bone Jt. Surg. 60B, 106. (1978).
17. op cit.
18. S. Sevitt. Bone repair and fracture healing in man. Churchill Livingstone, (1981).
19. A. M. Anderson, G. W. Hastings, T. R. Fisher, E. R. S. Ross, A. Shuttleworth. Collagen types present at human fracture sites - a preliminary report. Injury in press (1985).
20. D. H. R. Jenkins, I. W. Forster, B. McKibbin, Z. A. Ralis. J. Bone Jt. Surg. 59B, 52-57. (1977)
21. R. J. Minns, D. S. Muckle, J. E. Donkin. The repair of osteo-chondral defects in osteoarthritic rabbit knees by the use of carbon fibre. Biomaterials, 3, 81-86. (1982).
22. J. Black. Systemic effects of biomaterials. Biomaterials, 5, 11-18. (1984).
23. Method for testing bending strength and stiffness of bone plates for use in orthopaedic surgery. Draft for Development. British Standards Institution, DD87/1983.
24. Endurance testing of orthopaedic implants. Draft standard British Standard Institution (BSI), BS3531 (part not yet assigned).
25. Implants for Surgery - Ultra high molecular weight polyethylene International Standards Organisation Draft International Standard. ISO/DIS5834/Parts 1 and 2.
26. Eldon Frisch. The cost dilemma: The high cost and low volume of biomaterials in "Contemporary Biomaterials". J. W. Boretos, M. Eden (Eds), Noyes Publications, 607-625 (1984).
27. M. Vert, P. Christel, F. Chabot, J. Leray. Bioresorbable plastic materials for bone surgery in "Macromolecular Biomaterials". G. W. Hastings, P. Ducheyne (Eds), CRC Press, Boca Raton, Chapter 6 (1984).
28. M. Spector, S. L. Harmon, J. T. Eldridge, R. J. Davis. Porous polymeric coatings for orthopaedic implants in "Mechanical Properties of Biomaterials". G. W. Hastings, D. F. Williams, (Eds), J. Wiley, Chichester, Chapter 23, (1980).
29. D. L. Leake, M. B. Habal, H. C. Schwartz, S. Michieli, A. Pizzoferrato. Urethane elastomer coated cloth in "Mechanical Properties of Biomaterials" G. W. Hastings,

D. F. Williams (Eds), J. Wiley, Chichester, Chapter 34, (1980).

30. W. Bonfield, M. D. Grynpas, A. E. Tully, J. Bowman, J. Abram. Hydroxyapatite reinforced polyethylene - a mechanically compatible material for bone replacement. Biomaterials 2, 185-186 (1981).

31. F. Hahn, V. Strunz, J. Boese-Landgraf. Adv. Biomater, 4, 95 (1982).

32. L. L. Hench, June Wilson. Surface active biomaterials Science, 226, 630-636 (1984).

33. P. Ferruti, A. Casini, F. Tempesti, R. Barbucci, E. Mastacchi., M. Sarret. Heparinisable materials III, Biomaterials, 5, 232-234 (1984).

34. A. E. Baier, Surface phenomena in in vivo environments. Proc. NATO Adv. Study Institute, Marbella. In press (1985)

35. M. J. Lydon, T. W. Minett, B. J. Tighe. Requirements for cell adhesion to synthetic polymer substrates in culture. Proc. Sympos. Interaction of Cells with Natural and Foreign Surfaces. Royal College of Surgeons of England In press (November 1984).

36. C. R. Masson, H. W. Melville. J. Polymer Sci., 4, 323, (1944).

37. Biopol - A unique biodegradable thermoplastic Steam; Imperial Chemical Industries, 14, (1985).

ESCA FOR THE STUDY OF BIOMATERIAL SURFACES

Buddy D. Ratner

Department of Chemical Engineering and Center
for Bioengineering, BF-10
University of Washington
Seattle, WA 98195

INTRODUCTION

Polymer surface analysis is essential for the study of biomaterials because the ultimate biological response to implanted materials depends on what proteins and cells "see" at the interface. Ideally, we would like to routinely correlate the characteristics of the surface structure with the events that comprise the observed biological response. Electron spectroscopy for chemical analysis (ESCA) has been found to be useful in enhancing our understanding of biomaterials surfaces and has demonstrated the potential to generate data that might be correlated with complex biological interactions.

Some of the primary concerns associated with the characterization of polymer surfaces for biomaterial applications are:

- the presence and causes of surface contamination (1-3),

- the mobility of surface molecules in response to changes in the environment (4-10), and

- the compositional inhomogeneity from the surface layer into the depth of a material (11-17).

In this paper, a brief review of the ESCA technique is presented and I discuss how the above concerns are examined using ESCA.

THE ESCA TECHNIQUE

ESCA, also called x-ray photoelectron spectroscopy (XPS), is perhaps the single most valuable tool presently available for studying biomaterial surfaces (18). However, the ESCA method, coupled with other contemporary surface analysis tools such as ATR-IR, SEM, contact angle methods, secondary ion mass spectrometry, etc., provides a more complete picture of the surface without the ambiguity associated with any one technique. The ESCA technique can provide the following information:

1. The elemental composition (except for hydrogen and helium) of the sample may be determined.
2. Quantitative analysis may be performed.
3. Specific details of the chemical structure can be obtained (bonding state and/or oxidation level of most atoms).
4. The technique, if used with care, is nondestructive.
5. Depth profiling may be performed to assess surface heterogeneity and to understand the relationship between surface and bulk structure.
6. Samples may be studied in a hydrated, frozen condition (4).
7. ESCA has allowed a direct correlation to be established between surface chemistry and in vivo blood interaction (19, 20).

The ESCA experiment makes use of the photoelectron effect -- matter bombarded with x-rays (electromagnetic radiation) will emit photo-electrons with an energy:

$$K.E. = h\upsilon - \text{binding energy} - \phi$$

where $h\upsilon$ is the energy of the bombarding x-ray, ϕ is a work function that is established for each spectrometer, and K.E. is the kinetic energy of the photoelectron which is measured by the ESCA instrument. Thus, the binding energy of the ejected electron can be solved for. This binding energy is a sensitive function of the atomic environment, which is defined by the nature of the atom with which the ejected electron was associated and the atoms bound to that atom. A diagram schematically illustrating the components of an ESCA spectrometer is presented in Figure 1. Review articles and books are available which describe the ESCA technique (21-26) and specifically address applications for polymeric systems (27-32).

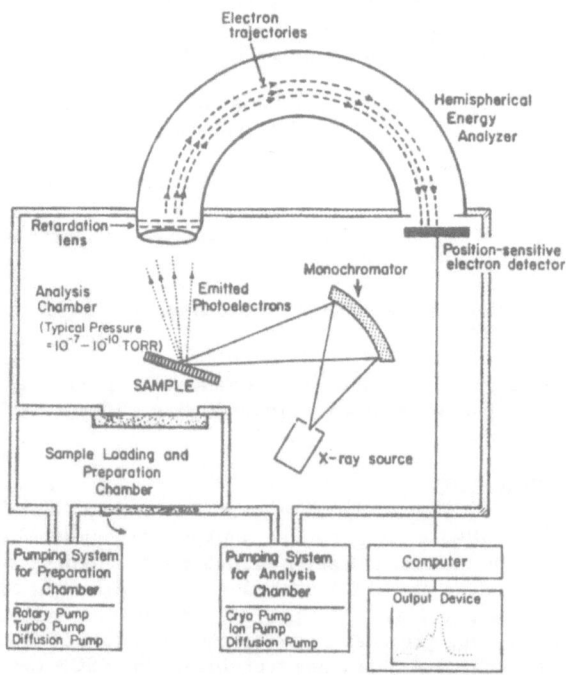

Figure 1. Schematic diagram of an ESCA system.

Measuring the intensity and energy distribution of the photoemission yields information with which to determine the surface chemistry. Each peak in a survey spectrum (usually covering a 1000 eV range) is indicative of the core level energy of an element. Chemical groups attached to an atom undergoing photoionization will alter the binding energy of core level electrons. Thus, at high resolution (a spectral range of typically 20 eV), for a given core level of an element, a number of different subpeaks may be observed, with each subpeak representing a different molecular environment. Resolving these subpeaks from a complex curve envelope with the assistance of tabulated peak shifts can be used to establish the chemical concentrations of various species and functional groups within a surface.

To illustrate how peak shifts reflect the molecular environment, the spectrum of the C1s core level photoemission of low-density polyethylene is presented in Figure 2a. The single peak at 285.0 eV is representative of carbons bound only to hydrogen atoms and to other carbon atoms. When one fluorine atom is added to each unit of the polyethylene backbone, i.e. poly(vinyl fluoride), the peak produced by carbons bound directly to the fluorine shifts to a higher binding energy (Figure 2b). Because the fluorine atom is highly electron-withdrawing, it draws electron density off of the carbon to which it is attached; this attraction increases the difficulty of extracting core level electrons from the carbon, causing the binding energy to increase. As shown in the spectrum for Teflon, poly(tetrafluoroethylene), in Figure 2c, this chemical shift effect is amplified with the addition of more electron-withdrawing fluorine atoms.

As an example of resolving subpeaks from a complex photoemission curve envelope in order to isolate different bonding environments, we examine the C1s and O1s spectra of poly(methyl methacrylate) (PMMA), a polymer often used for ophthalmologic and orthopedic applications. In the resolved C1s spectrum shown in Figure 3a, the subpeaks represent hydrocarbon-like environments (55%), carbons singly bonded to oxygen (25%), and carbons bound to oxygen with both single and double bonds (ester carbons) (20%). The well-defined doublet of the O1s spectrum (Figure 3b) indicates equal portions of single-bonded and double-bonded linkages.

This introduction to the ESCA technique is intended to present only some of the most basic concepts. The reader is referred to any of many fine books and articles expanding upon analysis by ESCA and data interpretation (21-36).

ESCA APPLICATIONS IN BIOMATERIAL SURFACE ANALYSIS

This section will review specific ESCA studies that deal with the biomaterial surface analysis concerns presented in the Introduction section.

Contamination

Because of its ability to describe in a semi-quantitative fashion the chemistry of a biomaterial surface, the ESCA technique can effectively monitor the extent of contamination caused by factors such as cleaning solution residues and atmospheric organics.

In a study of PMMA intraocular lenses (IOLs), indications were found of significant sodium dodecyl sulfate (SDS) residues on the lens surfaces (2). Specifically, an increase in the C/O ratio, unsymmetrical peaks in the oxygen spectrum, the presence of S and Na in a ratio of approximately 1:1, and the S2p peak shifted to a position suggestive of sulfate, were observed (see Table I). SDS solutions are common lens cleaners because of their effectiveness in

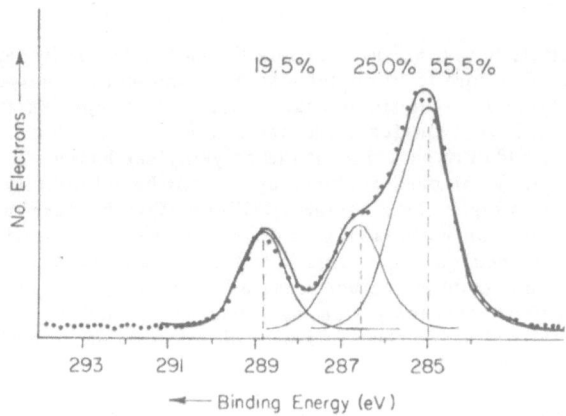

Figure 3(a). C1s spectrum of poly(methyl methacrylate).

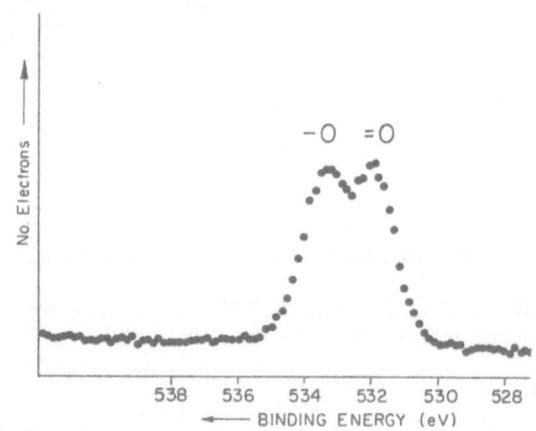

Figure 3(b). O1s spectrum of poly(methyl methacrylate).

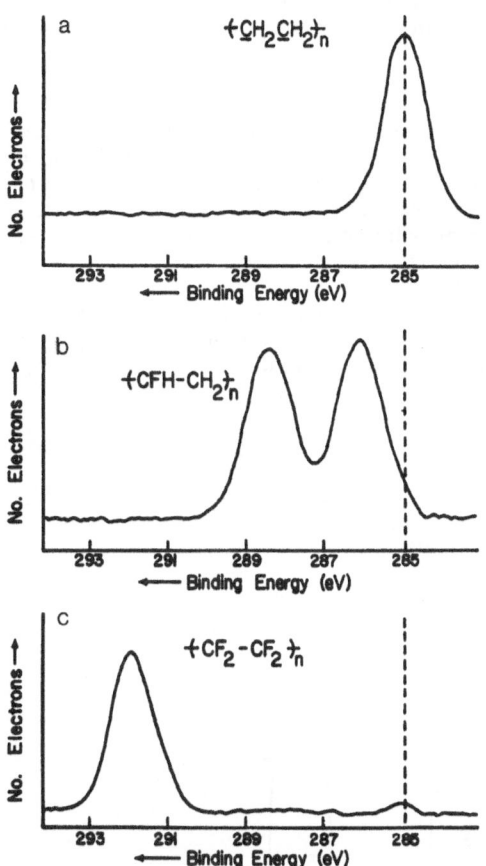

Figure 2(a).   C1s spectrum of polyethylene.
Figure 2(b).   C1s spectrum of poly(vinyl fluoride).
Figure 2(c).   C1s spectrum of polytetrafluoroethylene.

removing greaselike materials; however, in this case, the cleaning agent itself contaminated the device. This can have potentially serious consequences since SDS adherence to the IOL surface may induce denaturation of the proteins absorbed by the implanted IOL or lysis of cell walls contacting the lens. As an alternative to the SDS solutions, a wash based on NaOH and $NaHCO_3$ was developed that, based upon ESCA analysis, produced no surface residue on the PMMA lenses (see Table I).

Polytetrafluoroethylene vascular grafts are relatively "clean" materials; however, with the ESCA technique, evidence has been found of hydrocarbon contamination on the graft surfaces. As shown in the C1s spectrum in Figure 4, the bulge in the baseline at 285.0 eV indicates the presence of hydrocarbon contaminants. The amount of hydrocarbon contamination is small and the effect of this material on biological interactions is not known.

Molecular mobility

The ability of molecular groups in polymers to move to and from the surface in response to a change in their environment has been noted in several recent studies (4-10). In order to examine the chemical nature of the surface in environments relevant to biomaterial applications, the ESCA technique can be used not only on dehydrated samples at room temperature, but also on hydrated samples in a frozen state (4).

An example of the application of this technique in the analysis of molecular mobility is the examination of a poly(2-hydroxyethyl methacrylate) (HEMA) hydrogel radiation-grafted to silicone rubber (4). As shown in Table II,

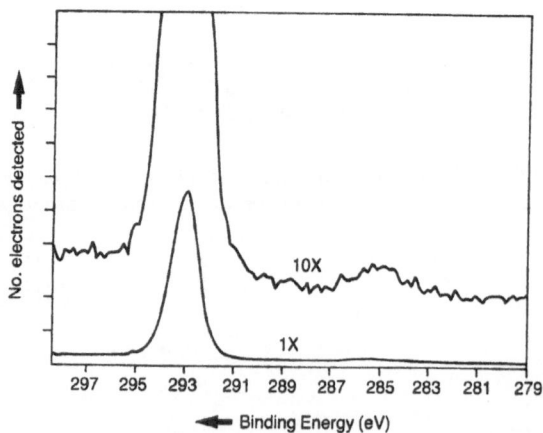

Figure 4. C1s spectrum of Goretex graft material. The 10x expansion of the y-axis clearly shows the presence of hydrocarbon contamination.

18

Table I

ESCA Data for PMMA Intraocular Lenses (IOLs) Treated by Various Methods[†]

| Lens Description | % Area Under Subpeaks in the C_{1s} Spectrum | | | C/O | O_{1s} Doublet | C_{1s} / S_{2p} |
	—O—C=O	—O—C—	—C—C—			
Poly(methyl methacrylate)—theory	20	20	60	2.50	Symmetric	∞
Poly(methyl methacrylate)—precipitated	20	25	55	2.72	Symmetric	∞
IOL (SDS—no soak)	15	21	64	3.35	Asymmetric	79.2[*]
IOL (SDS soak)	16	22	61	2.93	Asymmetric	181[*]
IOL (NaOH,SDS)	13	19	68	3.70	Asymmetric	146[*]
IOL (NaOH,NaHCO$_3$)	18	26	56	2.72	Symmetric	∞

[†] From reference 2

[*] S in —SO$_3^-$ position, S/Na~1

the ESCA data for the dry HEMA hydrogel grafted on silicone rubber reveal a chemistry indicative of silicone rubber; however, the hydrated (-160°K) hydrogel data show the poly(HEMA) chemistry present on the surface. Therefore, when the HEMA grafts are hydrated, the polar groups orient themselves outward, hydrogen-bonding to water; when the grafts are dehydrated, the polar groups apparently penetrate into the silicone rubber substrate, probably strongly hydrogen bonding to each other.

Non-Destructive Depth Profiling:
Compositional Variation as a Function of Depth

Clearly, the outermost layer of atoms of a surface will influence the surface properties and the biological response to the surface. However, the degree of influence that the second atomic layer and the layer below that has on biological reactions is still an area for speculation. The measurement of compositional changes as a function of depth is potentially important for understanding certain factors which might influence biocompatibility, e.g. the structural details of overlayers, the influence of bulk chemistry on surface structure, and the domain structures of phase-separated polymers. Therefore, in order to better understand the structure and properties at varying depths, the angular-dependent ESCA technique can be exploited to create a depth-concentration profile from the surface down into the bulk. As shown in Figure 5, the photoelectron emission from the sample is measured as the angle of the sample is varied, with the detector and the x-ray source in fixed positions. For any "layer" of material, the intensity of the signal is a function of the distance into the sample and of the angle with respect to the detector, and is proportional to the concentration of the atom of interest. The intensity is attenuated by an exponential factor which indicates the absorption undergone by a photoelectron attempting to emerge from the surface. Hence, an algorithm can be developed that, assuming a smooth surface, predicts the ESCA signal for any concentration profile (12,37). A comparison of this predicted ESCA data with actual ESCA data can indicate how closely the compositional variation with depth of the sample resembles the constructed model.

Table II
Elemental Ratios for Graft Polymers as Determined by ESCA

	C/Si		C/N	
	160°K	303°K	160°K	303°K
HEMA on Silicone Rubber	5.52	1.64	-	-
Acrylamide on Silicone Rubber	7.70	2.05	3.41	∞

160°K = hydrated
303°K = dehydrated

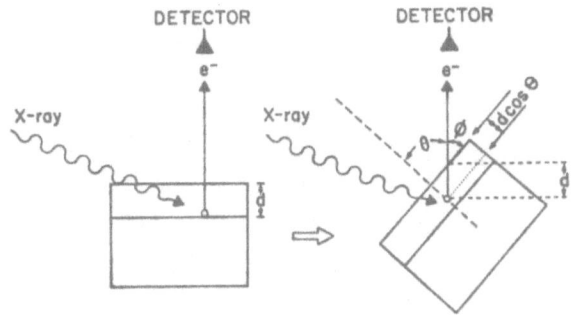

d = Sampling depth
d cos Θ = Effective depth sampled

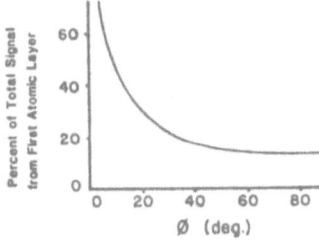

Figure 5. Diagram of the angular-dependent ESCA experiment.

One system in which this modelling technique was applied is a polyurethane containing a silicone oil contaminant. The model ESCA data set generated using the algorithm described in reference 37 from Figure 6 closely resembles the actual ESCA data set obtained using an angular dependent ESCA study on the contaminated polyurethane specimen (compare the experimental and calculated data sets at the bottom of Figure 6). The suggestion from Figure 6 is that the silicone strongly localizes itself at the surface.

Because this ESCA technique allows characterization of a system in the depth direction, it is quite useful in describing the bonding of extremely thin films to substrates. Cohn, et al., have described the structure of a thin fluoropolymer overlayer bonded to a poly(ethylene terephthalate) substrate (38).

The angular-dependent ESCA technique can also help us to appreciate the surface structure of phase-separated block copolymers. A number of earlier studies have laid the groundwork for studying these complex systems (13,39,40). In a study performed in our laboratory to examine the surface domain structures of segmented polyurethanes and polyurethane-ureas, "labelled" hard segments

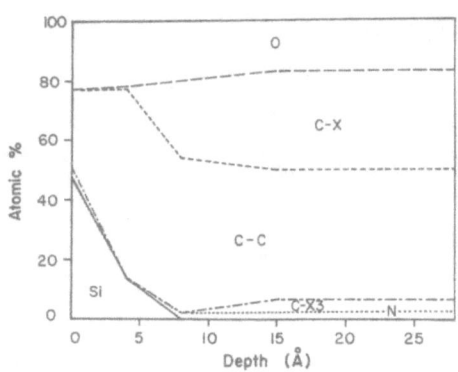

ATOMIC%

	EXPERIMENTAL						CALCULATED					
θ	C-C	C-X	C-X3	N	O	Si	C-C	C-X	C-X3	N	O	Si
80	47.8	8.5	2.0	0.0	21.7	20.0	47.5	9.0	1.4	0.6	21.8	19.7
72	49.3	13.4	1.8	0.0	21.3	14.2	48.1	15.4	1.7	1.1	20.8	12.9
62	43.6	24.0	1.8	1.1	20.9	8.7	47.5	19.8	2.1	1.4	20.0	9.2
52	46.0	23.9	2.3	1.1	19.8	6.7	47.0	22.3	2.4	1.6	19.4	7.2
31	46.6	25.0	2.7	2.0	18.7	5.0	46.3	24.8	2.7	1.8	18.9	5.4

Figure 6. Postulated depth profile obtained from Biomer centrifugally cast onto glass from a 2% solution in hexafluoroisopropanol. C-C represents hydrocarbon-like functionalities resolved from the C1s spectrum. C-X represents carbon species singly bound to oxygen resolved from the C1s spectrum. C-X3 represents carbon species bound with two or more bonds to oxygen or nitrogen from the C1s spectrum.

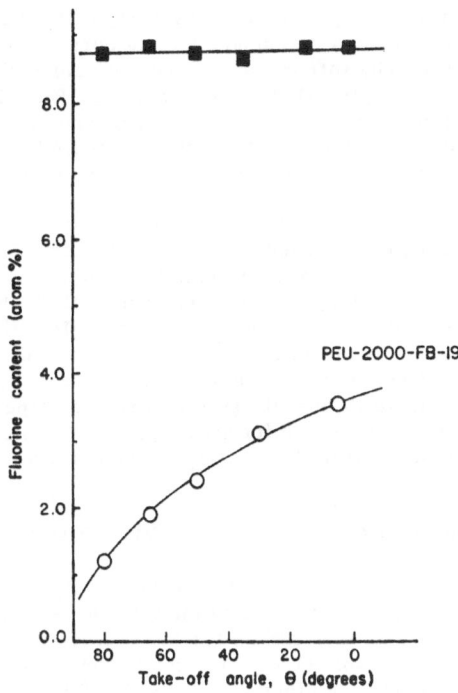

Figure 7. The angular dependence of the fluorine F_{1s} ESCA signal of two of
polyurethanes with fluorine-labelled hard segments. PEU-1000-FP24
contains fluorinated pentanediol (an "odd" diol) while PEU-2000-
FB14 contains fluorinated butanediol with an even number of
carbons.

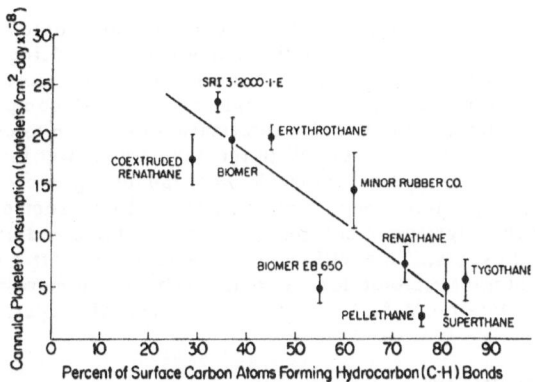

Figure 8. Platelet consumption by polyurethanes plotted against a ratio of
the fraction of the C_{1s} in a hydrocarbon-type environment to the
fraction in an ether-like environment.

based on 4,4'-methylene bis(phenylene isocyanate) with fluorinated diol chain extenders containing even and odd numbers of carbons, were synthesized, purified and cast as films (41). The soft segments used were molecular weight 1000 and 2000 poly(tetramethylene glycol)s (PTMO). For the PTMO 2000 samples, the odd diol chain-extended polymers reveal more hard segment at the surface than the even diol chain-extended polymers. As shown in Figure 7, the fluorine content increases monotonically to 80-100% of the bulk value as the sampling location drops from the surface down to the maximum depth of ~100A. A decrease in the soft-segment molecular weight from 2000 to 1000 produces an increase in the fluorine content (hard-segment concentration) in the surface region. For the PTMO 1000 samples, the odd diol chain-extended polymers show no angular dependence at all and have the same fluorine content throughout the surface and bulk regions. The extent of phase separation is strongly influenced by the number of carbons in the chain extender, with chain extenders containing even numbers of carbons demonstrating pronounced phase separation (42-44). Thus, these depth profiles illustrate that the surface structure depends strongly on the extent of phase separation, and for highly phase-separated polyurethanes, the polyether phase may uniformly "overlayer" the domain structure.

CORRELATION OF SURFACE STRUCTURE WITH BIOLOGICAL EVENTS

Utilizing data from ESCA and other surface analysis techniques, and data generated from studies on biological interactions with materials, investigators are beginning to correlate surface characteristics with biological responses (20,45-50). Hanson, et al., have established a strong relationship between surface chemistry and blood reactivity (19, 20). Using a series of polyurethanes in a baboon arterio-venous shunt system, they correlated the hydrocarbon/ether linkage ratio with platelet consumption (see Figure 8). As the ratio decreases, the platelet consumption significantly increases. In addition, the ESCA angular-dependent measurements and other chemistry measurements confirm that polyurethanes of the type that were noted in reference 19 to be highly blood-reactive are phase-separated with probable surface overlayers of the polyether soft segment (12,14,51,52). Thus, the polyether groups are implicated in this blood reaction assuming that the polyurethane surface does not undergo radical reorganization at the interface with an aqueous solution.

CONCLUSION

Composition-biological interaction correlations should be helpful in designing and perfecting biomaterials for implantation. Using the ESCA technique to study particular aspects of the surface (e.g., structure, contamination, molecular mobility, the influence of bulk composition, and phase separation) can increase the body of information available to correlate with present or future biological data. With the advent of newer techniques of surface analysis, such as quantum mechanical tunnelling microscopy (also called scanning tunnelling microscopy) (53), the potential may someday exist to image on surfaces at the angstrom resolution protein structures and other molecular-level biological events. This will enable researchers to better understand the biomaterial-biological interaction. Perhaps then we can more meaningfully and accurately correlate biological events with biomaterial properties.

ACKNOWLEDGEMENT

Generous support from N.I.H. grants HL25951 and RR 01296 has been received during the preparation of this manuscript and for some of the studies described in this work.

REFERENCES

1. B. Ratner, in: <u>Treatise on Clean Surface Technology</u> (K.L., Mittal, ed.), p., Plenum Press, New York (1985). (in press)

2. B.D. Ratner, Analysis of surface contaminants on intra-ocular lenses, <u>Arch. Ophthal.</u> 101, 1434-1438(1983).

3. B.D. Ratner, J.J. Rosen, A.S. Hoffman and L.H. Scharpen, in: <u>Surface Contamination</u> (K.L., Mittal, ed.), Vol. 2, pp.669-686, Plenum Publishing Corp., New York (1979).

4. B.D. Ratner, P.K. Weathersby, A.S. Hoffman, M.A. Kelly and L.H. Scharpen, Radiation grafted hydrogels for biomaterial applications as studied by the ESCA technique, <u>J. Appl. Polym. Sci.</u> 22, 643-664(1978).

5. F.J. Holly and M.F. Refojo, in: <u>Hydrogels for Medical and Related Applications, ACS Symposium Series</u> (J.D., Andrade, ed.), Vol. 31, pp.252-266, American Chemical Society, Washington, DC (1976).

6. D.S. Everhart and C.N. Reilley, The effects of functional group mobility on quantitative ESCA of plasma modified polymer surfaces, <u>Surf. Interf. Anal.</u> 3, 126-133(1981).

7. D.S. Everhart and C.N. Reilley, Polymer functional group mobility II. Partition of ion pairs between hydrophobic and hydrophilic phases of plasma oxidized polyethylene, <u>Surf. Interf. Anal.</u> 3, 258-268(1981).

8. J.D. Andrade, S.M. Ma, R.N. King and D.E. Gregonis, Contact angles at the solid-water interface, <u>J. Coll. Interf. Sci.</u> 72, 488-494(1979).

9. Y.C. Ko, B.D. Ratner and A.S. Hoffman, Characterization of hydrophilic-hydrophobic polymeric surfaces by contact angle measurements, <u>J. Coll. Interf. Sci.</u> 82, 25-37(1981).

10. R.G. Azrak, Surface property variations in melt-formed thermoplastics, <u>J. Coll. Interf. Sci.</u> 47, 779-794(1974).

11. B.D. Ratner, in: <u>Surface and Interfacial Aspects of Biomedical Polymers</u>(J.D., Andrade, ed.), Vol. 1, pp.373-394, Plenum Publishing Corp., New York (1985).

12. R.W. Paynter, B.D. Ratner and H.R. Thomas, in: <u>Polymers as Biomaterials</u>(S.W., Shalaby, A.S., Hoffman, B.D., Ratner and T.A., Horbett, eds.), pp.121-133, Plenum Publishing Co., New York (1984).

13. J.J. O'Malley, H.R. Thomas and G.M. Lee, Surface studies on multicomponent polymer systems by X-ray photoelectron spectroscopy. Polystyrene/poly(ethylene oxide) triblock copolymers, <u>Macromolecules</u> 12, 996-1001(1979).

14. C.B. Hu and C.S.P. Sung, Surface chemical composition - depth profile of polyether polyurethaneureas as studied by FTIR and ESCA, <u>Polym. Prepr., Am. Chem. Soc., Div. Polym. Chem.</u> 21, 156-158(1980).

15. D. Briggs, D.M. Brewis and M.B. Konieczko, X-ray photoelectron spectroscopy studies of polymer surfaces, <u>J. Mater. Sci.</u> 14, 1344-1348(1979).

16. D.T. Clark, W.J. Feast, W.K.R. Musgrave and I. Ritchie, Applications of ESCA to polymer chemistry. Part VI. Surface fluorination of polyethylene. Application of ESCA to the examination of structure as a function of depth, <u>J. Polym. Sci., Polym. Chem. Ed.</u> 13, 857-890(1975).

17. R. Chujo, T. Nishi, Y. Sumi, T. Adachi, H. Naito and H. Frentzel, Vertical distribution of components in a polymer blend with the aid of the secondary ion mass spectroscopy, J. Polym. Sci., Polym. Lett. Ed. 21, 487-494(1983).

18. B.D. Ratner, Surface characterization of biomaterials by electron spectroscopy for chemical analysis, Ann. Biomed. Eng. 11, 313-336(1983).

19. S.R. Hanson, L.A. Harker, B.D. Ratner and A.S. Hoffman, in: Biomaterials 1980; Advances in Biomaterials(G.D., Winter, D.F., Gibbons and H., Plenk Jr., eds.), Vol. 3, pp.519-530, John Wiley and Sons Ltd., Chichester, England (1982).

20. S.R. Hanson, L.A. Harker, B.D. Ratner and A.S. Hoffman, In vivo evaluation of artificial surfaces with a nonhuman primate model of arterial thrombosis, J. Lab. Clin. Med. 95, 289-304(1980).

21. D. Briggs(ed.), Handbook of X-ray and Ultraviolet Photo-electron Spectroscopy, Heyden & Sons, Ltd., London (1977).

22. T.A. Carlson, Photoelectron and Auger Spectroscopy, Plenum Press, New York (1975).

23. P.K. Ghosh, Introduction to Photoelectron Spectroscopy, John Wiley & Sons, New York (1983).

24. K. Siegbahn, C. Nordling, A. Fahlman, R. Nordberg, K. Hamrin, J. Hedman, G. Johansson, T. Bergmark, S.E. Karlsson, I. Lindgren and B. Lindberg, ESCA: Atomic, molecular and solid state structure studied by means of electron spectroscopy, Nova Acta Regiae Societatis Scientiarum Upsaliensis, Ser.IV 20, 5-282(1967).

25. K. Siegbahn, Electron spectroscopy for atoms, molecules, and condensed matter, Science 217, 111-121(1982).

26. N. Winograd and S.W. Gaarenstroom, in: Physical Methods in Modern Chemical Analysis(T., Kuwana, ed.), Vol. 2, pp.115-169, Academic Press, New York (1980).

27. D. Briggs, New developments in polymer surface analysis, Polymer 25, 1379-1391(1985).

28. D.T. Clark, in: Polymer Surfaces(D.T., Clark and W.J., Feast, eds.), pp.309-351, Wiley, J. & Sons, Chichester (1978).

29. A. Dilks, in: Electron Spectroscopy: Theory, Techniques, and Applications(A.D., Baker and C.R., Brundle, eds.), Vol. 4, pp.277-359, Academic Press, London (1981).

30. B.D. Ratner and B.J. McElroy, in: Spectroscopy in the Biomedical Sciences(R.M., Gendreau, ed.), p., CRC Press, Boca Raton, Fl (1985). (in press)

31. J.D. Andrade(ed.), Surface and Interfacial Aspects of Biomedical Polymers, Plenum Press, New York (1985).

32. D. Briggs and M.P. Seah(eds.), Practical Surface Analysis, John Wiley & Sons, Chichester (1983).

33. D.T. Clark, Advances in ESCA applied to polymer characterization, Pure & Appl. Chem. 54(2), 415-438(1982).

34.   M.P. Seah, The quantitative analysis of surfaces by XPS: A review, <u>Surf. Interf. Anal.</u> <u>2</u>, 222-239(1980).

35.   R.S. Swingle and W.M. Riggs, ESCA, <u>Crit. Rev. Anal. Chem.</u> <u>5</u>, 267-321(1975).

36.   C.D. Wagner, W.M. Riggs, L.E. Davis and J.F. Moulder, <u>Handbook of X-Ray Photoelectron Spectroscopy</u>, Perkin-Elmer Corporation, Eden Prairie, MN (1979).

37.   R.W. Paynter, Modification of the Beer-Lambert equation for application to concentration gradients, <u>Surf. Interf. Anal.</u> <u>3</u>, 186-187(1981).

38.   D. Cohn, A.S. Hoffman, B.D. Ratner and Y. Haque, Plasma-treated surfaces for biomedical applications: compositional analysis, Abstracts of the 2nd International Conference on Polymers in Medicine, Capri, Italy, June 3-7, 1985, C9.

39.   D.T. Clark, J. Peeling and J.J. O'Malley, Application of ESCA to polymer chemistry. VIII. Surface structures of AB block copolymers of polydimethylsiloxane and polystyrene, <u>J. Polym. Sci., Polym. Chem Ed.</u> <u>14</u>, 543-551(1976).

40.   H.R. Thomas and O'Malley, Surface studies on multicomponent polymer systems by X-ray photoelectron spectroscopy: Polystyrene/poly(ethylene oxide) homopolymer blends, <u>Macromolecules</u> <u>14</u>, 1316-1320(1981).

41.   S.C. Yoon and B.D. Ratner, Surface structure of segmented polyetherurethanes and polyetherurethaneureas with various perfluoro chain extenders. An x-ray photoelectron spectroscopic investigation, <u>Macromolecules</u>. (submitted)

42.   J. Blackwell, M.R. Nagarajan and T.B. Hoitink, Structure of polyurethane elastomers: effect of chain extender length on the structure of MDI/diol hard segments, <u>Polymer</u> <u>23</u>, 950-956(1982).

43.   J. Blackwell and M.R. Nagarajan, Conformational analysis of poly(MDI-butandiol) hard segment in polyurethane elastomers, <u>Polymer</u> <u>22</u>, 202-208(1981).

44.   J. Blackwell, J.R. Quay, M.R. Nagarajan, L. Born and H. Hespe, Molecular parameters for the prediction of polyurethane structures, <u>J. Polym. Sci., Polym. Phys. Ed.</u> <u>22</u>, 1247-1259(1984).

45.   V. Sa Da Costa, D. Brier-Russell, E.W. Salzman and E.W. Merrill, ESCA studies of polyurethanes: blood platelet activation in relation to surface composition, <u>J. Coll. Interf. Sci.</u> <u>80</u>, 445-452(1981).

46.   J.P. Fischer, P. Fuhge, K. Burg and N. Heimburger, Methoden zur Herstellung und Charakterisierung von Kunststoffen mit verbesserter Blutvertraglichkeit, <u>Angew. Makromol. Chem.</u> <u>105</u>, 131-165(1982).

47.   S.K. Chang, O.S. Hum, M.A. Moscarello, A.W. Neumann, W. Zingg, M.J. Leutheusser and B. Ruegsegger, Platelet adhesion to solid surfaces. The effect of plasma proteins and substrate wettability, <u>Med. Progr. Technol.</u> <u>5</u>, 57-66(1977).

48.   D.J. Lyman, W.M. Muir and I.J. Lee, The effect of chemical structure and surface properties of polymers on the coagulation of blood. I. Surface free energy efects, <u>Trans. Am. Soc. Artif. Int. Organs</u> <u>11</u>, 301-306(1965).

49.   N. Mohandas, R.M. Hochmuth and E.E. Spaeth, Adhesion of red cell to foreign surfaces in the presence of flow, J. Biomed. Mater. Res. 8, 119-136(1974).

50.   H. Yasuda, B.S. Yamanashi and D.P. Devito, The rate of adhesion of melanoma cells onto nonionic polymer surfaces, J. Biomed. Mater. Res. 12, 701-706(1978).

51.   B.D. Ratner and R.W. Paynter, in: Polyurethanes in Biomedical Engineering, Progress in Biomedical Engineering(H., Planck, G., Egbers and I., Syre, eds.), Vol. 1, pp.41-68, Elsevier, Amsterdam (1984).

52.   M.D. Lelah, L.K. Lambrecht, B.R. Young and S.L. Cooper, Physiochemical characterization and in vivo blood tolerability of cast and extruded biomer, J. Biomed. Mater. Res. 17, 1-22(1983).

53.   G. Binning and H. Rohrer, Scanning tunneling microscopy, Physica 127B, 37-45(1984).

CONTACT ANGLE ANALYSIS OF BIOMEDICAL POLYMERS: FROM AIR TO WATER TO ELECTROLYTES

J.D. Andrade

College of Engineering, University of Utah

Salt Lake City, Utah 84112

I. The Early Years Prior to 1960

The surface characterization of biomedical materials prior to the early 1960's consisted primarily of relatively qualitative observations as to whether surfaces were hydrophobic or hydrophilic (1). Contact angle techniques were well-known and were widely applied in industry, and general correlations had evolved between blood interactions in glass as opposed to siliconized glass tubes. Vroman was just beginning his studies on protein interactions with surfaces (2). There was considerable interest during this time period on the interaction of cells in culture with solid substrates, and there was some attempt to correlate and quantitate that interaction through contact angle measurements (3).

II. The 1960's

The situation improved considerably with the development of Zisman's critical surface tension (γ_c) concept in which the advancing contact angle of a series of probe liquids was measured, the data plotted as indicated in Figure 1, extrapolated to cosine $\Theta=1$, and the intercept identified as the critical surface tension for wetting. This concept is very well-known and is described in all basic textbooks and reviews on this subject (4,5), therefore will not be discussed further.

Detailed studies with a variety of model compounds, primarily Langmuir/Blodgett monolayers, allowed Zisman and coworkers to relate the critical surface tension to the nature of the functional group or groups present in the surface (5). It was clearly documented (using carefully prepared control surfaces) that the advancing contact angle was highly reproducible and that the surfaces showed minimum contact angle hysteresis. This method was applied by Dr. Robert Baier, who was a research fellow in Zisman's laboratory, to the biomedical materials problem in the late 60's and early 70's (6). The availability of a reproducible and reasonably precise measure of surface properties provided by the critical surface tension encouraged many investigators to attempt to develop correlations between that variable and various biological responses, including blood coagulation, cell adhesion, in vitro cell culture, platelet interactions, and protein adsorption.

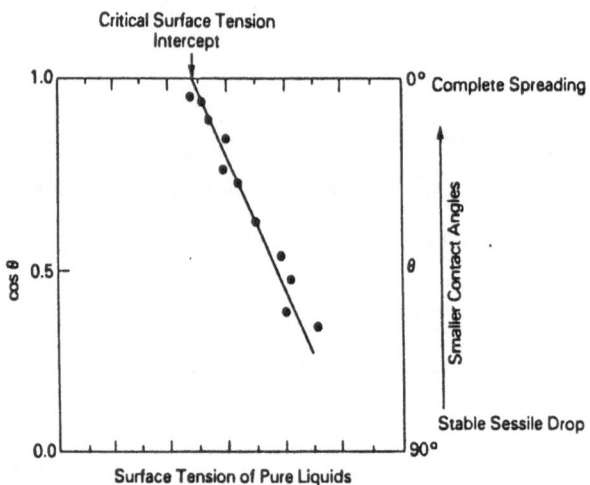

FIGURE 1: A typical Zisman γ_c plot. The cosines of the contact
angle for a range of pure liquids on a given solid are
plotted against the liquid surface tensions. The
critical surface tension is given by the intercept at
$\cos \theta = 1$ and is defined as the surface tension of that
liquid which would just totally spread on the solid
surface. This is an empirical measure related to the
surface free energy of the solid and is called the
critical surface tension for wetting of that particular
solid.

At about the time Zisman was refining the γ_c concept, Fowkes, Good, and coworkers were developing means to estimate surface and interfacial energetics using approximations of the work of adhesion at interfaces based in part, on intermolecular force considerations (7,8). The interactions at interfaces were basically considered to be due to two sources: London dispersion interactions and everything else generally called polar interactions. Fowkes showed that the London dispersion interactions could be deduced for various liquids and approximated for interfaces, using the geometric mean hypothesis commonly used in intermolecular forces (7) (Figure 2).

Simultaneously, Robert Good and coworkers were calculating interfacial interactions from first principles, using the expressions available for intermolecular forces (9). Given appropriate summation and/or integration and assuming intermolecular force additivity among molecules, expressions were developed which basically complimented the Fowkes treatment. Using Good and coworkers methods, it was possible to deduce the surface-free energy of polymer surfaces to compare or correlate that with the experimentally derived critical surface tension and generally to begin to develop a mechanistic understanding of the nature of interfacial processes and biological systems.

Lyman applied the methods of Fowkes and Good to the problem of coagulation and platelet adhesion on polymer surfaces and drew a correlation between the surface free energy of a polymer and its propensity to activate the coagulation of blood (10). Lyman's pioneering work stimulated a great deal of interest in relating the surface properties of polymers to their blood responses.

It is important to point out that all through this period, the contact angle data used was derived by advancing contact angle measurements of a series of highly purified model liquids (6,11).

A number of individuals extended the Fowkes geometric mean approximation for dispersion interactions to nondispersive or polar interactions (12), even though Fowkes clearly warned that such an approximation for polar interactions was not warranted and probably inaccurate. Nevertheless, many workers in the basic polymer surface science community (12) and in the biomedical community (13,14) showed that by such an approximation one could deduce not only the dispersion component of the surface energy or the surface tension of the solid, but its polar component as well. Given these components, one could approximate the interfacial-free energy, including its polar and dispersion component for a set of interfaces (10-14). There is considerable activity even today in estimating dispersion and polar components at interfaces and attempting to relate them to biological events. It is particularly pronounced in the field of bacterial adhesion and in the area of dental materials.

III. The Hydrogel Years

With the invention and development of hydrogels for soft contact lens application by Wichterle and Lim in Prague (15) and with the success and growth of the soft contact lens industry, considerable attention was focused on the role of water and hydrophilicity on biological

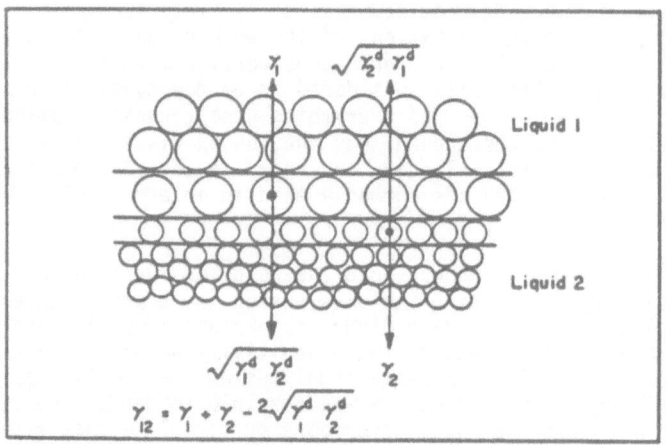

FIGURE 2: Fowkes' model of an interface showing the overall surface tensions (surface-free energies) γ_1 and γ_2 and the geometric mean approximations to the work of adhesion, $\sqrt{\gamma_1^d \gamma_2^d}$ (from Ref. 7).

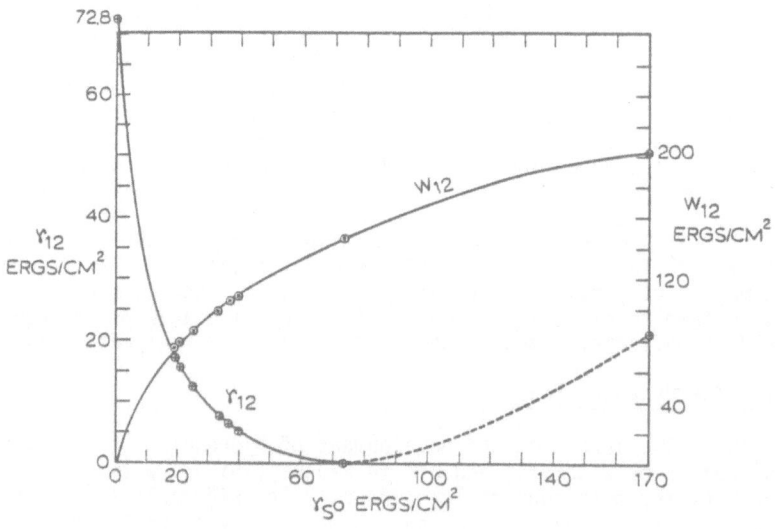

FIGURE 3: Plot of interfacial-free energy, γ_{12}, and work of adhesion, W_{12}, in water as a function of the surface-free energy, $\gamma_S o$, of various solid surfaces (from Ref. 14).

interactions. There were debates as to whether or not the surface of the endothelium was indeed hydrophilic or hydrophobic (6,16,17). There was considerable effort spent on the development of hydrophilic coatings and grafts for polymers. Radiation grafting of hydrophilic monomers to various substrates was a very active field.

Holly and Refojo observed that advancing water contact angles on these highly hydrophilic surfaces showed that the surfaces were hydrophobic, whereas receding water angles indeed showed that they were hydrophilic (18). Clearly the materials were hydrophilic under water, but outside of water some of the polymers appeared to be hydrophobic. This anomaly was explained by Holly and Refojo in terms of the restructuring or reorientation of polymer surfaces in response to their local environment, such as to minimize their interfacial energy (18,19).

In the early 70's, Andrade argued that the critical surface variable for biocompatibility is not the critical surface tension nor the surface-free energy, but rather the interfacial free energy of the polymer against water and showed that highly hydrophilic materials have very low interfacial-free energies (14,20) (Figure 3).

Although this was difficult to demonstrate experimentally, at about the same time W. C. Hamilton had developed a simple technique for measuring the interfacial energetics of polymers by the use of an octane droplet introduced at the polymer water interface (21). Because the surface tension of octane was exactly the same as the dispersion component of the surface tension of water (21.8 dynes/cm), it allowed a number of cancellations in the equations to occur, greatly simplifying the treatment and permitting one to deduce rather straight forwardly the interface energetics (22). (Holly and Refojo were performing similar experiments at about the same time.) Andrade and coworkers applied this method to attempt to deduce the interfacial energetics at gel-water interfaces, showing that indeed within the errors and uncertainties of the methods that the gel water interface has as low an interfacial-free energy as could be measured (20,22-23).

There is now considerable interest in measuring both advancing and receding water contact angles (24). It is well-known that the advancing angle tends to represent or sense the hydrophobic character of the surface, while the receding angle tends to sense primarily its hydrophilic character, and thus both angles are highly useful in getting a fuller understanding of the surface properties.

It has also become evident in the last ten years or so that polymer surfaces can be highly dynamic and can indeed show very different surfaces in air or vacuum as opposed to underwater. The question of the dynamics of polymer surfaces has been recently reviewed (19) and is the subject of a major symposium in June, 1986 (25).

IV. Where do we stand?

There are a number of major, unresolved problems:

1. We have no good way of estimating the polar interactions present at polymer surfaces. For the last ten years or so, Fowkes has treated these interactions as electron donor/acceptor or partial acid/base interactions and showed quite conclusively that, at least in nonaqueous systems, classical dipole/dipole interactions are relatively unimportant if partial acid/base character is present. He has demonstrated contact angle methods of probing these effects and has shown that IR analysis of polymer surfaces is probably the most useful

and direct way to characterize hydrogen bonding and electron donor/acceptor tendencies (26). No one has to my knowledge seriously applied these concepts to the study of biomedical polymer interactions with water and electrolyte solutions.

2. Although we have means to measure advancing and receding angles, the dynamics of such angles, and their hysteresis, it is somewhat difficult to interpret the results. This is because contact angle hysteresis is due to a number of different sources (Table 1). Although water is the solvent of choice due to its biological relevance, water is a difficult liquid to use for contact angle measurements because of its very small molecular volume; it readily penetrates into solid surfaces and is therefore not an inert probe of the surface energetics (27). The role of water penetration into the surface, subsequent water-induced plasticization of the surface region, and finally the inherent surface dynamics of the polymer chains and side chains all influence the contact angle results. The use of larger probe liquids, such as ethylene glycol, glycerol, and perhaps others may minimize the penetration effect (27). The surface dynamics may be partially sorted out by making measurements as a function of temperature. The dynamics can be probed in the absence of hydrophilicity by, for example, the use of model polymers with different alkyl side chain lengths (28,29). Molecular surface dynamics can in principle be probed directly by NMR methods (30).

Although considerable progress was made going from non-aqueous probe liquids to water, water is not the physiologic environment. The physiologic environment is a buffered ionic solution. Preliminary measurements of ionic surfaces with sodium chloride solutions of different ionic strengths suggest that we can no longer continue to ignore the role of ions in the interface energetics of biomedical polymers (31,32). Indeed it has been known for many years from electrokinetic measurements, using streaming potential and related techniques, that neutral hydrophobic polymers are highly negatively-charged based on electrokinetic measurements (33). This was considered an anamoly for a time, but it is now generally understood to result from ion adsorption at the polymer electrolyte solution interface (34). The adsorbed layer of ions basically sets up an interface potential, which is what the electrokinetic method measures. So even for "neutral" polymers, contact angle measurements in ionic solutions may be desirable. They are mandatory for polymers which we know are charged, such as the various sulfate and sulfonic acid containing polymers and surface films with synthetic heparin activity (32,35).

V. What do we know?

We have some current correlations. We know that the dispersion and polar contributions to the surface energy of a polymer and the consequent interfacial-free energies against water do indeed correlate with protein adsorption, cell adhesion, and coagulation time measurements, albeit not always in the direction in which we initially expected (29,36-38). We know that there are also correlations with γ_c and bioadhesion and related phenomena. Baier has postulated that optimum biocompatibility occurs in the γ_c range of about 25 ± 5 dynes/cm (6) because this is the range of minimum bioadhesion, which would be interpreted as minimum protein adsorption or minimum cell adhesion. Although this hypothesis is highly controversial, there is data in the literature to support it, and many groups are following that line of reasoning.

TABLE 1 (Ref. 24)

Sources of Contact Angle Hysteresis

General assumption	Specific assumption	Effect on hysteresis	Time dependent
Surface is smooth	Surface must be smooth at the 0.1 to 0.5 μ level	$\Delta\theta$ increases with increasing roughness (θ_{adv} increases and θ_{rec} decreases with increasing roughness)	No
Surface is heterogeneous	Surface must be homogeneous at the 0.1 μ level and above	θ_{adv} dependent on low-energy phase; θ_{rec} dependent on high-energy phase	No
Surface is nondeformable	Modulus of elasticity in surface $> \sim 3 \times 10^5$ dyn/cm^2	Not known	Yes—due to surface deformation/relaxation effects
Wetting liquid does not penetrate surface	Liquid molecular volume $>$ 60–70 cc/g-mole	Increased liquid penetration leads to increased hysteresis	Yes—due mainly to diffusion
Surface does not reorient	Reorientation time $\gg$ time of measurement	Increased tendency to orient leads to increased hysteresis	Yes
Surface immobile, therefore, surface entropy is constant	Configurational entropy independent of local environment	Unknown—but probably increase in hysteresis as surface mobility increases	Yes

VI. What do we need to consider?

Consider a hypothetical protein adsorption experiment (Figure 4). Figure 4a shows the polymer surface in air. We assume that this is a highly mobile polymer surface with both polar and apolar constituents. In air, it is an apolar material with its surface-free energy minimized. The material is introduced into water or electrolyte solution. It quickly restructures its surface and rearranges to show its polar constituent to the aqueous phase, thereby minimizing the interfacial-free energy (19). A protein diffuses by; the protein collides with the surface (Figure 4). Clearly the analysis of that initial contact must be done, respecting the fact that the surface is fully equilibrated in water just prior to and at the moment of contact. The surface the protein "sees" can therefore only be characterized or measured by underwater or receding water contact angle methods (24) (Figure 4b).

In order for the protein to contact the surface, water must be displaced from both the protein contact area as well as the surface contact area. Clearly, it is unlikely for that water to be displaced if it is tightly bound to either the protein or the surface. If that is the case, the protein will simply diffuse away, and the collision would have resulted in no net adsorption. If however, the protein now collides in a different orientation, perhaps one exposing a hydrophobic patch and perhaps if it hits the region of the surface which is partially hydrophobic and from which the surrounding waters can be removed, then there will be a transient interaction or adhesion. If the interaction energy is large enough, the protein will have a residence time at the surface. Given the various principles of protein adsorption (39), a number of things may happen. In light of the new local environment and of the statistics and dynamics of the process, another region or part of the protein may contact the surface leading to a second attachment point or "foot." If the polymer surface is itself in motion, a portion of the polymer surface may statistically approach and contact the protein. Once we have two or three contacts we have a cooperative binding process, and the protein is essentially adsorbed.

Now begins the process of long-term conformational change and accommodation--both of the protein, which is now said to be undergoing "denaturation," and the surface, which is restructuring or modifying in light of its new microenvironment. If the surface is binding to the protein through hydrophobic associations, then it may well be that characterization of the polymer surface in air (through advancing contact angle measurements) is important in estimating the hydrophobic "potential" of the surface. Although those hydrophobic polymer surface residues would not have been exposed during the initial phases of protein contact, they may well be exposed and participating in the interaction later on in the process as the protein and polymer surface maximally accommodate to one another (Figure 4d).

So measurements of polymer surface properties in air are neither good nor bad, just as measurements in water are neither good nor bad. Both are important and both are necessary in order to optimally correlate and understand the processes. In fact, ideally we would like to know the surface relaxation or restructuring time to get an idea as to what is the probability for a surface to accommodate to or "denature" in response to an adsorbed protein.

VII. Conclusions

The surface properties of biomedical polymers are indeed important in the adsorption of proteins and in subsequent biological interactions. It

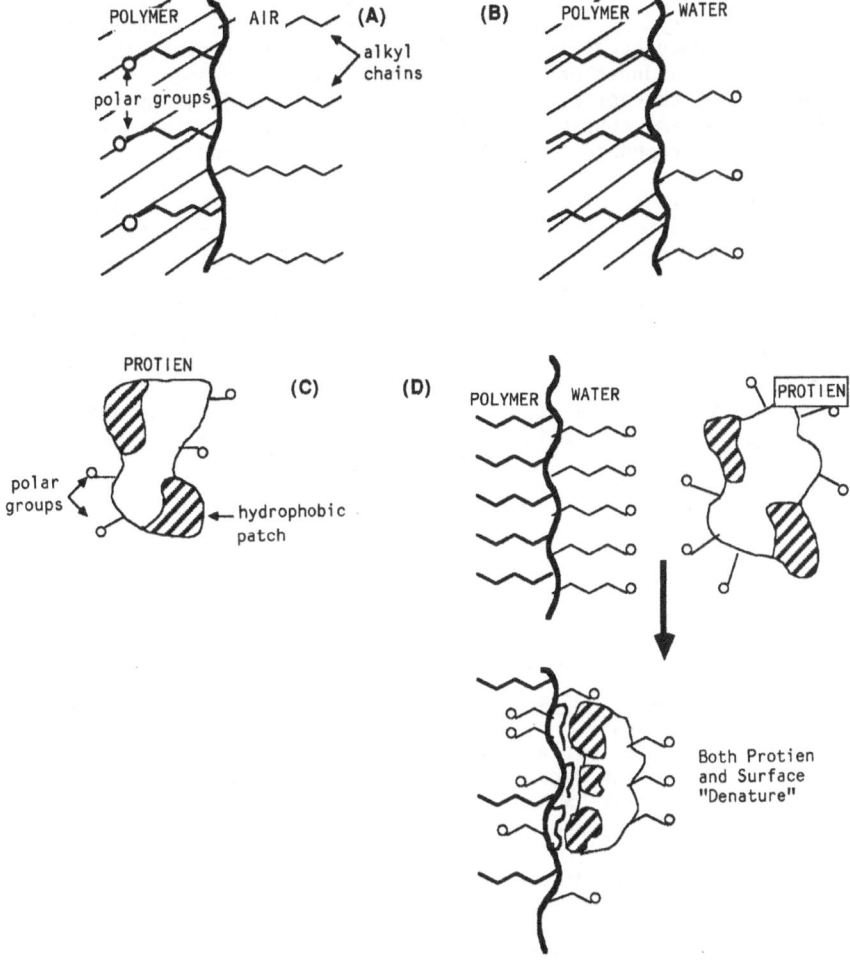

FIGURE 4: A schematic view of a dynamic polymer surface in
           different environments.
           a) in air;
           b) in water;
           c) a protein molecule in water exhibiting its
              equilibrium "surface";
           d) the polymer-adsorbed protein system with the protein
              shown interacting mainly via hydrophobic
              interactions and denatured to expose a third
              hydrophobic patch to the surface. The polymer
              surface has restructured to hydrophobically interact
              with the protein and, where appropriate,
              hydrophobically interact with the aqueous phase.

is important to characterize polymer surfaces with full awareness that they may be highly dynamic and somewhat amphoteric in character. Measurements should be made under conditions which carefully characterize both the hydrophobic and hydrophilic nature of the surface, which at this stage is best performed by advancing angle measurements to probe the hydrophobic character and under water or receding water measurements which probe the hydrophilic character. These measurements should, if possible, attempt to deduce the time course of surface restructuring or reorientation in going from the hydrophobic to hydrophilic environment and vice versa.

In addition, we must begin to develop standardized methods to probe the partial acid/base or electron donor/acceptor properties of polymer surfaces and their interfaces with aqueous electrolyte solutions. This can in part be done by electrokinetic (zeta potential) methods, as well as perhaps by infrared and Raman spectroscopy, preferably in the electrolyte solution of interest. The dynamics of polymer surfaces can indeed be directly probed by nuclear magnetic resonance and electron spin resonance spectroscopy, as well as perhaps by fluorescence spectroscopy. It is expected that these methods will continue to develop and will begin to be applied by the biomaterials community in the very near future.

IX. Acknowledgements

Discussions with R. Baier, T. Matsuda, F. M. Fowkes, and V. Hlady have been very helpful.

X. References

1. R. E. Baier, "Surface Properties Influencing Biological Adhesion," in Adhesion in Biological Systems (R. S. Manley, ed.), Academic Press, 1970, pp. 15-48.
2. L. Vroman, "Effect of Adsorbed Proteins on the Wettability of Hydrophilic and Hydrophobic Solids," Nature 196 (1962) 476
3. P. van der Valk, A. W. J. van Pelt, H. J. Busscher, H. P. de Jong, Ch. R. H. Wildevuur, and J. Arends, "Interaction of Fibroblasts and Polymer Surfaces: Relationship between Surface Free Energy and Fibroblast Spreading." J. Biomed. Mat. Res. 17 (1983) 807-817.
4. A. W. Adamson, Physical Chemistry of Surfaces, 4th ed., Wiley, New York (1983).
5. W. A. Zisman, "Relation of the Equilibrium Contact Angle to Liquid and Solid Constitution," in Contact Angle, Wettability, and Adhesion (F. M. Fowkes, ed), Adv. Chem. Ser. 43, 1-51 (1964).
6. R. E. Baier, "Role of Surface Energy in Thrombogenesis," Bull. New York Acad. Med. 48 (1972) 257-272.
7. F. M. Fowkes, "Attractive Forces at Interfaces," Chem. and Physics of Interfaces (S. Ross, ed.), Amer. Chem. Soc. (1965) 1-12.
8. R. J. Good and E. Elbing, "Theory for Estimation of Interfacial Energies," in Chem. and Physics of Interfaces 2 (S. Ross, ed.), Amer. Chem. Soc., (1968) 71-96.
9. R. J. Good, "Intermolecular Forces," in Treatise on Adhesion and Adhesives 1 (R. L. Patrick, ed.) Dekker, 1966 pp. 9-68.
10. D. J. Lyman, W. M. Muir, and I. J. Lee, "The Effect of Chemical Structure and Surface Properties of Polymers on the Coagulation of Blood. I. Surface Free Energy Effects," Trans. Am. Soc. Artif. Int. Organs 11 (1965) 301-306.
11. R. E. Baier, V. L. Gott, and A. Feruse, "Surface Chemical Evaluation...., Trans. Am. Soc. Art. Int. Org. 16 (1979) 50-57.
12. D. H. Kaelble, Physical Chemistry of Adhesion, Wiley, New York (1971).
13. E. Nyilas, W. A. Morton, D. M. Lederman, T. H. Chin, R. D. Cumming, "Interdependence of Hemodynamic and Surface Parameters in Thrombosis," Trans. Amer. Soc. Artif. Internal Organs 21 (1975) 55-63.

14. J. D. Andrade, "Interfacial Phenomena and Biomaterials," Med. Inst. 7 (1973) 110-120.
15. O. Wichterle and D. Lim, "Hydrophilic Gels for Biological Use," Nature 185 (1960) 117-119.
16. L. Vroman, Blood, Natural History Press, 1966.
17. J. D. Andrade, H. B. Lee, M. S. Jhon, S. W. Kim, J. B. Hibbs, "Water as a Biomaterial," Trans. Amer. Soc. Artif. Internal Organs 19 (1973) 1-6.
18. F. J. Holly and M. F. Refojo, "Wettability of Hydrogels", J. Biomed. Materials Res. 9 (1975) 315-326.
19. J. D. Andrade, D. E. Gregonis, L. M. Smith, "Polymer Surface Dynamics," in Surface and Interfacial Aspects of Biomedical Polymers 1 (J. D. Andrade, ed.), Plenum Press, 1985, pp. 15-41.
20. R. N. King, S. M. Ma, D. E. Gregonis, L. Brostrom, "Interfacial Tensions at Acrylic Hydrogel-Water Interfaces," J. Colloid Interface Sci. 103 (1985) 62-75.
21. W. C. Hamilton, J. Colloid Interface Sci. 40 (1972) 219; 47 (1974) 672.
22. J. D. Andrade, S. M. Ma, R. N. King, D. E. Gregonis, "Contact Angles at the Solid-Water Interface," J. Coll. Interface Sci. 72 (1979) 488-496.
23. J. D. Andrade, R. N. King, D. E. Gregonis, and D. L. Coleman, "Surface Characterization of Poly(hydroxyethyl) Methacrylate and Related Polymers I. Contact Angle Methods in Water," J. Polymer Sci., Polymer Symp. 66 (1979) 383.
24. J. D. Andrade, L. M. Smith, D. E. Gregonis, "Contact Angle and Interface Energetics," in Surface and Interfacial Aspects of Biomedical Polymers 1 (J. D. Andrade, ed.), Plenum Press, 1985, pp 249-291.
25. Polymer Surface Dynamics Symposium at the Rocky Mountain Regional American Chemical Soc. Meeting, Denver, CO, June 9-11, 1986.
26. F. M. Fowkes, "Interface Acid-Base/Charge-Transfer Properties," in Surface and Interfacial Aspects of Biomedical Polymers 1 (J. D. Andrade, ed.), Plenum Press, 1985, pp 337-372.
27. C. O. Timmons and W. A. Zisman, "Effect of Liquid Structure on Contact Angle Hysteresis," J. Colloid Interface Sci. 22 (1966) 165-171.
28. A. H. Hogt, D. E. Gregonis, J. D. Andrade, S. W. Kim, J. DanKert, J. Feijen, "Wettability of a Series of Methacrylate Copolymers," J. Colloid Interface Sci. (1985) in press.
29. D. L. Coleman, "Blood -Materials Interactions--A Multiparameter Approach," Ph.D. Thesis, University of Utah, 1980.
30. S. Nagaoka, Y. Mori, H. Takiuchi, K. Yokota, H. Tanzawa, and S. Nishiumi, "Interaction Between Blood Components and Hydrogels with Poly(Oxyethylene) Chains," Polymers as Biomaterials (S. W. Shalaby, A. S. Hoffman, B. D. Ratner and T. A. Horbett, ed.) Plenum Publishing Corp. 1984, pp. 361-374.
31. R. N. King, "Surface Characterization of Synthetic Polymers for Biomedical Applications," Ph.D. Thesis, University of Utah, 1980.
32. W. Y. Chen and J. D. Andrade, "Surface Characteristics of Polysulfoalkyl Methacrylates," J. Colloid Interface Sci. (1985) in press.
33. R. A. VanWagenen, D. L. Coleman, R. N. King, P. Triolo, L. Brostrom, L. M. Smith, J. D. Andrade, "Streaming Potentials of Polymer Thin Films," J. Colloid Interface Sci. 84 (1981) 155-165.
34. J. Lyklema, "Interfacial Electrochemistry of Surfaces with Biomedical Relevance," in Surface and Interfacial Aspects of Biomedical Polymers 1 (J. D. Andrade, ed.), Plenum Press, 1985, pp. 293-335.
35. M. Jozefowicz and J. Jozefonvicz, "Antithrombogenic Polymers," Pure & Appl. Chem. 56 (1984) 1335-1344.
36. B. R. Young, Plasma Proteins: Their Role in Initiating Platelet and Fibrin Deposition on biomaterials," in Biomaterials: Interfacial Phenomena and Applications (S . Cooper and N. A. Peppas, ed.) Adv. in Chem. Series 199 (1982) pp.317-350.(Is this the right reference?)
37. L. M. Smith, "Cell Adhesion as Influenced by Substrate Surface Properties," Ph.D. Thesis, University of Utah, 1979.

38. S. Hattori, J. D. Andrade, J. B. Hibbs, D. E. Gregonis, R. N. King, "Fibroblast Cell Growth on Charged Hydroxyethyl Methacrylate Polymers, J. Colloid Interface Sci. 104 (1985) 72-78.
39. J. D. Andrade, ed. Protein Adsorption Plenum Press, 1985.

HEPARIN-CONTAINING AND HEPARIN-LIKE POLYMERS

Marcel Jozefowicz and Jacqueline Jozefonvicz

Laboratoire de Recherches sur les Macromolécules
Université Paris-Nord, GRECO 48, UA 502 CNRS
Avenue J.B. Clément, F 93430 Villetaneuse

INTRODUCTION

The average expectancy of life is to day limited by cardiovascular diseases. Thus, considerable effort has been devoted to the research about atherosclerosis and the various methods of its prevention and treatment. Indeed, a major effort has been made to develop cardiovascular prosthetic devices in order to treat cardiac valve deficiences.

Actually, prosthetic cardiovascular devices remain essentially unsatisfactory either because of the necessity to use an anticoagulant therapy or because of failure, as for the treatment of diseased coronary arteries. In fact, blood coagulation is the primary effect which occurs when blood contacts any surface except the blood vessels. Therefore, blood compatible materials, i.e. materials suitable for use as cardiovascular prosthetic devices, have to be primarily non thrombogenic or antithrombogenic ones. Indeed, blood coagulation will not be induced at the interfaces with non thrombogenic polymers while antithrombogenic polymers will prevent clot formation by use of the mechanisms which control the so called coagulation cascade.

In order to show how antithrombogenic polymers may be tailored, in the following, we shall analyse the coagulation process induced at the interface between blood and biomaterials focusing attention on the control mechanisms. Then, we shall review the different antithrombogenic polymers consisting of:

-the anticoagulant drugs releasing materials.
-heparin or other natural anticoagulant covalently bound to polymers.
-synthetic heparin-like polymers which mimic the anticoagulant activity of heparin, including insoluble and soluble materials.

THE COAGULATION OF BLOOD INDUCED AT THE INTERFACE WITH MATERIALS

The coagulation process is now relatively well understood. Blood is a suspension of cells in an aqueous saline solution called plasma, containing several hundreds of proteins (Tables 1 and 2). In fact, the clot formation is the final result of a catastrophic event, called the coagulation cascade, a sequence of enzymatically catalyzed biochemical reactions in which both blood cells and plasma proteins are involved. With regard to their biological activity, these proteins are different. Some of them, zymogens, are able to undergo an activation process, i.e. a catalytic cleavage leading to the formation of enzymes. This is the case with the so called coagulation factors with the exception of fibrinogen and factors VIII and V. Other proteins, such as fibrinogen, play the role of enzyme substrates. They are able to undergo an hydrolysis reaction by an enzymatically catalyzed process. A further category of proteins consists of the so called cofactors. The presence of these proteins in the plasma in inactivated form, is essential in so far as when activated they drastically increase the rate of enzymatically catalyzed reactions. To this category belong factors VIII and V.

Table 1. Average composition of normal human blood

Cell	Number/ml
Erythrocytes	$(5.1 \pm 0.7) \, 10^6$
Leucocytes	7500 ± 2500
Polynuclear neuthrophile	$60 \pm 8\%$
Polynuclear eosinophile	$2 \pm 1\%$
Polynuclear basophile	$0.5 \pm 0.25\%$
Lymphocytes	$32 \pm 6\%$
Monocytes	$6 \pm 2\%$
Platelets	300.000 ± 50.000

On the other hand, the coagulation process is under control of inhibiting factors. Some of them are plasma enzyme inhibitors (Table 3) which are able to form specific inactive complexes with given enzymes. Other involve both membrane cell receptors and plasma zymogens which undergo activation by an enzymatically catalyzed process. In turn, the resulting enzymes are potent coagulation inhibitors.

Table 2. Average concentration of some proteins in human plasma

Protein	g/l
Albumin	35 - 55
Fibrinogen	2 - 4
Immunoglobulins	
IgG	9.2 - 14.8
IgA	1.4 - 2.6
IgM	0.9 - 1.8
Coagulation factors	
II	0.100
V	0.005 - 0.100
VII	0.01
VIII	0.010
IX	0.005
X	0.010
XI	0.003
Prekallikrein	0.050
High Molecular Weight Kikinogen	0.070

Table 3. Average concentration of some coagulation inhibitors in human plasma

Inhibitor	g/l
α_1 antitrypsin	2 - 4
α_2 antiplasmin	0.085
C_1 inactivator	0.250
α_1 antichymotrypsin	0.480
α_2 macroglobulin	3
antithrombin III	0.150 - 0.300
second heparin cofactor	-

a) The coagulation cascade

When blood is placed in contact with any foreign surface, the following events appear: ·

Adsorption of proteins In one minute or less, a competitive adsorption of proteins and glycoproteins occurs at the surface and forms a complex protein coating on the surface. Some of these adsorption processes are partially or completely reversible (1-3). Depending on the nature of the surface, some of the deposited proteins may initiate different coagulation pathways (Fig. 1).

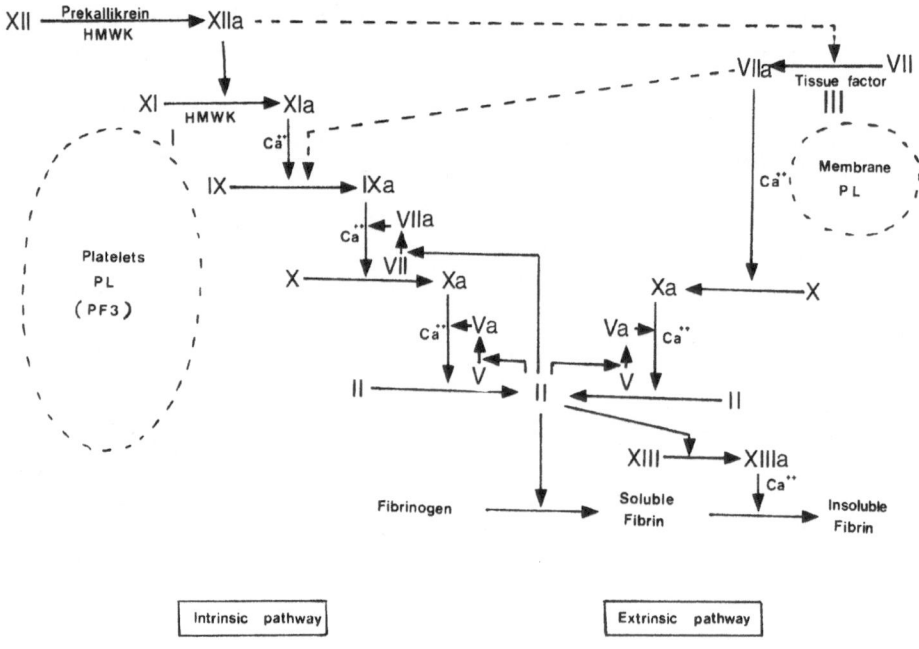

Figure 1. The coagulation cascade.

Platelet activation   First the von Willebrand factor (a polymer of
factor VIII) mediates platelet attachment onto the surface. Then, some
internal constituents are released from aggregated platelets into the
plasma. The   transformations, which are   under the control   of
prostaglandins, calcium   ions,   cyclic   adenyl   monophosphate   and
adenyldiphosphate concentration, are essential. They make available both
specific proteins and platelet membrane phospholipids .which are of prime
importance for the acceleration and control of the coagulation process.

    Platelet activation is also initiated by thrombin,adenyldiphosphate
and other reactants. It can also be produced by turbulences of the blood
flow and is a function of the flow rate.

Intrinsic pathway When artificial surfaces are placed in contact with
blood, the activation of the coagulation process occurs through the
so-called activation of the contact phase. This process involves several
proteins namely High Molecular Weight Kininogen (HMWK), prekallikrein,
kallikrein, factor XII and XIIa and factors XI and XIa and their
complexes and results in increasing amount of factor XIa attached to the
surface. It is self accelerated in that the formation of the enzymes
have feedback effects on the activation of the zymogens.

    The latter enzyme initiates a sequence of enzymatic reactions
which, in a short time, results in factor IIa (thrombin) formation

44

(Fig. 1). The rate of the latter reactions is slow and calcium dependent. In contrast, these reactions are speeded up when platelet factor 3 (PF3) from the platelet membranes and cofactors VIIIa and Va are present.

Extrinsic pathway The alternative pathway for the activation of blood coagulation, called " the extrinsic pathway", is involved when blood contacts natural surfaces excluding the normal vessel endothelium. In presence of tissue thromboplastin (factor III) and calcium ions, factor VII is activated and the resulting enzymatic activity causes factor IX and X activation. Then, thrombin formation occurs as in intrinsic pathway (Fig. 1).

Clot formation Thus, whatever the pathway, the coagulation cascade leads to the formation of thrombin, which plays numerous roles. First, the thrombin is able to induce platelet aggregation and release. Therefore, this enzyme has a feedback effect on its own formation. Second, thrombin is able to activate fibrinogen in an hydrolysis reaction which leads to fibrinopeptides A and B and the soluble fibrin polymer. The crosslinking of soluble fibrin which occurs is catalyzed by activated factor XIII forming the insoluble fibrin clot, and again thrombin is involved in the activation of factor XIII.

It is noteworthy that aggregated platelets, other blood cells and plasma proteins are entrapped in the crosslinked fibrin network.

b) Coagulation control systems

Considering the whole coagulation process, it appears that the clot formation should normally occur when blood contacts any surface including the normal vessel endothelium, resulting in general thrombosis of the blood circulation system. As this is not the case, it is obvious that blood is involved in some control system. Moreover, the normal healthy vessel endothelium itself should be able to control the coagulation cascade.

1. Endothelial cell control

The mechanism by wich endothelial cells achieve this control is not completely understood. However, two possible pathways have been recently proposed.

First, Weksler et al. (4) have shown that, when exposed to vessel endothelium, cells are able to release prostacyclin (PGI_2), a prostaglandin which is a very potent inhibitor of platelet aggregation.

Secondly, endothelial cell membranes contain a receptor called thrombomodulin (5). In the presence of the latter, thrombin enzymatically catalyzes the hydrolysis of protein C. The resulting activated protein C ($\overline{C}$) is able to bind to platelet membranes. It is suggested that, when fixed onto the platelet membrane, $\overline{C}$ is, in turn,

Table 4. Analyses of standard heparins

Iduronic acid % of total acid	Glucuronic acid % of total acid	Hexosamine %	Acetyl %	Sulfate %
45 – 75	30 – 60	20 – 26	0.4 – 2.5	25 – 34

able to catalyze the hydrolysis of factors Va (6) and VIIIa (7). As a result, the rate of the activation of factors X and II is decreased in the vicinity of endothelial cells.

Both these preceding pathways, and possibly others to be discovered in the future, are probably involved in the control of clot formation. Their failure probably contributes to atherosclerosis and, therefore, seems to be one of the possible way by which general thrombosis occurs as a consequence of this disease.

Other control plasmatic systems are involved in the prevention of general thrombosis.

2. Fibrinolytic system

The activation of the fibrinolytic system results from the hydrolysis of plasminogen, enzymatically catalyzed by factor XIIa. A potent enzyme, plasmin, formed during this reaction, promotes a sequence of reactions which lead to fibrinolysis and subsequent destruction of the fibrin clot. It is of interest to note that some foreign proteins, as for istance urokinase and spreptokinase have an enzymatic ability to activate plasminogen and to promote plasmin formation (8).

3. Antiproteases and heparin

Plasma contains enzyme inhibitors (Table 2) that are able to play an important role in the control of the coagulation process. These inhibitors are generally able to form complexes with several proteases, at various rates. For istance α_2 macroglobulin readily reacts with kallikrein and slowly reacts with thrombin.

The most potent plasma inhibitor is antithrombin III, which is able to form inactive stable complexes with serine-proteases including factors IIa, IXa, XIa and kallikrein. The reactions are irreversible and slow. They imply the formation of a chemical bond between the serin active site of the protease and an arginyl residue of the inhibitor, the exact nature of which is not completely clear. These reactions are more or less subject to catalysis by heparin and heparin analogs which might be present in some subendothelial or endothelial tissue. Commercially available heparin is a polysaccharide obtained from pig or beef mucosal linings of the lung and liver. In contrast with proteins, this copolymer is heterogeneous in molecular weight, chemical composition and sequences. It is mainly composed of alternating residues of sulfated

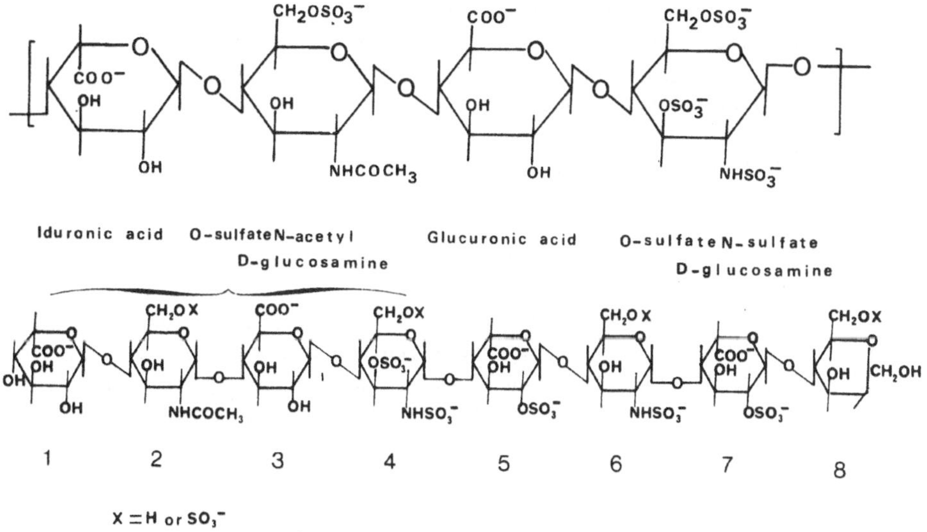

Figure 2. Tentative structures for the antithrombin-binding tetra-saccharide or octasaccharide derived from heparin.

glucuronic and iduronic acid and glucosamin derivatives (mainly N-sulfate and N-acetyl glucosamine) linked in the 1-4 position as shown in Fig. 2. Due to the large variability in the molecule of heparin (Table 4) and to difficulties in chemical and catalytic activity measurements, the mechanism of the catalytic activity and the exact chemical nature of the catalytic site are still investigated (9-12). Indeed, several mechanisms have been proposed. The first one is based on the assumption that antithrombin III binds first to heparin and that the complex formed readily reacts with the thrombin. The second model proposes that the thrombin binds first to heparin and then a fast irreversible reaction occurs. A third model is based on the hypothesis that heparin constitutes a bridge between enzyme and antithrombin III.

Moreover, it appears possible that the same mechanism should not apply for each of the inhibition reactions of the different enzymes by antithrombin III, as shown for istance in case of factors IXa, XIa and IIa on the one side and factors XIIa, Xa and kallikrein on the other side.

The nature of the catalytic sites also remains a matter of some controversy. It has been concluded that the binding site between the serine protease and antithrombin III are the serine active site of the thrombin and an arginyl residue of the antithrombin III. In contrast, disagreement exists about the nature and the number of the binding sites of heparin onto either inhibitor or serine protease. Indeed, lysyl residue of the antithrombin III is claimed to be a part of heparin-antithrombin III binding site. The number of these sites are discussed, some authors being in favour of one site, the other in favour of two.

The same situation exists with regard to serine proteases-heparin binding sites, which is even more complicated by the fact that the number of binding sites is claimed to be different and depending on the nature of the enzyme. Indeed, recently, hydrolytic cleavage of heparin has been performed and allowed to isolate an hexasaccharide and to hypothesize that the minimal sequence that binds to antithrombin III and elicits high anti-Xa activity was in fact contained in a pentasaccharide (9,13). These results have been confirmed by synthesis of the above mentioned pentasaccharide (14).

Relatively new investigations indicate that human plasma contains a substantial quantity of an unidentified heparin-dependent inhibitor of thrombin (15,16). This new inhibitor forms a complex with thrombin but is not identical to any of the other known plasma protease inhibitors.

ANTITHROMBOGENIC POLYMERS

Considering the whole process of thrombus formation, it appears that there are two concepts, and two only, on which the tailoring of blood compatible biomaterials can be based. First, the designed biomaterial should not activate any of the coagulation pathways. Unfortunately, no materials of this kind are known. Moreover it seems probable that even the endothelial cell membranes of the blood vessel wall are able to activate more or less the coagulation cascade. Second, the biomaterial surface, as any known material surface, will be blood coagulation activating, but in order to prevent thrombus formation at its blood interface, it should be able to inhibit the coagulation process.

Moreover, such biomaterials should not be inhibitors by themselves. Indeed, if it were the case, the biomaterial will be implied as a reagent in an irreversible inhibition reaction, the result of which will be an irreversible transformation of te biomaterial surface, which, in turn, will lose its inhibiting properties. Therefore, a suitable anticoagulant biomaterial should be a catalyst of inhibition reactions of the coagulation process. The surface of such a material should have a catalytic activity with regard to some of the control mechanisms of the coagulation of blood.

Indeed, several possibilities exist to design such biomaterial and each of them has been the basis of an active research development during the past two decades. Historically, the first attempt to achieve antithrombogenic materials was the ionic binding of eparin onto a polymeric surface by Gott (17) which resulted in a slow release of the anticoagulant drug into the blood stream. Since that time, the same strategy has been applied in numerous studies to different potent anticoagulant drugs.

a. Anticoagulant drugs releasing materials

It is the case, for istance, of biomaterials which release prostacyclin (PGI) entrapped in a macromolecular network (18,19). When, placed in contact with blood, these biomaterials prevent platelet aggregation and release, and thus, control the coagulation process. Such materials are potent as long as they are able to release PGI_2 . Unfortunately, PGI_2 is an expensive product and moreover unstable (its lifetime, after hydrolysis in biological conditions being less than one minute). Therefore biomaterials of this kind have found no practical uses at the moment. Other platelet anntiaggregating surfaces based on incorporation of antiaggregating groups such as dipyridamole for istance in polymer such as cellulose, cellulose triacetate, nylon and polyethyleneterephthalate have been prepared (20). The resulting materials have been claimed to be potent as antithrombogenic materials even when implanted in dogs.

b. Coating with heparin

Heparin is a negatively charged polysaccharide. This character allows its use to form macromolecular complexes in which heparin is bound ionically to a macromolecular polycation.

Since the early work of Gott (13) numerous studies have been devoted to the ionic attachment of heparin onto surfaces. In these studies, heparin was bound to various polycationic materials such as, graphite benzalkonium salts (13,21,22), tridodecylmethylammonium chloride (23,24), γ-aminopropyltriethoxysilane (25) as well as other derivatives of polyvinylchloride, cuprophane, polyester, nylon 6 and polyurethane. Cellulose films and tubing were also modified by treatment with ethyleneimine in order to allow ionic fixation of heparin onto the surfaces (26).

In other cases, copolymers derived from polyurethanes, polyacrylates, polystyrene have been prepared, in which aminated units were quaternized in order to form polysalts with heparin (27-34). Reviews of these techniques have been published by Leininger (35,36).

Whatever the techniques used, the resulting biomaterials have been proved potent as long as they release the ionically bounded heparin into the blood stream. During the release, the concentration of heparin near the surface is sufficiently high to prevent thrombus formation for periods of at least several days. Unfortunately, after that period of time, the release decreases. Therefore, such biomaterials are only suitable for use in devices of short term application as for instance extracorporeal circulation and catheters, but do not allow their use as permanent (37) implants. They were, indeed, developed for such temporary use by Japanese, American, Swedish and French manufacturers (38, 40-42), for the manufacture of catheters and peritoneovenous shunts.

An alternative approach to the ionic coating of heparin onto materials in order to get antithrombogenic surfaces involves the synthesis of polymeric gels in which heparin is entrapped. This can be achieved either by ionic binding of heparin to polymers in a first step followed by chemical (43-45) or radiation induced (46,47) crosslinking of the resulting network in a second step, or by trapping of heparin in a preformed hydrogel. The latter procedure was used in cases of polymeric gels prepared by radiation processing treatments involving vinylalcohol or vinylacetate and N-vinyl-2-pyrrolidone monomers (48-50). Cobalt-60 gamma rays were also used (51) to graft chloromethylstyrene on polysilicone samples followed by quaternization with pyridine. As in the preceeding case, uptake of heparin from solutions occurs by exposure of the polymeric gel.

All these gel procedures allow the preparation of antithrombogenic biomaterials. Indeed, the heparin release level appears to be decreased but sufficient to prevent blood clotting in ex-vivo assays; Nevertheless, the potency of these devices is still insufficient to allow their use as permanent implants.

Moreover, heparin appears to be an expensive starting material. Whatever the procedures, heparin coating may be of little practical interest for economic reasons.

c. Heparin covalently bound to polymer surfaces

The most promising way to design antithrombogenic biomaterials has appeared to be the covalent binding of heparin onto polymeric materials. In principle, such biomaterials will not release their heparin content by exposure to the blood stream. Therefore, if the covalently bound heparin remains active they will remain potent a longer time than it is the case for ionically bound heparin. Thus, it is not surprising that covalent binding of heparin onto surfaces has been extensively studied during the past decade.

Covalent binding of heparin onto polymers can be achieved by different ways. First, heparin has been coupled to preformed hydrogels by the use of various classical activatiuon techniques. Thus, Merrill et al. (52) prepared hydrogels derived from polyvinylalcohol on which heparin was fixed by acetal bridges. In other cases (53,59) heparin was bound to agarose and sepharose using the cyanogen bromide, the carbodiimide activation and so on. The same coupling reagents were used by Hoffman et al. (60-63) for the covalent binding of heparin either onto radiation grafted hydrogels derived from hydroxyethylmethacrylate and methacrylic acid or onto the same hydrogels radiation grafted on the surface of a silicone.

In a second general procedure pathway, a polymer is chemically or radiochemically activated and, then allowed to react chemically with heparin. Such a typical procedure was developed by Halpern et al. who fixed isocyanate groups onto polystyrene and then reacted the resulting polymer with heparin. Leininger et al. (64) also bound heparin to silicone in a two step procedure involving the synthesis of an amino derivative of the rubber followed by a reaction with heparin using the cyanuric chloride activation. Analog procedures were developed to bind heparin onto modified polyvinylalcohol hydrogels (65-67), elastomers, styrene-butadiene-stryrene elastomers (68), polyhydroxyethylmethacrylate-glycidyl methacrylate copolymers (69) or other polymers (70). Recently, H. Hasenfratz et al. (71) proposed the use of the same procedures to improve the blood compatibility of cellulosic membranes either by partial oxidation or by chemical amination of cellulose followed by binding heparin with suitable activation reagents. In turn, Chawla et al. (46) proposed a radiochemical grafting of heparin onto cellulose in order to get blood-compatible hemodialysis membranes.

Alternatively, heparin may be bound to suitably prepared substrates to produce thromboresistant materials. This approach has been used by Selfon et al.(68-72) to develop a highstrength styrene-butadiene-styrene block copolymer. The material was prepared by covalent coupling of an acetylated polyvinylalcohol-heparin mixture by an acetal bond to the hydroxylated surface of styrene-butadiene-styrene elastomer. Preliminary ex-vivo testing using a simple arterio-venous shunt in the leg of a rabbit showed good thromboresistance. These authors examined the mechanism of thromboresistance with polyvinylalcohol-heparin beads. Bound heparin appears to act as soluble heparin to promote thrombin-antithrombin III complex formation. The same approach was used by Larsson et al. (73) who adsorbed onto sulphated polyethylene colloidal particles composed of heparin and cetylamine hydrochloride and then reacted the mixture with glutaraldehyde. The authors claimed that in the resulting biomaterial heparin was covalently bound and that despite the fact that no release occured the biomaterial was thromboresistant.

In a third general procedure, a chemical or radiochemical treatment of heparin induces the formation of a macroradical which, in turn, induces the polymerization of a monomer resulting in a copolymer in which an heparin moiety is covalently bound to a synthetic polymeric sequence. This procedure was first achieved by Labarre et al. by use of Cerium (IV) peroxidation of heparin followed by reaction with acrylic and methacrylic monomers (74,75). Baquey et al. (76,77) described a similar method, using the radicals resulting from gamma ray irradiation of heparin instead of Cerium(IV) oxidation and used this method for the production of acrylic acid or dacron copolymer.

The properties and antithrombogenic activity of such covalently blonded heparin surfaces have been proved (78,79) to be similar to those of soluble heparin. An alternate pathway consisting in modification of heparin so that it should be able to polymerize or to copolymerize has been proposed by Mester et al. (80) and Plate et al. (81,82). According to these authors heparin is first acrylated and still remains a potent anticoagulant reagent. These acrylated heparin may be homopolymerized as well as copolymeryzed with suitable hydrophilic monomers or immobilized

in a polyacrylamide network. These biomaterials, as it was the case for
the preceding one, were shown to be anticoagulant.

The above methods allow the preparation of biomaterials where
heparin is covalently bound onto the surface. It is of interest to note
that they are generally able to bind antithrombin III when exposed to
blood or plasma. However, depending on the method used to fix heparin to
the polymer matrix they may or not have an anticoagulant heparin-like
activity (78,79,83).

It is also noteworthy that surfaces on which heparin is covalently
bound activate platelets when exposed to blood or platelets suspension
(44,59,84). However preincubation of these surfaces with either
antithrombin II aqueous solutions or plasma passivated the surface with
regard to platelet activation (44,84). This passivation seems to be
independent on the length of the spacer between the polymer matrix and
the bound heparin while in contrast there is a strong dependence of the
affinity of the coagulation factors toward the surface with the spacer
length (59).

As a conlcusion it was suggested (73,83,84) that biomaterials on
the surfaces of which heparin is covalently bound might be
thromboresistant beacuse of their heparin-like anticoagulant activity
and/or because of their ability to be passivated by binding antithrombin
III toward platelet activation. Nevertheless, the use of such
biomaterials may be limited for two reasons. The first one may be
economic as pointed out above; the second one may result from the fact
that an exposure to blood, may readily degrade heparin under the action
of heparinases. This is probably the reason why scientists started to
synthetize anticoagulant and heparin-like biomaterials.

d. Heparin-like materials

Natural heparin and heparin analogs of lower anticoagulant activity
such as heparan or chondroitin-6 sulfate are polyanionic derivatives of
polysaccharides. Thus, it is not surprising that a number of efforts
were devoted to the preparation of synthetic polyanions having
antithrombogenic or heparin-like properties.

As early as 1951, Lovelock and Porterfield (85) described te
anticoagulant properties of polystyrene sulfonates,, reacting a
polystyrene tubing with sulfuric acid. Then they observed that the blood
clotting time was longer (41 min) in the treated tubing than in the
starting one (29 min) or in glass tubing (9 min). Since that time, the
anticoagulant properties of sulfonated soluble polymers, specially
polystyrene and polyethylene sulphonates, have been described by Gregor
(86).

Furthermore Machovich (87) has shown that polymethacrylic acid has
an anticoagulant activity in aqueous solutions. He demonstrated that a
sample of this polymer at the concentration of 2µg/ml increases by a

factor of five the rate of the thrombin-antithrombin III reaction.

Later on, anticoagulant materials were obtained by reaction of N-chlorosulphonyl isocyanate with an unsaturated polymer followed by a suitable basic hydrolysis (88,89). For istance, a soluble polyelectrolyte derived from 1,4-cis polyisoprene:

$$
\begin{array}{c}
CH_3 \qquad\qquad\qquad\qquad\qquad CH_3 \\
| \qquad\qquad\qquad\qquad\qquad\qquad | \\
-CH_2 - C \!=\! CH - CH_{2n} - CH_2 - C - CH - CH_{2m} - \\
| \qquad | \\
NH \quad COONa \\
| \\
SO_3Na
\end{array}
$$

was shown to possess an anticoagulant activity which was 10% of that of heparin.

Such compounds were used to make ionic coatings covering thrombogenic hydrophobic surfaces of polyvinylchloride, for istance, by use of a coupling agent, the tridodecylmethyl ammonium chloride (TDMAC). (90) These coatings resulted in a drastic decrease of the platelet activation (adhesion and/or aggregation) on the surfaces. It appears to be stable since the polyelectrolyte seems to be not or only slightly eluted in presence of plasma or of an aqueous 0.15M sodium chloride solution.

Other methods were employed by the same authors in order to coat insoluble substrates with the same anticoagulant material (91). In a first approach, a block styrene-isoprene copolymer is reacted with N-chlorosulphonyl isocyanate. In the second approach, the soluble polyelectrolyte is partially reacted with a diol, the resulting cross-linked network being insoluble.

Other anticoagulant materials were developed by copolymerization of acrylic acid with ethylene (92,93) or ethylene with vinyl sulphonates (92) or again vinyl acetate and crotonic acid (92). In vivo testing by means of the Gott test showed that the blood compatibility dependes on the net negative charge of the macromolecular chains and the hydrophilicity of materials (92). In these series, the best materials appear to be polyvinylacetate-2% crotonic acid-60% sodium ionomer. The nature of the counter ions influences also the antithrombogenicity of the materials, calcium or magnesium ions being less favorable than sodium ions (93). Several works (45,94,95) were devoted to polyelectrolyte and polyelectrolyte complexes, i.e., insoluble materials formed by complexation of positively and negatively charged polymers. Generally, polystyrene derivatives were under consideration, as, for istance, the Ioplex complexes of sodium poly(styrene sulphonate) and poly(vinylbenzyltrimethyl ammonium) chloride (94,95), in variable proportions with regard to their net charge. These complexes can be either positive, negative, or neutral.

With regard to their thrombogenicity, the best materials seem to be those bearing a low negative net charge, the neutral or highly positively or negatively charged polymers being highly thrombogenic. Besides the net charge the charge distribution on the surfaces seems to determine the thrombogenicity some of them being favorable, other not (97).

Other polyelectrolyte complexes have been prepared by mixing chitosan either with carboxymethyldextran or with a mixture of carboxymethyldextran and sodium dextran sulfate (98,99,100). All these insoluble complexes whose composition depend on the preparation conditions have anticoagulant properties when prepared at pH between 3 and 6.5.

Some polyelectrolytes complexes have been prepared by mixing polyanions and polycations derived from acrylonitrile-sodium methallyl sulfonate copolymer and acrylonitrile-2-methyl 5-vinyl pyridine copolymer.

By varying the polyanion-polycation ratio, a large number of polyelectrolyte complexes can be prepared with an excess of anionic or cationic sites. Platelet adherence and protein adsorption in such materials were measured using the blood flow system in vitro described by Baumgartner.

Finally, Jozefowicz et al. (101-110) have obtained resins with significant antithrombogenic activity by binding sulfonate, carboxylic, aminoacid sulfamide or amide groups onto crosslinked polystyrene, polysaccharides or polystyrene-polyethylene graft copolymers. They have proved this activity to be antithrombic and anti Xa and to involve a plasmatic component by use of tests performed on platelets-poor plasma suspensions of the polymers.

This activity has also been proved to be located on the surface of the polymers. Moreover, by direct kinetic studies performed on suspensions of the polymers in aqueous solution of antithrombin III and proteases, they were able to show that the anticoagulant activity is heparin-like, i.e., the polymers act as heterogeneous catalyst with regard to the AT III-proteases inhibiting reaction. Thermodynamic studies of the adsorption of proteins on the surface of the resins allowed the authors to propose a mechanism for these catalytic reactions (107).

The anticoagulant activities of these materials are strongly dependent upon the content and the nature of the substituting groups which were shown to have either a cooperative or an additive effect. Given these observations, it was possible to correlate anticoagulant activities and compositions of the heparin-like resins as shown in Table 5 anf Fig. 3.

Based on these results, it was suggested that the isolated functional groups borne by the polysaccharide chain of heparin, rather than the secondary or tertiary structure of this mucopolysaccharide, are responsible for its interaction with AT III or protease. Moreover, the comparison and discussion of the estimates of activity parameters substantiate the hypothesis that carboxylic functions are essential for heparin-like activity.

Ex vivo animal experiments were performed on surface-treated small-diameter tubings made of polystyrene-polyethylene copolymers. They showed that no significant platelet adhesion and aggregation could be observed on the wall of the tubings, when pretreated with either plasma or AT III.

Recently, heparin-like soluble and biodegradable polysaccharides have been developed (110-112). The antithrombic heparin-like activity of these materials is strongly dependent upon the percentage of the monomer units bearing carboxyliuc and sulfonate groups. This activit may be the result of a cooperative effect between the functional groups. A comparison of the effect of some resulting products and heparin has been made by different clotting assays.

Table 5. Anticoagulant activity coefficients of some substituents linked to polystyrene resins

Substituent	Activity coefficient (meq^{-1}) for 1 NIH thrombin unit
$-SO_3^-$	50 – 70
$-SO_2$ NH butyl	0
$-SO_2$ 11-amino undecanoic acid	40 – 80
$-SO_2$ alanine	60 – 80
$-SO_2$ glycine	100 – 120
$-SO_2$ hydroxyproline	120 – 150
$-SO_2$ proline	120 – 150
$-SO_2$ methionine	120 – 150
$-SO_2$ threonine	120 – 150
$-SO_2$ N-benzyloxycarbonyl lysine	120 – 150
$-SO_2$ β-alanine	140 – 200
$-SO_2$ ε-amino caproic acid	300 – 350
$-SO_2$ glutamic acid	350 – 400
$-SO_2$ δ-amino valeric acid	400 – 450
$-SO_2$ γ-amino butyric acid	500 – 600
$-SO_2$ aspartic acid	500 – 600

The results prove that these materials exert their effects on the coagulation cascade by the same mechanism than heparin but are less effective (about 10 times) (110). In reverse, the inhibition of the alternative pathway of complement by these materials is very high and similar to the anticomplementary activity of heparin (112). Therefore such biomaterials can be used as heparin substituts of plasma expanders.

IV. CONCLUSION

The biomaterials used until now are only suitable for the making of some cardiovascular devices, for istance, aorta or cardiac valves. Their use implies that a permanent adjuvant anticoagulant therapy is delivered to the host patient.

Improvements to the blood compatibility of the devices can be achieved if the mechanical compliance of the biomaterial used for the making of the device is comparable to that of the natural, normal healthy vessel. This can also be a result of surface treatments performed in order to endow the materials surfaces with antithrombogenic properties. Discussions of the published results suggest that fundamental studies of the competitive protein adsorption that occurs

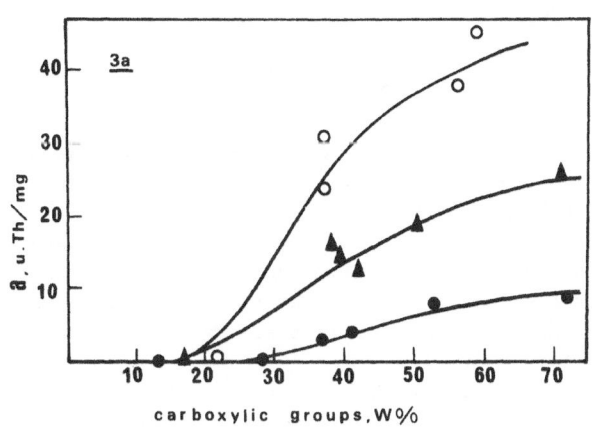

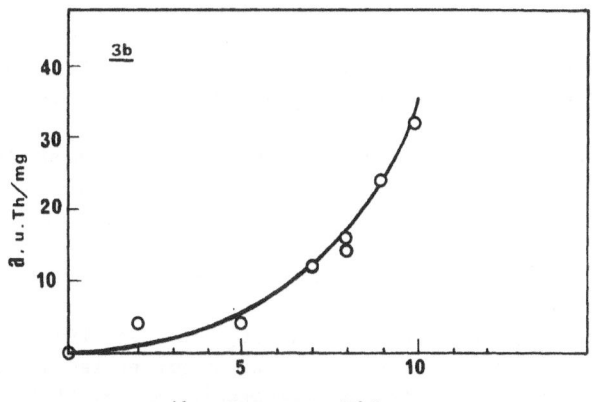

ANTITHROMBIC ACTIVITY OF THE SUBSTITUTED SEPHADEX.
3a : Influence of the units bearing carboxylic acid group($W\%$) :
  O $Y\% = 10 \pm 1$; ▲ $Y\% = 7 \pm 1$; ● $Y\% = 4 \pm 1$.
3b : Influence of the units bearing benzylsulfonate group($W\% = 40 \pm 2$).

Figure 3. Antithrombic activity of the substituted sephadex.

at the blood materials interface are necessary in order to make clear the materials structure, coagulation factors, and adsorption relationships responsible for the observed overall thrombogenicities of biomaterials. Moreover they are also necessary to analyze the antithrombogenic properties of heparin coated, heparin coupled and heparin-like materials which are to day available.

REFERENCES

1. J.L. Brash and D.J. Lyman, Adsorption of plasma proteins in solution to uncharged, hydrophobic polymer surfaces, J. Biomed. Mater. Res., 3:175 (1969).
2. D.J. Lyman, J.L. Brash, S.W. Chaikin, K.G. Klein and M. Carini, The effect of chemical structure and surface properties of synthetic polymers on the coagulation of blood. II. Protein and platelet interaction with polymer surfaces, Trans. Am. Soc. Artif. Intern. Organs, 14:250 (1968).
3. G.A. Boffa, N. Lucien, A. Faure, M.C. Boffa, J. Jozefonvicz, A. Szubarga, M.J. Larrieu and P. Mandon, Polytetrafluoroethylene-N-vinyl-pyrolidone graft copolymers: affinity with plasma proteins, J. Biomed. Mater. Res., 11:317 (1977).
4. B.B. Weksler, A.J. Marcus and E.A. Jaffe, Synthesis of prostaglandin I (prostacyclin) by cultured human and bovine endothelial cells, Proc. Natl. Acad. Sci. U.S.A., 74:3922 (1977).
5. C.T. Esmon, N.L. Esmon, J. Saugstab, and W.G. Owen, Activation of protein C by a complex between thrombin and endothelial cell surface protein, "Pathobiology and the endothelial cell", N.L. Wossel and H.J. Vogel, eds., 121 (1982).
6. W. Kisiel, W.N. Canfield, L.H. Ericsson, and E.W. Davie, Anticoagulant properties of bovine plasma protein C following activation by thrombin, Biochemistry, 16:5824 (1977).
7. G. Vehar and E. Davie, Preparation and properties of bovine factor VIII (antihemophilic factor), Biochemistry, 19:401 (1980).
8. E.J.P. Bromer, P. Brakman, F. Haverkate, C. Kluft, D. Traas, and G. Wijngaards, Progress in fibrinolysis, "Recent advances in blood coagulation", Vol. 3, L. Poller, ed., Churchill Livingston, 125 (1981).
9. J. Choay, J.C. Lormeau, M. Petitou, P. Siany, and J. Fareed, Structural studies on a biologically active hexasaccharide obtained from heparin, Ann. N.Y. Acad. Sci., 370:644 (1981).
10. G.M. Oosta, W.T. Gardner, D.L. Beeler, and R.D. Rosenberg, Multiple functional domains of the heparin molecule, Proc. Natl. Acad. Sci. U.S.A., 78(2):829 (1981).
11. L. Thunberg, G. Backstrom, and U. Lindahl, Further characterization of the antithrombin-binding sequence in heparin, Carbohyd. Res., 100:393 (1982).
12. R. Machovich, E. Regoeczi, and M.W.C. Hatton, The influence of heparin, NaCl and CaCl on the rate of the thrombin-antithrombin III reaction, Thromb. Res., 17:383 (1980).

13. J. Choay, J.C. Lormeau, M. Petitou, J. Fareed, and P. Sinay, Oligo saccharides de faible pods moleculaire présentant une activité inhibitrice du facteur Xa en milieu plasmatique, Ann. Pharmacent. Fran., 39 N.3:267 (1981).

14. J. Choay, M. Petitou, J.C. Lormeau, P. Sinay, B. Casu, G. Gatti, Structure-activity relationship in heparin: a synthetic pentasaccharide with high affinity for antithrombin III and eliciting high anti-factor Xa activity, Biochem. Biophys. Res. Comm., 116:492 (1983).

15. D.M. Tollefsen and M.K. Blank, Detection of a new heparin-dependent inhibitor of thrombin in human plasma, J. Clin. Invest., Am. Soc. Clin. Inv. Inc., 68:589 (1981).

16. D.M. Tollefsen, D.W. Majerus and M.K. Blank, Heparin cofactor II-Purification and properties of heparin-dependent inhibitor of thrombin in human plasma, J. Biol. Chem., 257:2162 (1982).

17. V.L. Gott, J.D. Whiffen, and R.C. Dutton, Heparin bonding on colloidal graphite surfaces, Sciences, 142:1297 (1963).

18. J.C. Mac Rea and S.W. Kim, Characterization of controlled release of prostaglandin from polymer matrices for thrombus prevention, Trans. Am. Soc. Artif. Inter. Organs, 24:746 (1978).

19. G.A. Grode, J. Pitman, J.P. Crowley, R.I. Leininger, and R.D. Falb, Surface-immobilized prostaglandin as a platelet protective agent, Trans. Am. Soc. Artif Intern. Organs, 20:38 (1974).

20. W. Marconi, New thrombogenic polymer compositions, Makromol. Chem. Suppl., 5:15 (1981).

21. Y. Imai and E. Masuhara, Preparation of nonthrombogenic plastic materials, Rep. Inst. Med. Dent. Engi., 3:72 (1969).

22. S. Uy and K. Kammermeyer, Nonthrombogenic surface preparation for silicone rubber, J. Biomed. Mater. Res., 3:587 (1969).

23. G.A. Grode, S.J. Anderson, H.M. Grotta, and R.D. Falb, Biocompatible materials for use in the vascular system, Trans. Am. Soc. Artif. Intern. Organs, 15:1 (1969).

24. R.I. Leininger, C.W. Cooper, R.D. Falb, and G.A. Grode, Preparation of nonthrombogenic plastic surfaces, Trans. Amer. Soc. Artif. Intern. Organs, 12:151 (1966).

25. R.L. Merker, L.J. Elyash, S.H. Mayhew, and J.Y.C. Wang, in Artificial Heart Program Conference Proceedings, R.J. Hegyeli, ed., National Heart Institute Artificial Heart Program, Washington D.C., 29 (1969).

26. R.A. Britton, E.W. Merrill, E.R. Gilliand, E.W. Salzman, W.G. Austen, and D.S. Kemp, Antithrombogenic cellulose film, J. Biomed. Mater. Res., 2(4):429 (1968).

27. S.P.S. Yen and P. Rembaum, Complexes of heparin with elastomeric positive polyelectrolytes, J. Biomed. Mater. Res. Symp., 1:83 (1971).

28. C. Pusineri, Polymers for use in medical apparatus, French Patent, 2, 264, 939 (1978).

29. Y. Idezuki, H. Watanabe, M. Hagiwara, K. Kasanugi, Y. Mori, S. Nagaoka, M. Hagio, K. Yamamoto, and H. Tanzawa, Mechanism of antithrombogenicity of a new heparinized hydrophilic polymer. Chronic in vivo studies and clinical application, Trans. Am.Soc. Artif. Intern. Organs, 21:436 (1975).

30. J.M. Courtney, G.B. Park, I.A. Fairwether, and R.M. Linday, Polymer structure and blood compatibility evaluation-application of an acrylonitrile copolymer, Biomater. Med. Devices. Artif. Organs, 4:263 (1976).

31. P. Ferruti and L. Provenzale, A new nonthrombogenic polymeric material, Transplant. Proc., 8(1):103 (1976).

32. C.S. Paik Sung, J. Bush, D.B. Mac Kie, E.W. Merrill, Copolymers containing aminohexyl residues in side chains for attaching heparin, J. Appl. Polym. Sci., 20:2603 (1976).

33. F. Tempesti, G. Casini, R. Barbucci, P. Ferruti, and L. Sprovier, Developments of surface grafted materials by heparin in new complexing polymers, presented at 4th Congress of Intern. Soc. Artif. Organs, Kyoto, Japan, 37 (1983).

34. B.J. Dudley and J.L. Williams, Nonthrombogenic articles, U.S. Patent, 4, 116, 898 (1978).

35. R.I. Leininger, Polymeric materials that don't clot blood, Chem. Technol., 5:172 (1975).

36. R.I. Leininger, J.P. Crowley, R.D. Falb, and G.A. Grode, Three years experience in vivo and in vitro with surfaces and devices treated by the heparin complex method, Trans. Am. Soc. Artif. Intern. Organs, 18:312 (1972).

37. Ch. Baquey, B. Masson, B. Basse-Cathalinat, G. Janvier, and J.M. Seris, Etude directe par traceurs radioactifs de l'interaction du sang circulant avec les protèses vasculaires, Chirurgie, 191(107):418 (1981).

38. R. Eloy, C. Pusineri, J. Baguet, J. Paul, and S. Serafini, Bulk heparinized catheters do not generate fibrinopeptide A: an ex-vivo test in dogs, presented at 4th Congress of Intern. Soc. Artif. Organs, Kyoto, Japan, 37 (1983).

39. J.C. Mc Rea, J. Lin and S.W. Kim, Heparin relasing polymers: variations in biological and release activity, presented at 4th Congress of Intern. Soc. Artif. Organs, Kyoto, Japan, 37 (1983).

40. C. Pusineri, R. Eloy, J. Baguet, J. Paul, J. Belleville, P. Leconte, and J.P. Farges, In vivo evaluation of heparinized catheters in dogs, Artif. Organs, 5(suppl.):512 (1981).

41. D.T. Mangano, Heparin bonding and long-term protection against thrombogenesis, The New England J. of Medec., 894 (1982).

42. C. Pusineri, R. Eloy, J. Baguet, J. Paul, and S. Serafini, In vivo evaluation of heparinized catheters in dogs, presented at Europ. Symp. Artif. Organs, Bruxelles, Belgium (1982).

43. H.R. Lagergreen and J.C. Eriksson, Plastics with a stable surface monolayer of crosslinked heparin: preparation and evaluation, Trans. Am. Soc. Artif. Intern. Organs,, 17:10 (1971).

44. P. Olsson, H. Lagergreen, R. Larsson, and K. Radergran, Prevention of platelet adhesion and aggregation by a glutardialdehyde-stabilized heparin surface, Thrombos. Haemostas, 37:274 (1977).

45. I.O. Salyer, A.J. Blardinelli, G.L. Ball, W.E. Weesner, V.L. Gott, M.D. Ramos, and A. Furuse, New blood compatible polymers for artificial heart applications, J. Biomed. Mater. Res. Symp., 1:105 (1971).

46. A.S. Chawla and T.M.S. Chang, Nonthrombogenic surface by radiation grafting of heparin: preparation, in vitro and in vivo studies, Biomat. Med. Dev. Art. Org., 2:157 (1974).

47. A. Rembaum, S. Singer, and H. Keyzer, Ionene polymers III dicationic crosslinking agents, Polym. Letters, 7:395 (1969).

48. D.E. Gregonis, G.A. Russel, J.D. Andrade, and A.C. de Visser, Preparation and properties of stereoregular poly(hydroxyethyl/methacrylate) polymers and hydrogels, Polymer, 19(11):1279 (1978).

49. B.D. Ratner, T. Horbett, A.S. Hoffman, and S.D. Hauschka, Cell adhesion to polymeric materials. Implications with respect to biocompatibility, J. Biomed. Mater. Res., 9(5):407 (1975).

50. I.A. Donetski, Z.M. Belomestnaya, A.K. Chepurov, and V.I. Shumarov, Use of hydrophilic gels as thromboresistant coatings, Aktual. Probl. Transplantol. Iskusstv. Organov, 246 (1975).

51. J.E. Wilson, Synthesis of thromboresistant heparinized polysilicone using radiation grafting, J. Macromol. Sci. Chem., A 16(4):769 (1981).

52. E.W. Merril, E.W. Salzman, P.S.L. Wong, T.P. Ashford, A.H. Brown, and W.G. Austen, Polyvinyl alcohol-heparin hydrogel 'G', J. Applied Physiol., 29(5):723 (1970).

53. P.H. Iverius, Coupling of glysosaminoglycans to agarose beads, Biochem. J., 124:677 (1971).

54. G. Schmer, The biological activity of covalently immobilized heparin, Trans. Am. Soc. Artif. Intern. Organs, 18:321 (1972).

55. I. Danishefsky and F. Tzeng, Preparation of heparin-linked agarose and its interaction with plasma, Thromb. Res., 4:237 (1974).

56. B. Nordenman and I. Bjork, Purification of thrombin by affinity chromatography on immobilized heparin, Thromb. Res., 11:799 (1977).

57. A.S. Carlstrom, K. Lieden, and I. Bjork, Decreased binding of heparin to antithrombin following the interaction between antithrombin and thrombin, Thromb. Res., 11:785 (1977).

58. S.W. Kim, C.D. Ebert, and J.C. Mac Rea, Pharmacological modification of polymers for the enhancement of blood compatibility, presented at Third Annual Meeting of the Internat. Soc. for Artif. Organs, Paris, 70 (1981).

59. C.D. Ebert and S.W. kim, Immobilized heparin: spacer arm effects on biological interactions, Thromb. Res., 26:43 (1982).

60. A.S. Hoffman, G. Schmer, C. Harris, and W.G. Kraft, Covalent binding of biomolecules to radiation-grafted hydrogels on inert polymer surfaces, Trans. Am. Soc. Artif. Intern. Organs, 18:10 (1972).

61. A.S. Hoffman and G. Schmer, New approaches to nonthrombogenic materials, "Current topics in coagulation", G. Schmenr, ed., Academic Press, N.Y., 201 (1973).

62. A.S. Hoffman and G. Schmer, New biocompatible and biofunctional materials via radiation-chemical synthesis, Paroi Arterielle -Arterial Wall, 1:95 (1973).

63. A.S. Hoffman, Use of radiation technology in preparing materials for bioengineering and medical sciences, "Industrial application of

radioisotopes and radiation technology", Int. Nat. Atom. Energ. Agency, Vienna, 279 (1982).

64. R.D. Falb, R.I. Leininger, and J.P. Crowley, Materials with chemically active substituents, Am. N.Y. Acad. Sci., 283:396 (1977).

65. N.A. Peppas and E.W. Merrill, Development of semicrystalline PVA hydrogels for biomedical applications, J. Biomed. Mater. Res., 11:423 (1977).

66. M.V. Sefton, F.I. Wan, G. Rollason, M.W.C. Hatton, and W. Zingg, The thromboresistance of a heparin-polyvinyl alcohol hydrogel, Chem. Eng. Comm, in press

67. C.H. Cholakis and M.V. Sefton, Chemical characterization of an immobilized heparin:heparin PVA, Polym. Prepr., 24(1):347 (1979).

68. M.F.A. Goosen and M.V. Sefton, Heparinized styrene-butadiene-styrene elastomers, J. Biomed. Mater. Res;, 13:347 (1979).

69. T. Nakashima and K. Takakura, Antiblood clotting medical polymers, Japanese, Kokai, 75 03 494 (1975).

70. F. Tempesti, G. Casini, R. Barbucci, P. Ferruti, and L. Sprovier, Developments of surface grafted materials by heparin in new complexing polymers, presented at 4th Congress of Intern. Soc. Artif. Organs, Kyoto, Japan, Conf. Proc., 37 (1983).

71. H. Hasenfratz and G. Knaup, Improvement of the blood compatibility of cellulosic membranes through the immobilization of heparin and measurement of biological heparin activity, Artif. Organs, 5(Suppl.):507 (1981).

72. M.V. Sefton and M.F.A. Goosen, Irreversible immobilization of heparin for biomaterials, "Chemistry and biology of heparin", R.L. Lundblad, W.V. Brown, K.G. Mann, and H.R. Roberts eds., Elsevier North Holland, Inc., N.Y., 463 (1981).

73. R. Larsson, P. Olsson, and U. Lindahl, Inhibition of thrombin on surfaces coated with immobilized heparin and heparin-like polysaccharides: a crucial non-thrombogenic principle, Thromb. Res., 19:43 (1980).

74. D. Labarre, M.C. Boffa, and M. Jozefowicz, Preparation and properties of heparin-poly(methylmethacrylate) copolymers, J. Polym. Sci. Symp., 47:131 (1974).

75. M. Jozefowicz, J. Jozefonvicz, C. Fougnot, and D. Labarre, New heparin-like insoluble materials, "Chemistry and biology of heparin", R.L. Lundblad, W.V. Bown, K.G. Mann, and H.R. Roberts, Elsevier North Holland, Inc., N.Y., 475 (1981).

76. Ch. Baquey, P. Blanquet, and D. Ducassou, Reactivity of paramagnetic species created in heparin under rays action, Ann. Phys. Biol. Med., 9(2):131 (1975).

77. Ch. Baquey,, A. Beziade, D. Ducassou, and P. Blanquet, Intéret du greffage radio-chimique de monomère vinyliques pour améliorer l'hémocompatibilité de matériaux artificiels, Innov. Tech. Biol. Med., 2(4):379 (1981).

78. D. Labarre, M. Jozefowicz, and M.C. Boffa, Properties of heparin-poly(methylmethacrylate) copolymers-II, J. Biomed. Mater. Res., 11:283 (1977).

79. M.C. Boffa, D. Labarre, M. Jozefowicz, and G.A. Boffa, Interactions between human plasma proteins and heparin-poly(methylmethacrylate) copolymer, Thrombos. Haemostas, 41 (2):346 (1979).

80. L. Mester, A. Amit-Amaya, and M. Mester, Heparin derivatives of high molecular weight carbohydrate sulfates, Am. Chem. Soc. Symp. Ser., 77:113 (1978).

81. N.A. Plate, L.I. Valuev, F.K. Gumirova, Covalent immobilization of modified heparin on synthetic polymers, Dokl. Akad. Nauk. SSSR, 244(6):1505 (1979).

82. N.A. Plate, L.I. Valuev, and F.K. Gumirova, Covalant binding of modified heparin with synthetic polymer matrix, presented at 26th Int. Symp. on Macromolecules, Mainz, FRG, Conf, 3:1522 (1979).

83. E. Salzman, M. Silane, J. Lindon, K. Brier-Russel, A. Dincer, R. Rosenbert, D. Labarre, and E. Merrill, Thromboresistance of heparin coated surfaces, "Chemistry and biology of heparin", W.V. Brown, K.G. Mann, H.R. Roberts, and R. Lundblad, eds., Elsevier North Holland, Inc., N.Y., 435 (1980).

84. J. lindon, R. Rosenbert, E. Merrill, and E. Salzman, Interaction of human platelets aith heparinized agarose gels, J. Lab. Clin. Med., 91(1):47 (1978).

85. J.E. Lovelock and J.S. Porterfield, Blood coagulation: its prolongation in vessels with negatively charged surfaces, Nature, 167:39 (1951).

86. H.P. Gregor, Anticoagulant activity of sulfonate polymers and copolymers, Polym. Sci. Technol. U.S.A, 7:51 (1975).

87. R. Machovich and I. Horvath, Heparin-like effect of polymethacrylic acid on the reaction between thrombin and antithrombin III, Thromb. Res., 11:765 (1977).

88. L. van der Does, J. Hofman, T.E.C. van Utteren, Reaction of N-chlorosulfonylisocyanate with unsaturated polymers- route to a synthesis of polyampholytes, Polym. Lett., 11:169 (1973).

89. T. Beugeling, L. van der Does, B.V. Rejda, and A. Bantjes, Antithrombogenic polymers synthetized from polyisoprenes, "Biocompatibility of implant materials", D. Williams, ed., Sector, London, 187 (1976).

90. T. Beugeling, The interaction of polymer surfaces with blood, J. Polym. Sci. Polym. Symp., 66:419 (1979).

91. T. Beugeling, L. van der Does, A. Bantjes, and W.L. Sederel, Antithrombin activity of a polyelectrolyte synthetized from cis-1,4-polyisoprene, J. Biomed. Mater. Res., 8:375 (1974).

92. J.S. Byck, G.T. Kwiatkowski, L.S. Gonsior, W.S. Creasy, D.D. Stewart, and E.O. Lundell, Factors affecting the blood compatibility of polyelectrolytes, Soc. Plast. Eng. Techn. Pap., 21:563 (1975)

93. M. Costello, B. Stanczewski, P. Vriesman, T. Lucas, S. Srinivasan, and P.N. Sawyer, Correlations between electrochemical and antithrombogenic characteristics of polyelectrolyte materials, Trans. Am. Soc. Artif. Intenr. Organs, 16:1 (1970).

94. D. Marshall, R. Cross, and H. Bixler, An evaluation of polyelectrolyte complexes as biomedical materials, J. Biomed. Mater. Res., 4:357 (1970).

95. R. Larsson, J.C. Eriksson, and P. Olson, The effects of the surface localized sulfate groups on the clotting time in plasma, Thromb. Res., 14(6):941 (1979).

96. L. Nelsen, R.A. Cross, M.A. Vogel, V.L. Gott, and A.M. Fadali, Synthetic thromboresistant surfaces from sulfonated polyelectrolyte complexes, Surgery, 67(5):826 (1970).

97. F. Leonard, Rationale for the preparation of non-thrombogenic materials, Trans. Am. Soc. artif. Intern. Organs, 15:15 (1969).

98. H. Fukuda and Y. Kikuchi, In vitro clot formation on the polyelectrolyte complexes of sodium dextran sulfate with chitosan, J. Biomed. Mater. Res., 12:531 (1978).

99. H. Fukuda, Polyelectrolyte complexes of chitosan with sodium carboxymethylcellulose, Bull. Chem. Soc. Jpn., 53:837 (1980).

100. Y. Kikuchi and Y. Onishi, Polyelectrolyte complex consisting of carboxymethyldextran, sodium dextran sulfate and chitosan, Nippon Kagaku Kaishi, 1:127 (1979).

101. C. Fougnot, J. Jozenfovicz, M. Samama, and L. Bara, New heparin-like insoluble materials I, Ann. Biomed. Eng., 7:429 (1979).

102. C. Fougnot, M. Jozefowicz, M. Samama, and L. Bara, New heparin-like insoluble materials II, Ann. Biomed. Eng., 7:441 (1979).

103. M. Mauzac, C. Fougnot, and J. Jozefonvicz, New antithrombogenic polysaccharides, Artif. Organs, 5(suppl.):504 (1981).

104. M. Mauzac, N. Aubert, and J. Jozefonvicz, Antithrombic activity of some polysaccharide resins, Biomaterials, 3:221 (1982).

105. C. Fougnot, M. Jozefowicz, L. Bara, and M. Samama, Interactions of anticoagulant insoluble modified polystyrene resins with plasmatic proteins, Thromb. Res., 28:37 (1982).

106. C. Fougnot, M.P. Dupillier, and M. Jozefowicz, Anticoagulant activity of amino acid modified polystyrene resins: influence of the carboxylic acid function, Biomaterials, 4:101 (1983).

107. C. Fougnot, M. Jozefowicz, and R.D. Rosenberg, Affinity of purified thrombin or antithrombin III for two insoluble anticoagulant polystyrene derivatives:I. In vitro adsorption studies., Biomaterials, 4:294 (1983).

108. C. Fougnot, M. Jozefowicz, and R.D. Rosenberg, Catalysis of the generation of thrombin-antithrombin complex by insoluble anticoagulant polystyrene derivatives, Biomaterials, 5:94 (1984).

109. C. Fougnot, J. Lindon, L. Kirshner, E.W. Salzman, and M. Jozefowicz, Interactions of human platelets with insoluble anticoagulant modified polystyrene resins, Biomaterials, 5:169 (1984).

110. M. Mauzac and J. Jozefonvicz, Anticoagulant activity of dextran derivatives. Part I: Synthesis and characterization, Biomaterials, 5:301 (1984).

111. M. Mauzac, J. Jozefonvicz, and M. Jozefowicz, Nouveaux dérivés du dextran a activités anticoagulante et antiinflammatoire, leur procédé de préparation et utilisation de ces dérivés en tant qu'anticoagulants et en tant que substituts du plasm sanguin, French Patent, N. 83 19110 (1983).

112. M. Mauzac, F. Maillet, J. Jozefonvicz and M.D. Kazatchkine, Anticomplementary activity of dextran derivatives, Biomaterials, in press.

HEPARIN IMMOBILIZATION ON POLYURETHANE SURFACE; GRAFTING OF HEPARIN COMPLEXING POLY(AMIDO-AMINE)S

Rolando Barbucci, Manuela Benvenuti, Giuliana Casini, Paolo Ferruti, and Federica Tempesti

C.R.I.S.M.A. Università di Siena, Le Scotte

53100 Siena, Italy

Poly(amido-amine)s (PAA) are a relatively new family of tertiary amino polymers in which amido - and tertiary amino groups are regularly arranged along the macromolecular chain. They may be obtained by polyaddition of primary or secondary amines to bis acrylamides (1-4):

a) $x CH_2=CHCON-R^2-N-CO-CH=CH_2 + xH_2NR^4$
       $\quad\quad\quad | R^1 \quad | R^3$

$$\left[CH_2-CH_2-CO-N-R^2-N-CO-CH_2CH_2-N- \right]_x$$
$$\quad\quad\quad\quad | R^1 \quad | R^3 \quad\quad\quad\quad\quad | R^4$$

b) $x CH_2=CH-CO-N-R^2-N-CO-CH=CH_2 + xHN-R^5-NH$
       $\quad\quad\quad\quad\quad | R^1 \quad | R^3 \quad\quad\quad\quad\quad | R^4 \quad | R^6$

$$\left[CH_2-CH_2-CO-N-R^2-N-CO-CH_2CH_2-N-R^5-N- \right]_x$$
$$\quad\quad\quad\quad | R^1 \quad | R^3 \quad\quad\quad\quad\quad | R^4 \quad | R^6$$

The polymerization takes place under very mild conditions, usually in water or alcohols, at room temperature, and without any added catalysts. The number average molecular weights of the polymers usually range between 5.000 and 20.000. They are influenced by the molecular proportions of the monomers and by the nature of the polymerization solvent. As a rule, higher molecular weights are obtained, with the same monomeric mixture, in water followed by water/alcohols mixtures, alcohols, pirydine/alcohols mixtures and pirydine in the order.

Poly(amido-amine)s are usually soluble in, or at least, swollen by, water, except when they carry large hydrophobic substituents. Most of them are also soluble in several organic solvents.

Poly(amido-amine)s, as a class, present some interesting properties. First of all, the above polymerization reaction is a general one. Pratically all aliphatic and cycloaliphatic primary monoamines and bis(secondary amine)s, and bis acrylamides, can be used as monomers, the only limit being that the presence of bulky substituents in the former, the other conditions being the same, results in lower polymerization rates and lower molecular weights (1,3). Therefore, a large variety of structures can be obtained, as shown in Table 1.

Furthermore a number of chemical functions, if present as side substituents, do not interfere with the polymerization process. As a consequence, poly(amido-amine)s bearing carboxyl-hydroxyl-allyl-, and additional tertiary amino groups can be easily obtained. Some functionalized poly(amido-amine)s are reported in Table 2.

Chemical Properties

Poly(amido-amine)s are basic polymers of medium strength. The protonation and complex formation of a number of poly(amido-amine)s in aqueous solution have been studied by potentiometric, calorimetric, viscosimetric, spectrophotometric, esr, ^{13}C nmr, and quantum chemical tecniques.

In PAA's, as a rule, the basicity of the aminic nitrogens of each repeating unit does not depend on the degree of protonation of the whole macromolecule. Consequently, "real" basicity constant can be determined. The number of basicity constants is in all cases equal to the number of the aminic nitrogens present in the repeating unit. On this respect, the behaviour of these polymers is very similar to that of their non macromolecular models, apart for some minor differences in basicity due to entropy effect(5,6,8,9,10).Calorimetric studies, in fact, demonstrate that the protonation enthalpies are always about the same for polymers and models, and a linear relationship can be obtained between the enthalpies and the net charges of nitrogens that undergo protonation (7)

Poly(amido-amine)s having more than one basic nitrogen in their repeating unit are also capable of forming in aqueous solution stable complexes with heavy metal ions, e. g., with the Cu^{2+} ion, with the exception of those which the aminic part has a piperazinic structure (8,10). When complexes are formed, sharp stability constants may be determined. Furthermore, solid Cu^{2+} complexes have been isolated in several instances, having a well defined composition, namely one Cu^{2+} ion per unit. Thus, in complex formation with metal ions poly(amido-amine)s also

Table 1. Some examples of poly(amido-amine)s

Syst. N°	Structure of the repeating unit[a]	Ref.[b]

1

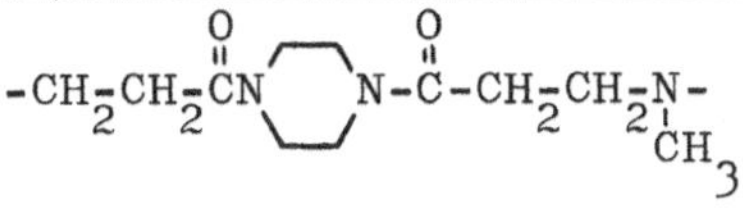

$$-CH_2-CH_2-\overset{\overset{O}{\parallel}}{C}N\bigcirc N-\overset{\overset{O}{\parallel}}{C}-CH_2-CH_2-\underset{\underset{CH_3}{\mid}}{N}-$$

1,4

2

$$(1)-\underset{\underset{(CH_2)_3\ CH_3}{\mid}}{N}-$$

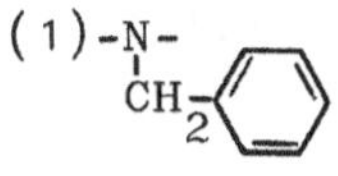

1,4

3

$$(1)-\underset{\underset{CH_2}{\mid}}{N}-\bigcirc$$

1,4

4

$$-CH_2-CH_2-\overset{\overset{O}{\parallel}}{C}-\underset{\underset{C_2H_5}{\mid}}{N}-CH_2-CH_2-\underset{\underset{C_2H_5}{\mid}}{N}-\overset{\overset{O}{\parallel}}{C}-CH_2-CH_2-\underset{\underset{CH_3}{\mid}}{N}-$$

1,4

5

$$(1)-\underset{\underset{CH_3}{\mid}}{N}(CH_2)_2\underset{\underset{CH_3}{\mid}}{N}-$$

1,3

6

$$(1)-\underset{\underset{CH_3}{\mid}}{N}(CH_2)_3\underset{\underset{CH_3}{\mid}}{N}-$$

5

7

$$(1)-\underset{\underset{CH_3}{\mid}}{N}(CH_2)_4\underset{\underset{CH_3}{\mid}}{N}-$$

5

8

$$(1)-\underset{\underset{CH_3}{\mid}}{N}(CH_2)_6\underset{\underset{CH_3}{\mid}}{N}-$$

6

9

$$(1)-\underset{\underset{CH_3}{\mid}}{N}(CH_2)_8\underset{\underset{CH_3}{\mid}}{N}-$$

6

10

$$(1)-\underset{\underset{CH(CH_3)_2}{\mid}}{N}-(\ CH_2\ )-\underset{\underset{CH(CH_3)_2}{\mid}}{N}-$$

1,3

11

$$(1)-N\bigcirc N-$$

1,2

12

$$(1)-N\overset{\overset{CH_3}{\diagup}}{\bigcirc} N-$$

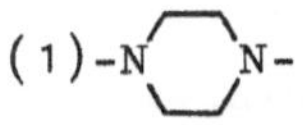

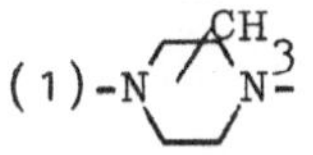

1,2

13

$$(1)-\underset{\underset{CH_3}{\mid}}{N}(CH_2)_2\underset{\underset{CH_3}{\mid}}{N}(CH_2)_2\underset{\underset{CH_3}{\mid}}{N}-$$

7,8

(continued)

Table 1. Continued

14 $-CH_2CH_2\overset{O}{\overset{\|}{C}}NH(CH_2)_2NH\overset{O}{\overset{\|}{C}}CH_2CH_2N\underset{}{\bigcirc}N-$ 1,3

15 $-CH_2CH_2\overset{O}{\overset{\|}{C}}NH(CH_2)_{12}NH\overset{O}{\overset{\|}{C}}CH_2CH_2\underset{CH_3}{N}(CH_2)_2\underset{CH_3}{N}-$ 6

16 $(15)-\underset{CH_3}{N}(CH_2)_4\underset{CH_3}{N}-$ 6

17 $(15)-\underset{CH_3}{N}(CH_2)_6\underset{CH_3}{N}-$ 6

18 $(15)-\underset{CH_3}{N}(CH_2)_8\underset{CH_3}{N}-$ 6

19 $-CH_2CH_2\overset{O}{\overset{\|}{C}}\underset{CH(CH_3)_2}{N}CH_2CH_2\underset{CH(CH_3)_2}{N}\overset{O}{\overset{\|}{C}}CH_2CH_2\underset{CH_3}{N}CH_2CH_2\underset{CH_3}{N}-$ 1,3

---

[a]When the amidic portion is the same as in an already reported PAA, it is no longer displayed; instead, the Syst. N° of that PAA is given in parentheses.

[b]Where the synthesis has been reported for the first time.

70

Table 2. Some examples of poly(amido-amine)s carrying functional groups as side substituents

Syst. N°	Structure of the repeating unit[a]	Ref.[b]
1	$-CH_2CH_2\overset{O}{\overset{\|}{C}}N\bigcirc N\overset{O}{\overset{\|}{C}}CH_2CH_2\overset{\|}{\underset{CH_2COOH}{N}}-$	1,13
2	$(1)-\underset{(CH_2)_2COOH}{N}-$	1,13
3	$(1)-\underset{(CH_2)_4COOH}{N}-$	1,13
4	$(1)-\underset{(CH_2)_5COOH}{N}-$	1,13
5	$(1)-\underset{(CH_2)_2CO(D\text{-Alanine})}{N}-$	1[c]
6	$(1)-\underset{(CH_2)_2CO(L\text{-Alanine})}{N}-$	1[d]
7	$(1)-N\bigcirc N-$ with COOH	1
8	$(1)-N\bigcirc N-$ with HOOC COOH	1
9	$(1)-\underset{CH_2CH=CH_2}{N}-$	(e)
10	$(1)-\underset{(CH_2)_2N(CH_3)_2}{N}-$	5,9

(continued)

Table 2. Continued

11 $(1)-N-$
$(CH_2)_3N(CH_3)_2$
5

12 $(1)-N-CH_2CH_2N-$
$(CH_2)_2 \quad (CH_2)_2$
$N(CH_3)_2 \quad N(CH_3)_2$
9

13 $(1)-NCH_2CH_2-N-$
$(CH_2)_2OH \quad (CH_2)_2OH$
2.1(f)

14 $(1)-N-CH_2CH_2 \quad N-$
$CH_2 \qquad CH_2$
22(f)

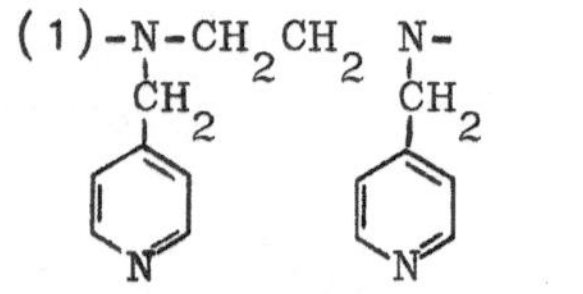

---

a,b See the corresponding footnotes at Table 1.

c $[\alpha]_D^{25} = + 10.4°$

d $[\alpha]_D^{25} = - 9.8°$

e P. Ferruti and M.A.Marchisio, unpublished.

f As crosslinked resin.

exhibit a quite unusual behaviour among polyelectrolytes.
The behaviour of poly(amido-amine)s bearing carboxyl groups
as side substituents (e.g., N⁇s 1-4 of Table 2) is more complex (11,12).
Both protonation constants of each polymer follow the modified
Henderson-Hasselbach equation in the $0.2 < \alpha < 0.8$ region.
The n values relative to $\log K_1$ corresponding to the protonation
equilibrium of the nitrogens, are always >1 and are very similar in all cases indicating that the protonation of a single
basic nitrogen becomes more and more difficult as the degree
of protonation (α) of the whole macromolecule increases.
The addition of the second proton to the polymer aminoacids
in their zwitterionic form leads to the neutralization of the
negative $-COO^-$ charges. $\log K_2$ values increase with the number
of methylene groups between the nitrogen and $-COO^-$ groups.
With the exception of N⁇ 1 of Table 2 the n values relative to
$\log K_2$ are > 1, and decrease from N⁇ 2 to N⁇ 4 (Table 2)
reaching a value very close to 1 for the last polymer. Of all
the aminoacid polymers only N⁇ 1 and N⁇ 2 (Table 2) are able
to form Cu^{2+} complexes in aqueous solution. The equilibrium
constants relative to the Cu^{2+} complex formation for each point
of the titration curve were calculated by taking into account
the dependence of logK's on pH via the Henderson-Hasselbach
equation (11 ,12).

Heparin-PAA interaction in aqueous solution

In previuos work (13) we have found that several poly(amido-amine)s are able, in a linear form, to neutralize the anticoagulant ability of heparin, much as protamine sulfate does.
This is apparently due to their ability to form complexes with
heparin in aqueous solution.
A calorimetric study on the complex formation between heparin
and poly(amido-amine)s (N⁇s 5-7 of Table 1) has been performed
in water. The measurements were rather difficult because the
complex has a tendency to precipitate unless the complexation
reaction has carried out in very dilute solution ~ 10^{-2}M (14),
and with a heparin/polymer molar ratio above about 10. Polymer
N° 5 of Table 1 gives ampler opportunities on this respect because its complex with heparin is less insoluble. The results
obtained with this polymer are reported in Fig. 1. It may be
observed that the heat effect is exothermic, and its value decreases by increasing the amount of poly(amido-amine). In other
words it follows the same trend observed during the neutralization with NaOH of same ionizing groups present in insulin.
The heat effect also increases by decreasing the ionic strength
of the medium, and also by decreasing pH. All these results
clearly indicate that electrostatic interaction takes place

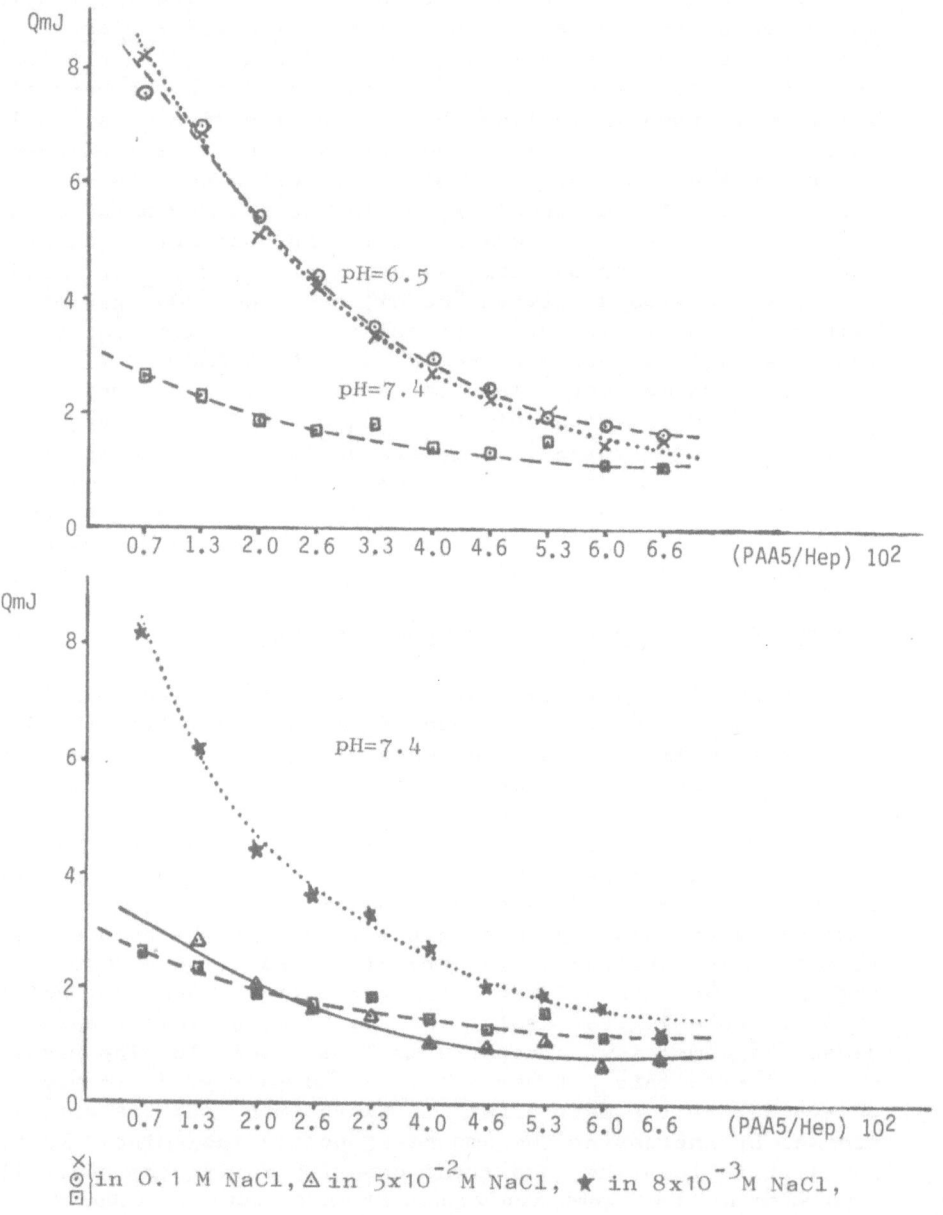

Fig. 1 Heat determined for PAA-heparin complex formation
at different PAA5/heparin ratios.

betweeen heparin and poly(amido -amine). In fact the intensity
of electrostatic interaction is lowered at high ionic strength
because of shielding effect; on the other hand lowering pH increases the charge density on poly(amido-amine), while has no
effect in the range considered on the charge density of heparin.
The results obtained with polymer N° 7 (Table 1) at the same
pH shows that the heat effect is more exothermic. This showed
be related with the higher basicity of the latter poly(amido-
amine).
A solid complex between heparin and polymer N° 5 (Table 1) could be isolated. This complex precipitated out when complexation reaction was performed at pH 6.
The elemental analysis corresponds to a 1:1 complex, its FTIR
spectrum compared with the spectra of both heparin and N° 5
(Table 1) shows the appearance of new bands indicating that the
product is not a simple mixture of the two polymers.

Surface grafting of poly(amido-amine)s on polyurethane

It is well known that most non-physiological materials
induce thrombus formation when put in contact with blood.
Surface heparinization of biomedical articles is presently recognized as a method of rendering them nonthrombogenic (15,16).
Surface-grafting of heparin-complexing poly(amido-amine)s on
a given material should result in a heparinizable surface, thus
leading either to a potentially non-thrombogenic material, or
to ammaterial able to deheparinize plasma or blood, as the crosslinked resin do. In this respect, synthetic methods for grafting poly(amido-amine)s on to the surface of glass (17,18),
plasticized poly(vinyl chloride) (PVC)(19, 20), poly(dimethyl-
siloxane) (Silastic) (21)and polyethyleneterephtalate (22),
have been developed.
It may be observed that this approach has the advantage, in
principle, of maintaining the size, shape, and mechanical per-
formances of the starting articles, thus avoiding any problem
connected with the manufacture of entirely new products.
It is well known that segmented polyurethane are _per se_ endowed
with interesting properties as prosthetic materials. Therefore
as a further development of our studies on heparinizable materials, we thought it interesting to study a simple method for
surface grafting of poly(amido-amine)s on polyurethane articles
(23).
Experiments were performed with poly(amido-amine) N° 7 of Table 1 which by previous results (see above) appears to be par-
ticurarly effective as heparin complexing agent. The following
reaction sequence applies in principle to every type of polyurethane. Most of our experiments, however, were performed on
a polyurethane of the following structure synthetized by us(PUS)

$$PUS \equiv \left[\left(OCH_2CH_2 \right)_x OCNH \bigcirc CH_2 \bigcirc NHCNHCH_2CH_2NHCNH \bigcirc CH_2 \bigcirc NHC \right]_n$$

and another commercial PU Pellethane 2363-80A (PUC).

The first step of the grafting reaction was the addition of diisocyanates to PU with the formation of an ureic bond

$$
\begin{array}{c}
\wr \\
NH \\
| \\
C=O \\
| \\
\wr \\
PU
\end{array}
\quad + \quad
O=C=N-(CH_2)_6-N=C=O
\quad \longrightarrow \quad
\begin{array}{c}
\wr \quad O \\
\quad \parallel \\
N-C-NH-(CH_2)_6-N=C=O \\
| \\
C=O \\
\wr \\
PU
\end{array}
$$

$$\text{excess}$$

The presence of the diisocyanates on the surface was determinated by FTIR spectroscopy (see below).

Contrary to what expected, the amount of isocyanate groups which can be introduced on the PU surface, the other conditions being equal, is higher when some moisture is present in the atmosphere above the reaction medium. This might be related with the change of the hexamethylene diisocyanate configuration at the polyurethane surface (see below).

A further reaction of a secondary amino and-capped poly(amido-amine) N° 7 of Table 1 with the terminal isocyanate group led to grafting of this polymer:

$$
\begin{array}{c}
\wr \quad O \\
\quad \parallel \\
N-CNH(CH_2)_6NCO \\
| \\
C=O \\
\wr \\
PU
\end{array}
\; + \;
HN(CH_2)_4N\!-\!\!\left[\!CH_2CH_2\overset{O}{\overset{\parallel}{C}}N\bigcirc N\overset{O}{\overset{\parallel}{C}}CH_2CH_2N(CH_2)_4N\!-\!\right]_n
$$
$$
\begin{array}{cc}
CH_3 & CH_3 \qquad\qquad\qquad\qquad\qquad\qquad CH_3 \quad CH_3
\end{array}
$$

$$\Big\downarrow$$

$$
\begin{array}{c}
\wr \quad O \\
\quad \parallel \\
N-CNH(CH_2)_6NHCN(CH_2)_4N\!-\!\!\left[\!CH_2CH_2\overset{O}{\overset{\parallel}{C}}N\bigcirc N\overset{O}{\overset{\parallel}{C}}CH_2CH_2N(CH_2)_4N\!-\!\right]_n \\
| \qquad\qquad\quad | \quad\quad | \\
C=O \qquad\quad CH_3 \quad CH_3 \qquad\qquad\qquad\qquad\qquad CH_3 \quad CH_3 \\
\wr \\
PU
\end{array}
$$

Also the presence of N° 7 (Table 1) chains on the PU surface was revealed by the FTIR method, while the amount was evaluated by a titrimetric method.

It may be observed that this treatment has the advantage of maintaining the size, shape and mechanical performance of the starting articles.

Mechanical properties was determined by stress-strain analysis; Fig. 2 shows stress-strain curves of native PUC and treated PUC.

(Measurements carried out at Istituto di Ricerche su Tecnologia dei polymeri e Reologia, C.N.R., Arco Felice, Napoli, Italy).

The poly(amido-amine) N° 7 (Table 1) grafted surfaces were heparinized and amount of heparin adsorbed on the surfaces was evaluated by biological test (see below).

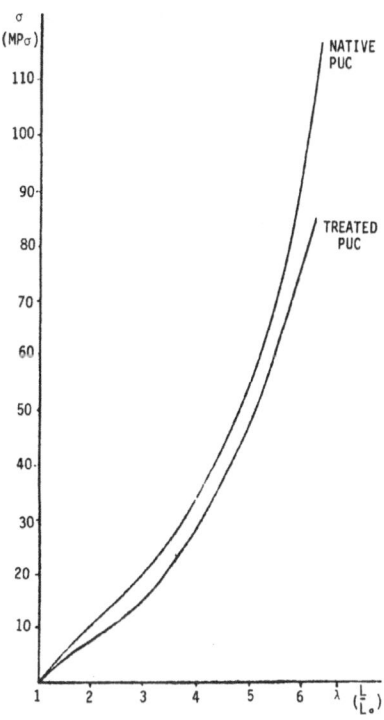

Fig. 2 Stress-strain measurements of native and treated PUC

FTIR Transmittance spectra

Fig. 3 shows the transmittance spectrum of native PUS and the spectrum of the same sample after one day of diisocyanate treatment. The presence of a new band at 2270 cm^{-1}, assignable to the asym. stretching of the NCO group is immediately evident. This band shows that grafting of the diisocyanates occurs. Other new bands, or variations in either intensity or frequency of the PUS I.R. bands are not evident.(23)
As in the previous case, the appearance of a band at 2270 cm^{-1} indicates the presence of diisocyanate groups on the surface. In addition a new band at 1628 cm^{-1} is noted.

Difference spectra

Poly(amino-amine) treated PU minus diisocyanate treated PU

The presence of poly(amido-amine) N°7 of Table 1 chains on the surface of PU was demonstrated by susbtracting the di-

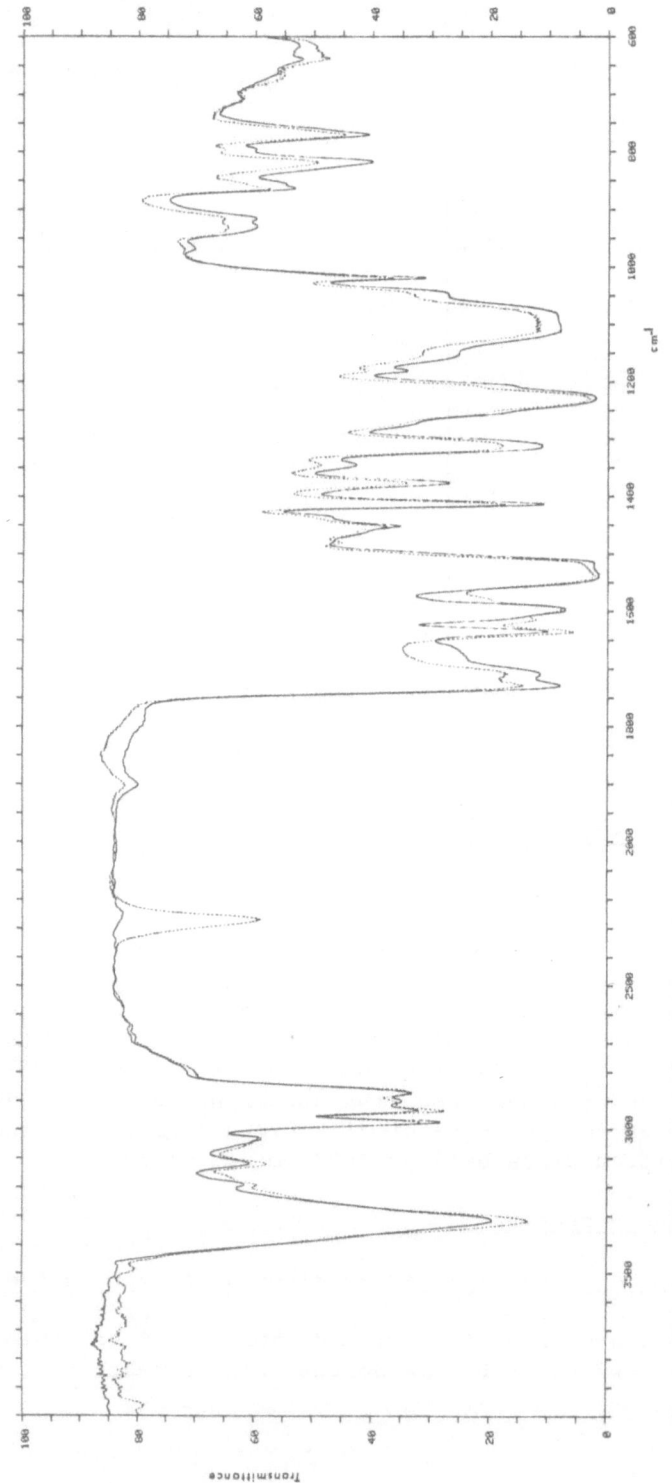

Fig. 3 FTIR spectra of native PUS (——) and diisocyanate treated PUS (....)

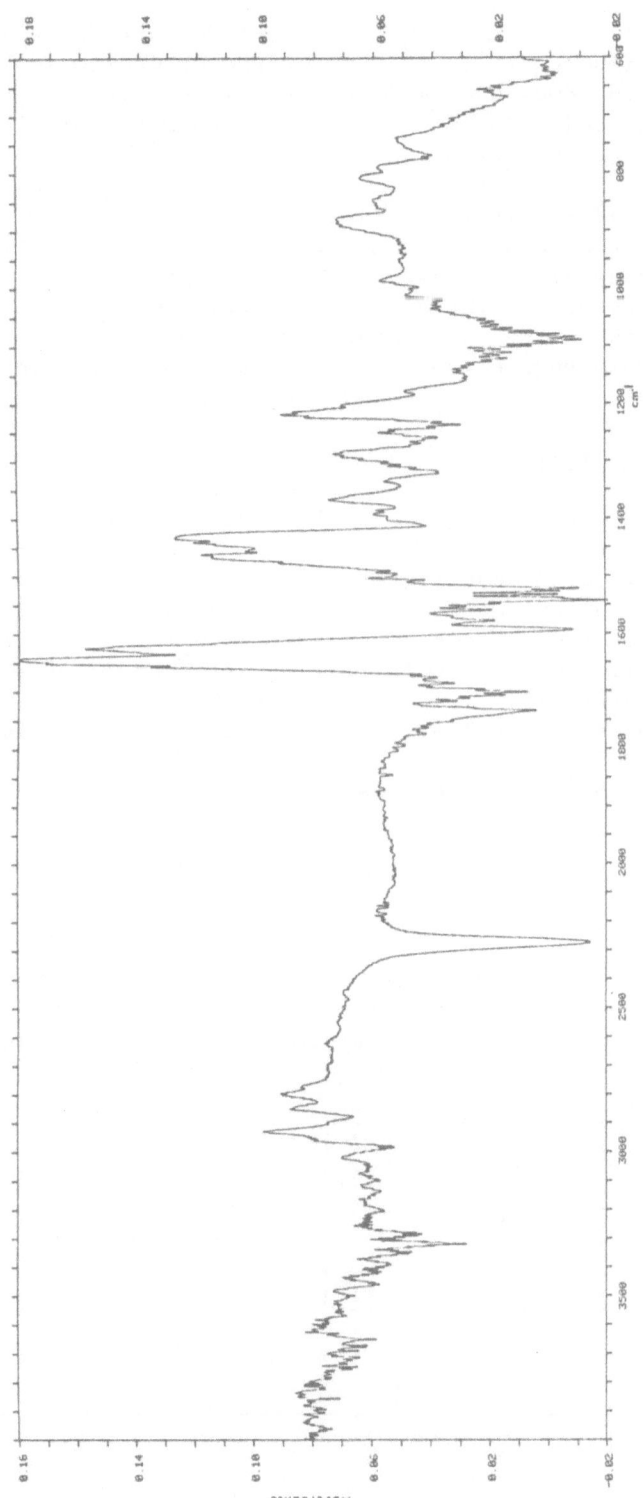

Fig 4 Difference spectra of PAA N° 7 treated PUS minus diisocyanate treated PUS

isocyanate-grafted PU spectrum from that N° 7 (Table 1) grafted PU.
The results obtained in the case of PUS are shown in Fig. 4.
The spectrum consists of three bands of medium intensity in
the region 2800-3000 cm^{-1} and four more intense bands at 1640,
1465,1435 and 1220 cm^{-1}. These are the most intense bands present in the polymer N° 7 (Table 1) spectrum. Moreover, the
general patterns of the difference spectrum and of polymer N°
7 (Table 1) spectrum are similar.
Similar results are obtained with PUC, even if only three bands
are evident at 1640, 1465, 1430 cm^{-1}. The absence of the other
bands, present in the difference spectrum of PUS, may be attributed both to the high intensity of the bands of the native
PUC in this region, and to the feeble intensity in the same
region of the N° 7 (Table 1) bands.(23)

Biological tests

      In the order to determine the stability of heparin in-
teraction with poly(amido-amine) N° 7-surface grafted samples,
and to evaluate quantitatively the amount of heparin adsorbed,
the heparinized samples were extracted with saline, and then
with aqueous 0.1M NaOH. The latter treatment had been found to
completely remove heparin from heparinized poly(amido-amine)
-based materials(20). Only the amount of heparin retained after rinsing with saline was quantitatively determined both in
PUC and PUS samples, and it was found to be much higher in the
case of PUS (about ten times) (Table 3).

Table 3. Quantity of complexed heparin on PUS and PUC samples

	Samples	UI Hep/cm^2PU	mg Hep/cm^2PU
	I	7,4	0,048
PUS	II	5,1	0,033
	III	5,3	0,034
	I	0,6	0,0042
PUC	II	1,1	0,0069
	III	1,2	0,0078

This data agrees very well with the fact that the quantity of N° 7 (Table 1) on the PUS samples is larger than that on PUC. A preliminar biological tests have been performed in parallel on films of native PU, and PAA grafted PU both as such and heparinized. The heparinized grafted PU had been exhaustively washed with phosphate buffer saline (PBS) until no heparine was present in the washing. These films have been dipped in platelet riched plasma (PRP) at 37°.

At the end of each experiment the samples were removed from plasma, washed in PBS and fixed with a 2.5% buffered glutaraldehyde.

Both faces of the samples were then examined by scanning electron microscopy (SEM).

The results show that on the PAA grafted PU considerably thick deposit of platelets and fibrinogen was present (Fig. 5a). On the contrary the heparinized film was pratically clear (Fig. 5b). By comparing the results obtained with the native PU (Fig. 6) it could be observed that grafting of PAA induced an unfavourable change. It may observed that this particular PU is known by literature data to be considerably non thrombogenic. The presence of stably adsorbed heparin i.e. heparin which can be washed out only by dilute NaOH, not only reserved this trend but gave a product with somewhat improved properties, within the limits of the test considered.

Surface characterization of Heparin complexing poly(amido-amine) chains grafted on polyurethane

The wetting properties of the surfaces of native polyurethane, and of the product of each reaction step reported in the previous chapter, as well as the corrisponding values of their surface free energies, solid-water adhesion energies (W_a), solid-water interfacial free energies (γ_{sw}) dispersive (W_a^d) and non-dispersive (I_{sw}) components of the solid-water adhesion energies, have been determined in the laboratory of Dr. Baszkin in Paris (24).

Contact angles measuraments were performed on films casted from 6% and 3% DMF solution.

The following conclusions can be drawn

1) Polyurethane films casted from 6% DMF solutions are more hydrophilic than those casted out of 3% DMF solutions.

2) Treatment of polyurethanes with hexamethylene diisocyanate produces hydrophobic surfaces. The hydrophobicity of 3% DMF casted films after this treatment is higher than of those casted from 6% DMF.

Treatment with hexamethylene diisocyanate seems to expose CH_2 groups in the uppermost position (high contact angles with water θ_A; some spots of examined samples exhibit contact angles even higher than 100°).

81

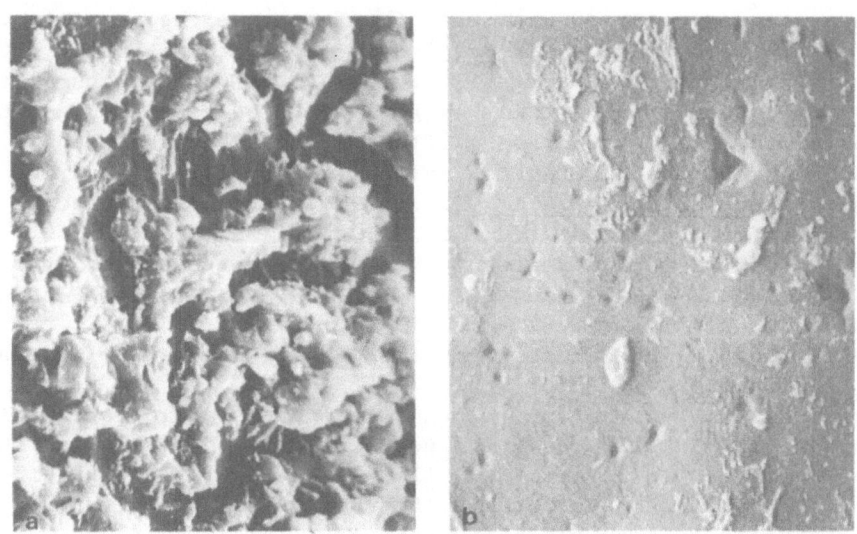

Fig. 5
SEM of the surface of N° 7 grafted PU(a) and heparinized
PU(b) after exposition to PRP.

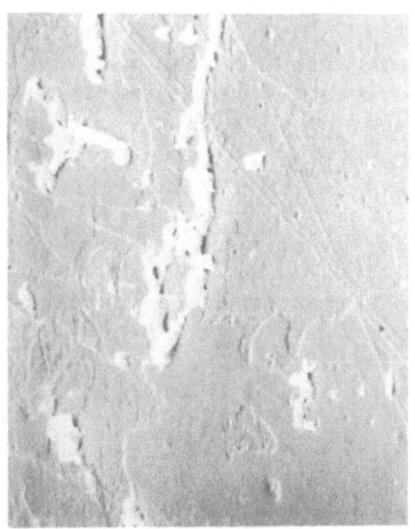

Fig. 6
SEM of the surface of native PU after exposition to PRP

When the hexamethylene diisocyanate treated surfaces are immersed in water and air bubble contact angles are measured, the receeding contact angles are low as compared to the advancing angles. Such pronounced hysteresis of advancing-receeding contact angles may be explained by the reorientation of polymer chains at the surface and the exposure of N=C=O hydrophilic groups in their uppermost position towards water. This is confirmed by the high values of contact angles measured with octane; these angles would have been much lower of the chain reorientation did not occur.

3) Treatment with poly(amido-amine) decreases the hydrophobicity of 3% and 6% DMF casted samples.

4) Heparinization of samples produces highly hydrophilic surfaces.

Furthermore, by considering the results reported in the literature (25), surface with high I_{sw}^p/W_a^d value ($\geqslant 1.0$) inhibit protein adsorption when put in contact with blood. Our heparinized-PAA grafted PU films show a I_{sw}^p/W_a^d value largely higher than 1 (≈ 2.0).

Acknowledgements

The authors wish to thank to Dr. V. Sforza of the Istituto di Anatomia e Istologia Patologica, University of Siena, Siena, Italy,for performing the electron scanning micrographs of the PU surfaces.

References

1) F. Danusso, P. Ferruti, Polymer ,11, 88 (1970) and references therefrom.

2) F. Danusso, P. Ferruti, G. Ferroni, Chim. Ind.Milan, 49; 271 (1967).

3) F. Danusso, P. Ferruti, G. Ferroni, Chim. Ind., Milan, 49, 453 (1967).

4) F.Danusso, P. Ferruti, G. Ferroni, Chim. Ind., Milan, 49, 587 (1967).

5) R. Barbucci, M. Casolaro, P. Ferruti, V. Barone, F. Lelj, L. Oliva, Macromolecules, 14, 1203 (1981).

6) R. Barbucci, P. Ferruti, C. Improta, M. Delfini, A.L. Segre, F. Conti, Polymer, 19, 1329 (1978).

7) R. Barbucci, P. Ferruti, C. Improta, M. La Torraca, L. Oliva, M.C. Tanzi, Polymer, 20, 1298 (1979).

8) R. Barbucci, P. Ferruti, Polymer, 20, 1061 (1979).

9) P. Ferruti, R. Barbucci, Advances in Polymer Science, 58, 55 (1984), and references therefrom.

10) P. Ferruti,L. Oliva, R. Barbucci, M.C. Tanzi, Inorganica Chim. Acta, 41, 25 (1980).

11) R. Barbucci, M. Casolaro, M. Nocentini, G. Reginato, P. Ferruti, Journal of Solution Chemestry, in press.

12) R. Barbucci, M. Casolaro, M. Nocentini, S. Corezzi, P.Ferruti, V. Barone, Macromolecules,in press.

13) M.A. Marchisio, P. Longo, P. Ferruti, F. Danusso, Europ. Surg. Research, 3, 240 (1971).

14) F. Tempesti, I. Wadso, Preliminary study.

15) D.J. Lyman, Rev. Macromol. Chem., 1, 355 (1966).

16) Akutsu, Artificial Heart, IgakuShoin Ltd, Tokyo,(1975).

17) P. Ferruti, I. Domini, R. Barbucci, M.C. Beni,E. Dispensa, S. Sancasciani, M.A. Marchisio, M.C. Tanzi, Biomaterials, 4, 218 (1983).

18) R. Barbucci, P. Ferruti, L. Provenzale, F. Conti, G. Delfini, G. Segre, It. Pat. Application N° 24638 A/78.

19) P.Ferruti, R. Barbucci, N. Danzo, A. Torrisi, O. Pugliesi, S. Pignataro, P. Spartano, Biomaterials,3, 33 (1982).

20) P. Ferruti, G. Casini, F. Tempesti, R. Barbucci, R. Mastacchi, M. Sarret, Biomaterials, 5, 234(1984).

21) R. Barbucci, P. Ferruti, L. Provenzale, It. Pat. Applica - tion N° 23610 A/80.

22) R. Barbucci, M. Benvenuti, G. Casini, P.Ferruti, F. Tempesti, Biomaterials, 6, 102 (1985).

23) R. Barbucci, M. Benvenuti, G. Casini, P. Ferruti, M. Nocentini, Makromol. Chem., in press.

24) A. Baszkin, M. Da Costa, R. Barbucci et all. in preparation.

25) A. Baszkin, D.J. Lyman, J.Biomed. Mat. Res.,14, 393 (1980).

THE DESIGN AND SYNTHESIS OF BIOABSORBABLE POLY(ESTER-AMIDES)

T. H. Barrows, J. D. Johnson, S. J. Gibson, and
D. M. Grussing

3M, 3M Center, St. Paul, Minnesota 55144-1000

INTRODUCTION

The use of synthetic fibers as surgical suture is a well recognized improvement over the use of natural materials for a number of reasons.[1] Although a wide variety of synthetic nonabsorbable sutures are commercially available, the choice of synthetic absorbable sutures is rather limited. Polyglycolic acid sutures in braided form have been available for 15 years whereas the first monofilament absorbable polyester, poly(p-dioxanone), has only recently been developed.[2]

In most applications, surgical sutures are not permanently required. Thus absorbability is an attractive feature provided that the fiber handling qualities are good and the rate of strength retention is appropriate for the healing rate of the tissue that is being repaired.

Other uses for synthetic bioabsorbable polymers include suture substitutes such as ligating clips[3], staples[4], vessel[5] and nerve[6] anastomosis devices, and a variety of drug-releasing implants[7].

In designing a new class of bioabsorbable polymers, we sought to combine the excellent fiber, film, and molding properties of nylon with the degradability of polyglycolic acid by introducing hydrolytically unstable ester linkages into the polyamide structure.[8,9] The synthesis of our poly(ester-amides) begins with a linear, aliphatic diamine which is condensed with two moles of glycolic acid to form a diamidediol monomer (Eq. 1). This diol with preformed amide linkages is then polyesterified, preferably with a diacid chloride under carefully controlled conditions in chlorobenzene (Eq. 2).

$$NH_2 \text{--}(CH_2)_x\text{--}NH_2 + 2\ HO\text{-}CH_2\text{-}COOH \xrightarrow[\Delta,\ -H_2O]{} HO\text{-}CH_2\text{-}\overset{\overset{O}{\|}}{C}\text{-}NH\text{--}(CH_2)_x\text{--}NH\text{-}\overset{\overset{O}{\|}}{C}\text{-}CH_2\text{-}OH$$

Eq. 1

$$HO-CH_2-\overset{O}{\overset{\|}{C}}-NH-(CH_2)_x-NH-\overset{O}{\overset{\|}{C}}-CH_2-OH \;+\; Cl-\overset{O}{\overset{\|}{C}}-(CH_2)_y-\overset{O}{\overset{\|}{C}}-Cl \xrightarrow{\;\emptyset-Cl,\; -HCl\;}$$

$$-[O-CH_2-\overset{O}{\overset{\|}{C}}-NH-(CH_2)_x-NH-\overset{O}{\overset{\|}{C}}-CH_2-O-\overset{O}{\overset{\|}{C}}-(CH_2)_y-\overset{O}{\overset{\|}{C}}]-$$

Eq. 2

This polymer structure offers considerable latitude for variation. For example, x and y can be varied independently or mixtures of monomers with different values for x and y can be copolymerized in different ratios. Aromatic or heteroatomic substituents also can be present in the diamine or the diacid. While all of these possibilities have been explored to some extent, simple variation of x and y has received the most attention.

EXPERIMENTAL

Detailed descriptions of the procedures for polymer synthesis have been published previously.[8] The preferred method, especially where y=2, is to react a diacid chloride in chlorobenzene with the diamidediol by heating the mixture under conditions that allow polyesterification. Excessive reaction time or temperature leads to crosslinked material.

The diamidediol with x=12 polymerizes most conveniently to high molec. wt. and was held constant for a series of different diacids. The effect of diacid methylene chain length on polymer melting temperature is shown in Figure 1.

A series of experiments was performed to determine gross bio-compatibility and relative degradation rates of various poly(ester-amide) fibers in vivo.[9] Monofilament fibers (2-0 U.S.P. suture size) prepared by melt spinning were implanted subcutaneously in mice and examined at various times post-implantation for up to 19 months. The chemical compositions of fibers tested are given in Table 1 and the strength loss profiles are shown in Figure 2.

The in vivo absorption times for various diamidediols (x=6,8,10, and 12) and the corresponding low molec. wt. polymers (y=2) were obtained for comparison by subcutaneous implantation of 20 mg pellets in mice. Based on the result that $x \leq 10$ polymers are faster absorbing than the x=12 polymer, the x=6 polymer was arbitrarily chosen to study the mechanism of in vivo poly(ester-amide) degradation further using radiolabeled polymer.[10] Diamidediol labeled with carbon-14 in the hydroxyacetamide moiety was polymerized with succinyl chloride to give radiochemically pure polymer with a specific activity of 6.07 μCi/mg. The carbon-14 polymer was melted, drawn by hand into thin filaments, and cut into 1.0 cm pieces. One piece of polymer was implanted in the right leg in each of 35 rats. The rats were sacrificed in groups of 5 at 7 days, 1,3,6, and 9 months post-implantation. At the time of necropsy, various tissues and organs (including the implant site) were separately collected and analyzed radiometrically. The remaining 10 rats were used to monitor radioactivity in the urine and feces for 12 months. The urine was also analyzed chromatographically. The rate of decrease in radioactivity at the implant site is shown in Figure 3.

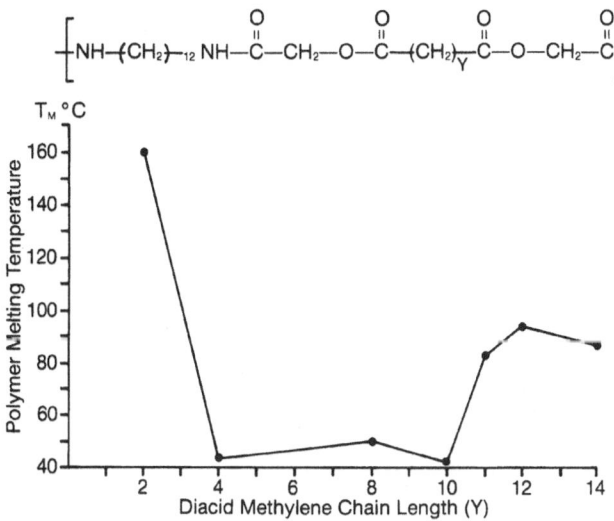

$$\left[\text{NH}-(\text{CH}_2)_{12}\text{NH}-\overset{\text{O}}{\underset{\|}{\text{C}}}-\text{CH}_2-\text{O}-\overset{\text{O}}{\underset{\|}{\text{C}}}-(\text{CH}_2)_Y\overset{\text{O}}{\underset{\|}{\text{C}}}-\text{O}-\text{CH}_2-\overset{\text{O}}{\underset{\|}{\text{C}}}\right]$$

Figure 1. The Effect of Diacid Methylene Chain
Length on Melting Temp. of Poly(ester-amides)

Table 1. Chemical Composition of Poly(ester-amide) Fibers Tested *In Vivo*

2-0 U.S.P. Suture Size
Poly(ester-amide) Fibers

$$\left[\text{O}-\text{CH}_2-\overset{\text{O}}{\underset{\|}{\text{C}}}-\text{NH}-\text{R}_1-\text{NH}-\overset{\text{O}}{\underset{\|}{\text{C}}}-\text{CH}_2-\text{O}-\overset{\text{O}}{\underset{\|}{\text{C}}}-\text{R}_2-\overset{\text{O}}{\underset{\|}{\text{C}}}\right]$$

Sample	R_1	R_2	Initial Tenacity (g/denier)	% Tensile Strength Retained 4 Weeks Post-Implantation
A.	$(CH_2)_{12}$	$(CH_2)_2$	3.15	87.4
B.	90% $(CH_2)_{12}$ 10% $(CH_2)_3$O-$(CH_2)_4$O-$(CH_2)_3$	$(CH_2)_2$	6.53	96.5
C.	80% $(CH_2)_{12}$ 20% $(CH_2)_3$O-$(CH_2)_4$O-$(CH_2)_3$	$(CH_2)_2$	4.67	76.6
D.	$(CH_2)_{12}$	90% $(CH_2)_2$ 10% $-CH_2-O-CH_2-$	4.11	79.4
E.	$(CH_2)_6$	$(CH_2)_2$	3.60	64.0
F.	$(CH_2)_{12}$	$(CH_2)_{12}$	2.50	0
G.	$(CH_2)_{12}$	$(CH_2)_8$	0.82	0
Control Samples		Dexon™	5.67	0
		Vicryl™	5.24	2.8
		Chromic Catgut	4.10	28.7

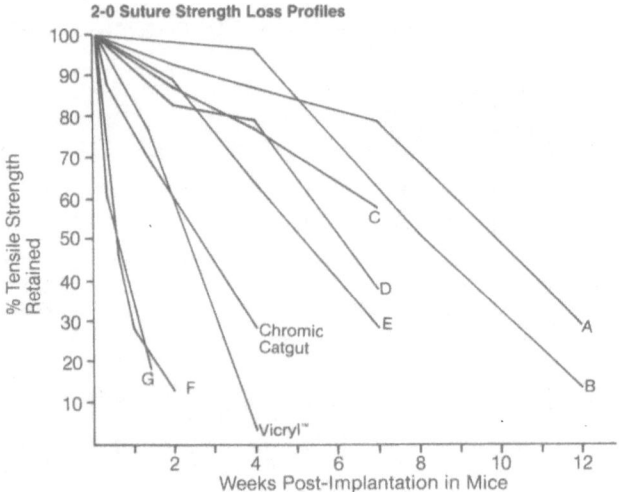

Figure 2. Poly(ester-amide) Fiber Strength
Retention In Vivo

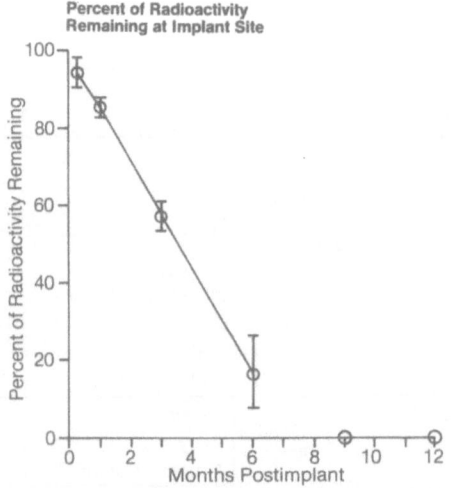

Figure 3. Rate of Decrease in Radio-
Activity at the Implant Site

RESULTS AND DISCUSSION

Poly(ester-amides) with y>2 are relatively low melting (Tm<100°C) and produce low strength fibers which lose strength rapidly in vivo. Poly(ester-amides) with y=2 (and x=6,8,10, or 12) are higher melting (Tm≅165°C) and yield high strength fiber. The rate of strength loss in vivo for a given poly(ester-amide) fiber varies somewhat depending on the quality of the fiber, higher tenacity samples generally exhibiting a more prolonged rate of strength retention.

The rate of poly(ester-amide) absorption in vivo is not related to the rate of fiber strength loss. For example, fiber samples A. and F. (Table 1) both required 19 months in vivo before significant mass loss could be visually confirmed. All of the implants appeared grossly bio-compatible at all time points, which was in contrast to chromic catgut controls where severe inflammation was occasionally observed.

The implantation of diamidediol and low molec. wt. polymer pellets for visual inspection of absorption time showed that the x=12 monomer and polymer are much slower absorbing than any of the other homologs. Since the x=12 monomer is water insoluble whereas the other homologs are at least sparingly. soluble, it appears that diamidediol insolubility prolongs the absorption time of poly(ester-amides).

The radiolabeled implant study shows that loss of radioactivity from the implant site is virtually linear (Figure 3) with no accumulation of radioactivity in any of the peripheral tissues. The primary route of excretion is via urine and the major radiochemical compound in the urine is unchanged diamidediol monomer.

In separate experiments, the x=6 diamidediol monomer and polymer were parenterally administered to groups of 10 rats each in doses of 5,000 mg/Kg to evaluate acute toxicity. Under these conditions there was no evidence of toxic effects for two weeks post-dose nor were any visible lesions detected during necropsy.

CONCLUSIONS

Poly(ester-amides) are a new class of fiber-forming bioabsorbable polymers with potential utility as monofilament absorbable suture. The general formula has been optimized further by requiring the use of a water soluble diamidediol to minimize polymer absorption time and the use of succinic acid as the comonomer to maximize polymer melting point and in vivo fiber strength retention.

In vivo absorption of radiolabeled poly(ester-amide) results in complete elimination of all radioactivity from the body. The principle mechanism of in vivo degradation appears to be hydrolysis of the ester bonds. This produces the diamidediol monomer as a major metabolite which is non-toxic as determined by the LD_{50} test in rats.

REFERENCES

1. D. E. Clark, Surgical Suture Materials, Contemporary Surgery, 17: 33 (1980).
2. J. A. Ray, N. Doddi, D. Regula, J. A. Williams, and A. Melveger, Polydioxanone (PDS), A Novel Monofilament Synthetic Absorbable Suture, Surg. Gynecol. Obstet., 153: 497 (1981).

3. C. J. Schaefer, P. M. Colombani, and G. W. Geelhoed, Absorbable Ligating Clips, Surg. Gynecol. Obstet., 154: 513 (1982).

4. D. G. Noiles, Surgical Fastening Method and Device Therefor, U.S. Patent 4,060,089, November 29, 1977.

5. R. K. Daniel and M. Olding, An Absorbable Anastomotic Device for Microvascular Surgery: Experimental Studies, Plast. Reconstr. Surg., 74: 329 (1984).

6. T. H. Barrows, Absorbable Nerve Repair Device and Method, U.S. Patent Pending.

7. H. J. Sanders, Special Report: Improved Drug Delivery, Chemical & Engineering News, April 1, 1985, p. 30.

8. T. H. Barrows, Synthetic Absorbable Surgical Devices of Poly(esteramides), U.S. Patent 4,343,931, Aug. 10, 1982.

9. T. H. Barrows, D. M. Grussing, and D. W. Hegdahl, Poly(ester-amides): A New Class of Synthetic Absorbable Polymers, Trans. Soc. Biomater., 6: 109 (1983).

10. T. H. Barrows, S. J. Gibson, and J. D. Johnson, Poly(ester-amides): In Vivo Analysis of Degradation and Metabolism using Radiolabeled Polymer, Trans. Soc. Biomater., 7: 210 (1984).

HEPARINIZABLE SEGMENTED POLYURETHANES FOR CARDIO-VASCULAR APPLICATIONS

M.C.Tanzi, P.Albonico, C.Barozzi, A.Bolognesi, R.Fumero°,
and G.Tieghi

Dipartimento di Chimica Industriale e Ingegneria Chimica
° Dipartimento di Energetica
Politecnico - Piazza L.Da Vinci, 32 - 20133 Milano (Italy)

INTRODUCTION

Segmented polyurethanes, introduced as new elastomeric materials for biomedical application in 1967[1], have found increasing attention as blood-compatible materials for vascular repair and cardiovascular devices[2-6]. These thermoplastic elastomers are block copolymers generally composed of short polyurethane sequences (from aromatic diisocyanates extended with low molecular weight diols or diamines, hard segments) connected to long and flexible polyether or polyester chains (soft segments). These materials show a combination of properties not easily available in other materials: high modulus of elasticity and resistance to flex-fatigue[6,7], together with a considerable degree of blood compatibility which has been the subject of various investigations[2,8-11].

This work deals with the synthesis and the properties of new segmented polyurethanes, containing poly-ether-urethane segments coupled with heparinizable poly-amido-amine (PAA) segments.

These new poly(ether-amidoamine-urea-urethane)s (PEAAUU) have been prepared in order to match the mechanical and processability properties of polyurethanes, with the ability to form a stable complex with heparin, characteristic of the PAA segments.

SYNTHESIS

The synthesis of the PEAAUU copolymers can be obtained from aromatic diisocyanates, macrodiols and amino terminated poly-amido-amines (PAA) following different pathways[12]. Two lines can be considered as shown in Scheme 1.

Scheme 1. Synthesis of PEAAUU copolymers

LINE 1

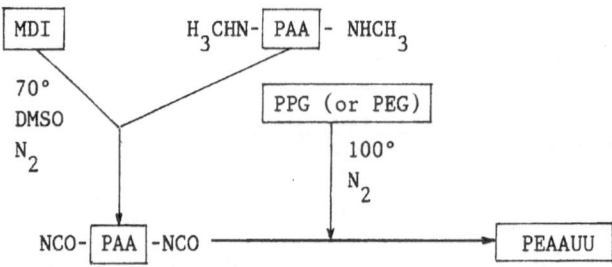

LINE 2

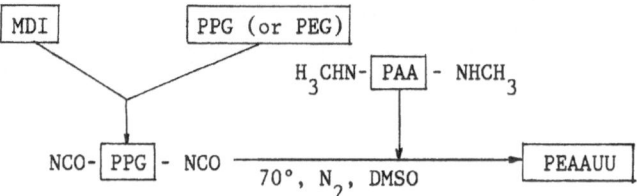

4,4'-Diphenylmethane (MDI) was chosen as the most suitable aromatic diisocyanate for the reaction with aliphatic diamines[13], and because of its rather low toxicity. The macrodiols were commercial low molecular weight (MW = 400 ÷ 2000) polyoxypropyleneglycols (PPG) or polyoxyethyleneglycols (PEG), while secondary - amino terminated PAAs were purposely synthesized. The PAAs utilized in the present work are reported in Table 1.

Table 1. Poly-amidoamine macromonomers utilized in the synthesis of PEAAUU

PAA TYPE-A

$$H-N(CH_2)_n-N-(CH_2CH_2C-N\underset{O}{\parallel}\bigcirc NC-CH_2CH_2N(CH_2)_n \ N)-H$$
$$\begin{array}{ccc} CH_3 & CH_3 & O \end{array} \qquad \begin{array}{ccc} O & CH_3 & CH_3 \end{array}$$

PAA-A$_1$	and	PAA-B$_1$	n = 2
PAA-A$_2$	"	PAA-B$_2$	n = 4
PAA-A$_3$	"	PAA-B$_3$	n = 6
PAA-A$_4$	"	PAA-B$_4$	n = 8

PAA TYPE-B

$$H-N(CH_2)_n-N-(CH_2CH_2C-NH(CH_2)_{12}NH-C-CH_2CH_2N(CH_2)_n \ N)-H$$
$$\begin{array}{ccc} CH_3 & CH_3 & O \end{array} \qquad\qquad \begin{array}{ccc} O & CH_3 & CH_3 \end{array}$$

It is well known that a rigorous control of purity and of reaction conditions is essential if high molecular weight linear polyurethanes are to be obtained. The presence of water and small variations of the stoichiometric ratio between isocyanate and hydroxyl or amino groups from equimolarity, cause a decrease of the intrinsic viscosities of the products, and the occurrence of side reactions, resulting in branching or crosslinking[13,14]. Since the order of reactivity with isocyanates is: RNH_2 > RCH_2OH > H_2O > secondary or tertiary amino groups, rigorous anhydrous conditions are requested in the case of secondary-amino terminated PAAs. Unfortunately PAAs are inherently hygroscopic and, even when essiccated, they retain a little amount of water.

As a matter of fact, line 1 has been discarded because of partial crosslinking occurring during the first synthetic step. In the case of line 2, good results have been obtained, especially when B type PAAs were used. A preliminary investigation on the properties and blood compatibility of the PU_{725}-B_3 copolymer (obtained from PPG 725, MDI and PAA-B_3) has already been reported[15].

All the PEAAUUs synthesized have been isolated from the reacting solutions, by precipitation in water, as white or pale yellow gummy or tough solids.

The structure of the copolymers obtained was confirmed by elemental analysis, infrared and H^1-NMR spectroscopy. Some I.R. and NMR spectra are reported in Fig. 1 and Fig. 2, together with the corresponding PAA spectra. The values of the intrinsic viscosity (g/100 ml), in DMF solution at 30°C, were as follows: 0.38-0.40 for standard PU; 0.25-0.35 for

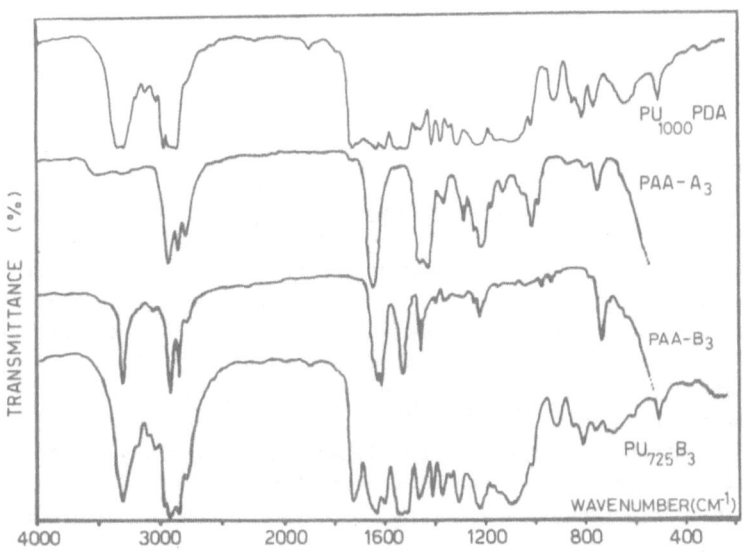

Fig. 1. Infrared spectra of segmented polyurethanes, and pure poly-amidoamines.

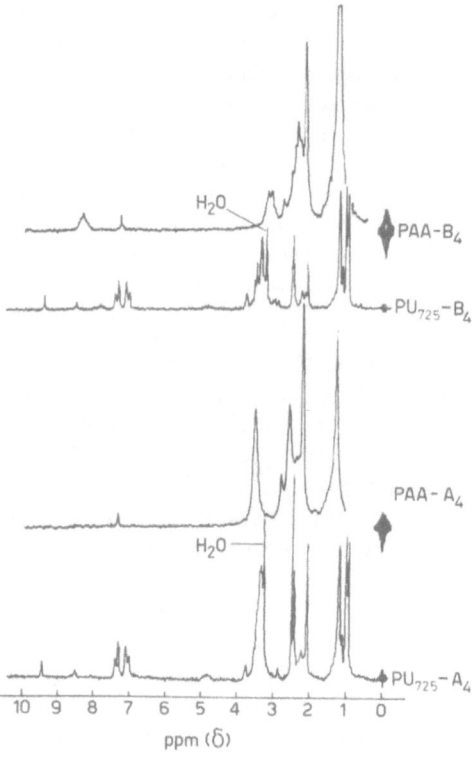

Fig. 2. N.M.R. spectra of two copolymers, and of corresponding
pure poly-amidoamines.

PEAAUUs from line 2 (like $PU_{725}-B_4$, $PU_{1000}-B_3$, $PU_{725}-A_4$ obtained from
MDI, PPG 725 or 1000, PAA-B_4 or PAA-B_3 or PAA-A_4).

PHYSICAL CHARACTERIZATION

The DSC traces, in a temperature range from -100°C to 200°C, of all
PEAAUU copolymers show two different glass transition temperatures (Tg),
attributable to PU and PAA segments. Some Tg values are reported in Table
2. The traces of copolymers containing B-type PAAs show melting effects in a temperature range between 75°C and 110°C, which are attributable to PAA segments, whereas those of copolymers containing A-type PAAs
indicate a very poor crystallinity. As an example, the DSC scans of two
copolymers containing PAA-B_3 and PAA-B_4 are reported in Fig. 3.

The crystallinity of the copolymers containing B-type PAAs was also
revealed by wide-angle X-ray diffraction as shown as an example in Fig.
4 while PEAAUUs containing A-type PAAs appear to be almost amorphous in
the conditions we obtained the samples.

Preliminary mechanical tensile tests on PEAAUU and standard PU specimens, before and after the blood compatibility test (see below), indic-

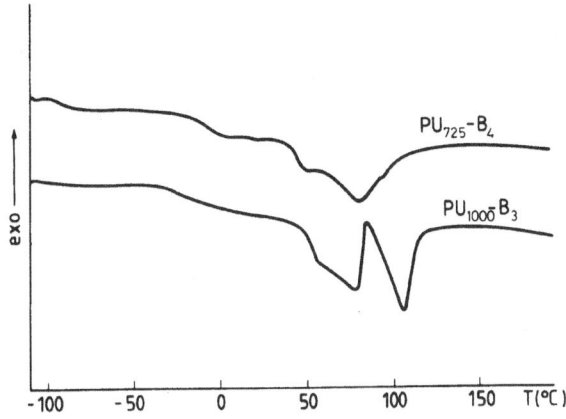

Fig. 3.   DSC traces of two copolymers.

ated that blood exposition did not affect the mechanical properties of all samples. PEAAUU and standard PU copolymers show quite similar properties.

Table 2. Tg values of PEAAUU and standard PU copolymers

Copolymer	Tg I (°C)	Tg II (°C)
PU_{1000}-PDA	-14	--
PU_{725}-ESA	-14	--
PU_{725}-A_4	-22	+39
PU_{725}-B_4	-18	+39
PU_{1000}-B_3	-34	+46

HEPARIN-ADSORBING ABILITY

The ability to form a stable complex with heparin has been previously proved for PU_{725}-B_3[15] by activated partial thromboplastin time (PTT) and thrombine time (TT) tests. The sample was heparinized with a 1% heparin solution for 24 h; the total heparin amount complexed resulted 68.7 mg/g (7% by weight). The formation of the complex with heparin was also investigated by FT-IR spectroscopy on the PU_{1000}-B_3 copolymer. In Fig. 5 the spectrum of the heparinized copolymer is compared with those of heparin and of not heparinized PU_{1000}-B_3.

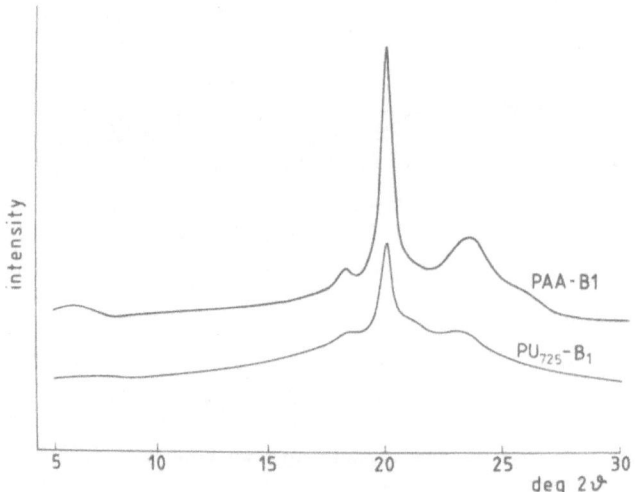

Fig. 4. Wide-angle X-Ray diffraction patterns of
PU_{725}-B_1, and of corresponding pure PAA-B_1

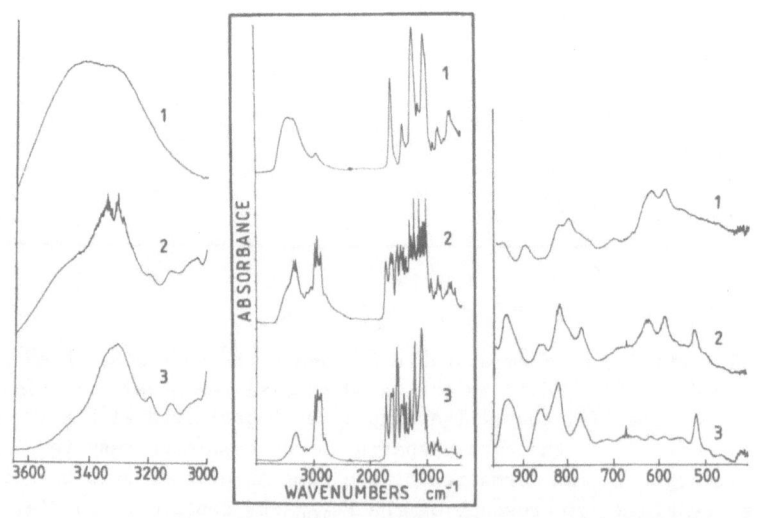

Fig. 5. Middle: FT-IR spectra of heparin (1), heparinized PU_{1000}-B_3 (2),
and PU_{1000}-B_3(3). Left and right: magnification of significant
regions.

BLOOD COMPATIBILITY

Sheets of PEAAUU copolymers, before and after heparinization, and
of standard PUs have been exposed to bovine blood in closed circuit for
6 h, then fixed in glutaraldehyde and repeatedly washed with water. The
surface of the sheets, before and after contact with blood, has been ex
amined with optical and scanning electron microscopy. The surface of
all samples appeared practically unchanged after the contact with blood,
save some slight alterations that could be detected on the PU-PDA sam-

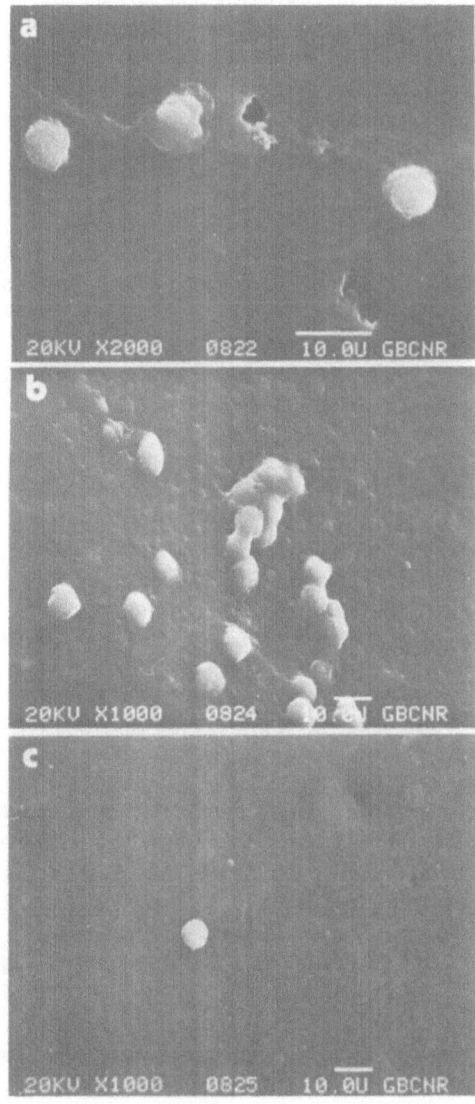

Fig. 6. S.E.M. micrographs of PU-PDA (a), PU_{1000}-B_3 (b) and heparinized
PU_{1000}-B_3 (c) surfaces after blood exposition.

ples. The nature of the surface deposit of blood elements was the same for all the examined samples, resulting of particles with a diameter about 5 μm. The deposit amount was rather poor on PU-PDA, slightly larger on PEAAUU copolymers and practically inappreciable on the heparinized PEAAUUs. Examples of the surface SEM micrographs are reported in Fig. 6.

CONCLUSIONS

The materials obtained following the synthetic pathway of line 2 in Scheme 1 retain the ability to form a stable complex with heparin, without noticeable variations of their mechanical properties, and show a satisfactory behaviour in the contact with blood. Thus, on the whole, the new copolymers appear promising materials for cardiovascular applications.

ACKNOWLEDGEMENTS

This work has been financially supported by C.N.R. (Progetto Finalizzato Chimica Fine e Secondaria) and Ministero Pubblica Istruzione, Italy. Many thanks are due to the Centro Gino Bozza (Politecnico, Milano) no) for SE microscopy, to Prof. G.Zerbi (Politecnico, Milano) for FT-IR spectroscopy, and to Mr. L.Bordogna and Dr. A.Daghetti (Politecnico, Milano) for assistance in physical characterization.

REFERENCES

1. J. W. Boretos and W. S. Pierce, Segmented polyurethane. A new elastomer for biomedical applications, Science, 158:1481 (1967).
2. International Colloquium "Polyurethane in Medical Technics", Fellbach (Germany), January 27-29, 1983, Polyurethanes in Biomedical Engineering, 1, H. Planck, G. Egbers, I.Syré, eds. Amsterdam, Oxford, New York, Tokyo: Elsevier (1984).
3. D. L. Lyman, W. J. Seare, Jr., D. Albo, Jr., S. Bergman, J. Lamb, L.C. Metcalf and K. Richards, Polyurethane Elastomers in Surgery, Int. J. Polymeric Mater. 5:211 (1977)
4. G. L. Wilkes, Necessary considerations for selecting a polymeric material for implantation with emphasis on polyurethanes, in: "Polymer Science and Technology (vol. 8): Polymers in Medicine and Surgery", R. L. Kronenthal, Z. Oser and C. Martin eds., Plenum Press, New York, London (1975).
5. J. W. Boretos, D. E. Detmer and J. H. Donachy, Segmented polyurethanes: a polyether polymer, II. Two years experience, J. Biomed. Mater. Res., 5:373 (1971).
6. E. Nyilas, Development of blood compatible elastomers. II. Performance ce of Avcothane blood contact surfaces in experimental animal implantations, J. Biomed. Mater. Res., 3:97 (1972)
7. B. D. Ratner and R. W. Paynter, Polyurethane surfaces: the importance of molecular weight distribution, bulk chemistry and casting conditions, in ref. 2, pag. 41.

8. S. K. Hunter, D. E. Gregoris, D. L. Coleman, B. Hanover, R. L. Stephen and S. C. Jacobsen, Surface modification of polyurethane to promote long-term potency of peritoneal access devices, <u>Trans. Am. Soc. Artif. Int. Organs</u>, 29:250 (1983).

9. P. M. Knight and D. J. Lyman, Evaluation of thrombogenicity of selected microporous oxigenator membranes, <u>Artificial Organs</u>, 9:28 (1985).

10. D. J. Lyman, L. C. Metcalf, D. Albo, Jr., K. F. Richards and J. Lamb, The effect of chemical structure and surface properties of synthetic polymers on the coagulation of blood. III. In vivo adsorption of proteins on polymer surface. <u>Trans. Am. Soc. Artif. Int. Organs</u>, 20:474 (1974).

11. D. J. Lyman, K. Kunston, B. Mc.Neill and K. Shibatani, The effects of chemical structure and surface properties of synthetic polymers on the coagulation of blood. IV. The relation between polymer morphology and protein adsorption. <u>Trans. Am. Soc. Artif. Int. Organs</u>, 21:49 (1975).

12. M. C. Tanzi, Italian Patent Appl. n° 20868 A/84.

13. H. Rinke, Elastomeric fibers based on polyurethanes, <u>Angew. Chem. Int. Ed.</u>, 8:419 (1962).

14. D. J. Lyman, Polyurethanes. I. The solution polymerization of diisocyanates with ethylene glycol. <u>J. Polymer Sci.</u>, 55:49 (1960).

15. M. C.Tanzi, C.Barozzi, G. Tieghi, L. Grassi and R. Fumero, New poly(ether-amidoamine-urea-urethane) block copolymers for biomedical use, International Workshop: "Future Trends on Polymer Science and Technology. Polymers: Commodities of Specialities?" Capri, October 8-12 (1984).

CELL ATTACHMENT TO VARIOUS POLYMER SURFACES

Yasushi Tamada and Yoshito Ikada*

Research Center for Medical Polymers and Biomaterials
Kyoto University
53 Kawahara-cho, Shogoin, Sakyo-ku, Kyoto 606, Japan

INTRODUCTION

Protein adsorption and the subsequent cell adhesion are thought to be the first occurrence when a biomedical material is implanted in body or comes in contact with blood. Recently a large number of investigations have been performed, focusing attention on the interaction of biomedical materials with plasma proteins[1] and cultured cells[2]. The observed results have been interpreted in terms of the DLVO theory[3,4], the critical surface tension[5], the free energy change[6,7], the adhesion kinetics[8,9], the biological behavior[10,11], and others. We also have published papers, for instance, on the dynamic, selective adsorption of proteins[12] and the dependence of protein adsorption on the surface characteristics of polymeric materials[13].

The objective of this work is to study the cell adhesion, that is, the cell attachment *in vitro* onto a variety of materials including conventional polymers and glass. We often would like to select a biomedical material that has specific properties, for instance, a bioinert material that invokes the minimum encapsulation and thrombus formation, or the bioadhesive material that yields strong bonding to the contacting tissue without giving any deadspace between the implanted material and the connective tissue. To do this, it is essential to have knowledge of cell behaviors to the foreign materials possessing different surface properties. For these purposes, polyethylene films grafted with synthetic and naturally occurring water-soluble polymers also have been employed in the present study.

EXPERIMENTAL

Substrates

Commercial films were used as the polymeric substrates for cell adhesion with the exception of poly(methyl methacrylate)(PMMA) film, which was prepared by casting from a benzene solution of PMMA. For purification, all films were subjected to Soxhlet extraction with methanol or immersed in methanol at room temperature for longer than

---

* To whom all correspondence should be addressed.

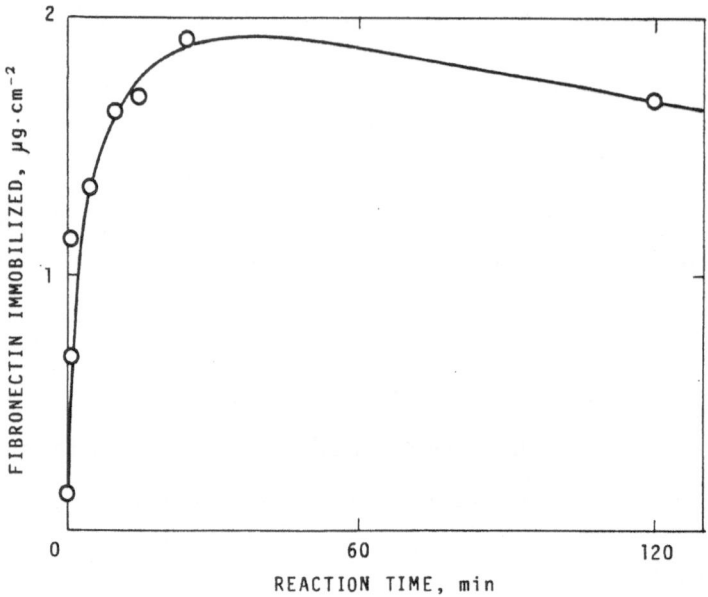

Fig.1. Amount of fibronectin immobilized as a function of
WSC pretreatment time for coupling reaction onto
polyethylene films grafted with poly(acrylic acid).
WSC pretreatment condition:[WSC]=1 mg·ml^{-1}, 0 °C

12 hours, followed by drying in vacuum at room temperature.

Grafting

Graft copolymerization of acrylamide(AAm) and acrylic acid(AA)
onto a high density polyethylene(PE) film was carried out using a
corona-pretreatment technique as follows. The PE film was pretreated
with corona discharge in air at 15 kV for 2 minutes at room temperature,
put in a 10 wt% aqueous solution of either of the monomers in a glass
ampoule, and then sealed after degasing. This sealed ampoule was
kept at 50 °C for one hour to allow the graft copolymerization to
proceed. The homopolymer formed was removed from the grafted film
by washing with water at 70°C for about 30 hours.

To effect the graft coupling of bovine serum albumin(BSA, crystallized), collagen (atelocollagen), and plasma fibronectin(FN) onto
the surface of PE film, AA was first graft copolymerized to the film
surface, leading to introduction of carboxyl groups onto the surface.
These carboxyl groups on the film were reacted with a water-soluble
carbodiimide(WSC) at 10 mg·ml^{-1} in a phosphate buffered saline(PBS)
for 30 minutes at 0 °C. After that, the film was put in the solution
of BSA (1 mg·ml^{-1}), collagen (1 mg·ml^{-1}) or FN (0.3 mg·ml^{-1}) at 4°C. BSA was
purchased from Miles Laboratories, Inc., USA, collagen from Nitta
Gelatine Co., Japan, and FN was isolated in our laboratory from frozen
human plasma by affinity chromatography with a gelatin-Sepharose
column[14]. An example of the coupling reaction is demonstrated in
Figure 1 for the grafting of FN onto the PE film previously graft-
copolymerized with AA.

Measurements of Contact Angle and Zeta Potential

Water contact angles were measured at 25 °C for dried films and glass with the sessile drop method. At least five readings on different parts were averaged. Streaming potentials were measured for films of 3 x 5 cm^2 at 25 °C using the cell unit described by Andrade and his co-workers[15] The electrode was made of platinum and the pH value and the ionic strength of the electrolyte solution used for the potential measurement were kept to 7.4 and 6 x 10^{-3}, respectively, by the use of HCl and NaOH. The zeta potential was calculated according to the Helmholtz-Smoluchowski equation.

Cell Culture

All cells were cultured in Eagle MEM containing antibiotics (Kanamycin, 60 mg·l^{-1}) and 10 % fetal calf serum(FCS, M.A.Bioproducts, Maryland, USA). Cell lines used were L cell and HeLa S$_3$ cell. Primary-cultured rat fibroblasts were taken from the skin of fetal rats using trypsin and ethylenediaminetetraacetic acid disodium salt(EDTA).

Cell Adhesion and Growth

The HeLa S$_3$ cells and the rat fibroblasts were trypsinized from cultured flasks, washed once in a medium with 10 % FCS and once in a serum-free medium. L cells were scraped out of the culture flasks and washed twice in a serum-free medium. The cell density was adjusted to 1.0 x 10^5 cells/ml for the cell adhesion test in a serum-free medium or to 1.76 x 10^4 cells/ml for the cell growth test in a medium with 10 % FCS. 1 ml of cell suspensions was plated on each of the polymer films of 15 mm diameter in a multidish 24 wells Nunclon Delta SI (Nunc, Denmark) and kept for predetermined periods in a humidified incubator conditioned to 37 °C, 5 % CO$_2$ and 95 % air atmosphere. After incubation, the unattached cells were removed from the film surface by washing with PBS free of Ca^{2+} and Mg^{2+}. The number of cells attached to the films was counted with the lactate dehydrogenase(LDH) activity method of Schnaar and his co-workers[16]. Films with the adherent cells were placed in 2 ml of lysis buffer (0.5 % Triton-X in PBS) and incubated for longer than 60 minutes at room temperature to ensure the complete cell lysis. The LDH activity of the lysate containing the disrupted cells was measured using Monotest® LDH opt. (Boehringer Mannheim, FRG) as the LDH assay buffer. The cell number was determined using the calibration curve which is shown in Figure 2. This cell counting method is useful, as there is a good linear relationship between the enzyme activity and the cell number measured by a hemocytometer.

RESULTS

Polymer surfaces can be classified according to various criteria. The most common is the classification with respect to the hydrophilicity or water wettability. More quantitatively, the surface property can be expressed by the surface free energy. However, as it is very difficult to determine the surface free energy of polymeric surfaces with good accuracy except for a few polymers[17], the surface properties of the materials used here were characterized simply by the contact angle against water which is directly related to the hydrophilicity.

On the other hand, as is well known, a surface potential is generally observed when a material is brought into contact with an aqueous solution of electrolytes even for materials without any fixed, ionic groups such as polyethylene. This electrostatic potential

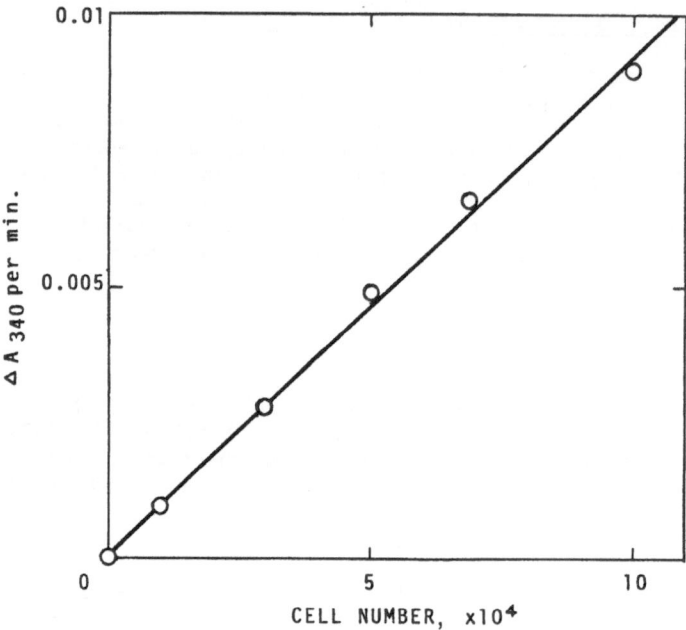

Fig.2. Determination of L cell number by lactate
dehydrogenase assay.

is expected to play a significant role in the interfacial phenomenon
like cell attachment to the polymer surface in contact with an aqueous
environment. Therefore, we determined also the zeta potential of
all the samples as a measure of the surface potential.

Effect of Surface Wettability on Cell Attachment

The number of L cells attached per cm^2 of the film surfaces after
one hour incubation at 37 °C is plotted in Figure 3 against the contact
angle of the films. The results obtained for the glass and the films
surface-grafted with proteins and synthetic polymers are also shown
in Fig.3. It appears that most of the experimental data fit on a
single curve. The distinct exceptions are the glass and the FN-grafted
PE where the cells adhered most remarkably. It is interesting to
note that the curve exhibits a maximum at the contact angle around
70°. The minimum cell attachment seems to occur for the surface
that has a contact angle as low as 40° except for the glass and the
FN-grafted PE. The collagen- and BSA-grafted PE belong to the surface
group characterized by high hydrophilicity and minimum cell attachment.

A similar relationship between the cell attachment and the contact
angle of the surfaces was found for the incubation study with HeLa
S3 cell. The results are illustrated in Figure 4. As can be seen,
very few cells attached to the surface with a contact angle of about
40° or less. However, as the film surface becomes more hydrophobic,
the number of cells attached increases to reach a maximum at the
contact angle around 70° and then decreases with the increasing
hydrophobicity. Grafting of FN onto the PE surface again results
in the largest cell attachment.

104

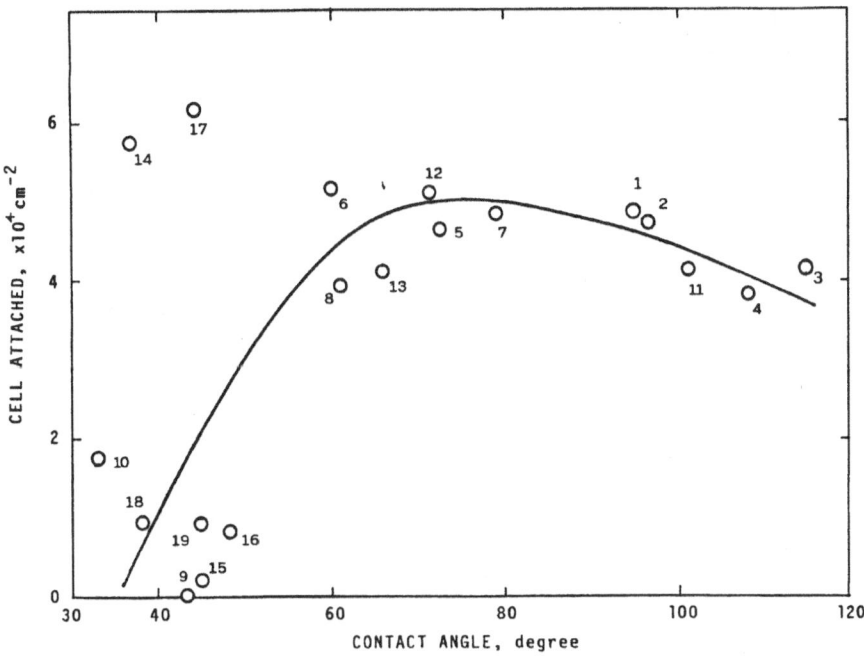

Fig.3. Effect of contact angle of surfaces on L cell adhesion at 37 °C. (60 min incubation using 1.14 x 10[5] L cells/cm^2 without FCS.)

Film Material: 1.polyethylene(PE) 2.polypropylene 3.polytetrafluoroethylene 4.tetrafluoroethylene-hexafluoropropylene copolymer 5.polyethylene terephthalate 6.PMMA 7.Nylon 6,6 8.vinylalcohol-ethylene copolymer 9.PVA 10.cellulose 11.silicone 12.polystyrene 13.commercial plastic sheet for cell culture 14.glass 15.PAAm-grafted PE 16.PAA-grafted PE 17.FN-grafted PE 18.collagen-grafted PE 19.BSA-grafted PE

It has been reported that the cell attachment to incubation substrates is dependent on the temperature[18]. Therefore, the incubation was attempted after lowering the temperature from 37 °C to 4 °C. Figure 5 shows the result obtained for L cells. Clearly, the number of cells attached is decreased to about 50 % as a whole by lowering the incubation temperature, but the characteristic dependence of the cell number on the surface wettability seems to be maintained even at 4 °C. It is interesting that the cell adhesion does not take place to the FN-grafted surface at such low temperatures, while adhesion remained almost the same level of incubation as at 37 °C for the glass surface.

In the incubation study described above, cell counting was performed after incubation only for one hour. This was because the cell interaction with the foreign surfaces in the initial stage of culture was the aim of the study. If the incubation duration is prolonged, the number of cells attached should be increased because of cell proliferation. It is highly probable that the cell proliferation rate will be also affected by the surface properties of the substrates. To

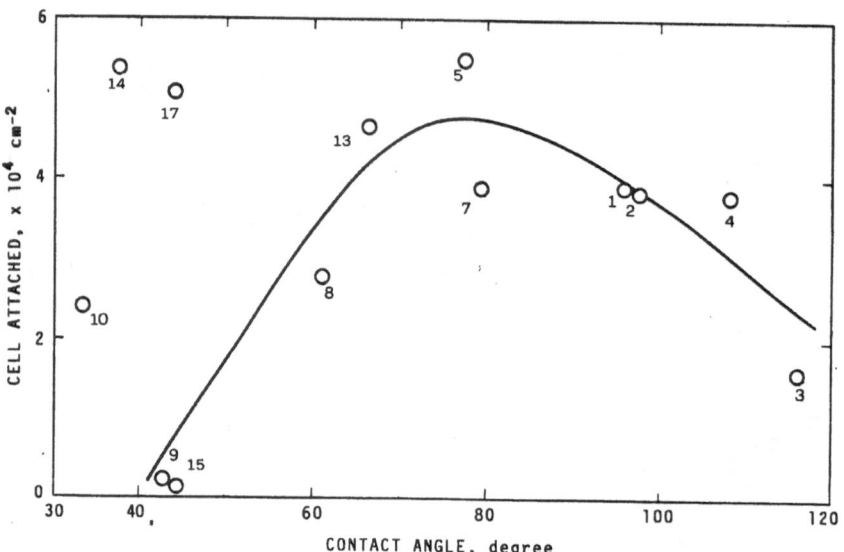

Fig.4. Effect of contact angle of surfaces on HeLa S_3 cell
adhesion at 37 °C. (60 min incubation using
1.14×10^5 HeLa S_3 cells/cm^2 without FCS)
(See Fig.3 for explanation of film material type code)

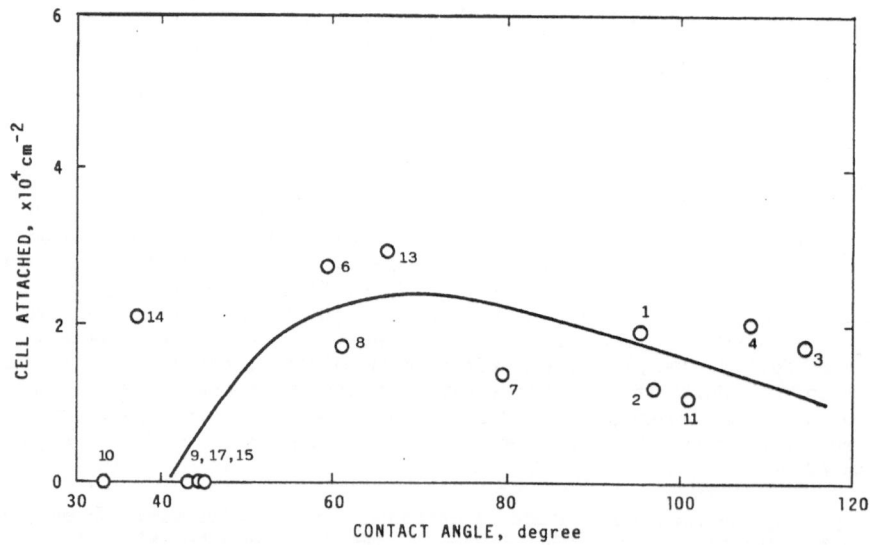

Fig.5. Effect of contact angle of surfaces on L cell adhesion at
4 °C. (60 min incubation using 1.14×10^5 L cells/cm^2
without FCS)
(See Fig.3 for explanation of film material type code)

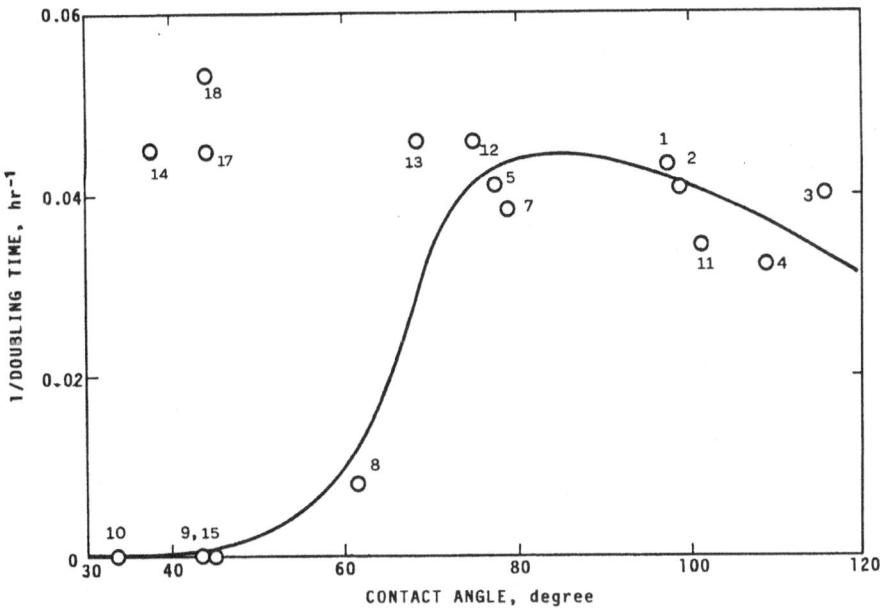

Fig.6. Effect of contact angle of surfaces on rat fibroblast
        proliferation.
        (See Fig.3 for explanation of film material type code)

examine this possibility, fibroblasts from the skin of fetal rat
were cultured on the various surfaces in the presence of 10 % FCS
at 37 °C. It was found that the fibroblasts cultured on these surfaces
showed a logarithmic proliferation rate after one day incubation,
similar to the usual cell culture. The reciprocal of the doubling
time in this logarithmic proliferation period is plotted against
the surface contact angle in Figure 6. In this case, the experimental
data scattered rather broadly, but it is likely that the highly hydrophilic surfaces such as PVA, cellulose, and PAAm-grafted PE do not
allow the cells to proliferate. On the contrary, although the glass
and the FN-grafted surface are very hydrophilic, the cells were able
to proliferate extensively.

Effect of Zeta Potential on Cell Attachment

       The number of L cells attached to the various films is plotted
in Figure 7 as a function of the zeta potential of the substrate
films. The incubation condition is identical with that in Fig.3.
Although the result in Fig.7 shows appreciable scatter, one can see
a trend that the number of cells attached, varies linearly to a first
approximation with the zeta potential of the surfaces. Any surfaces
used in the present work yield no positive potential. As is apparent
from Fig.7, the cell attachment is almost non-existent when the substrate
surface has a zeta potential close to zero except for the FN-grafted
film. In contrast to the case in Fig.3, the experimental result
for the glass seems to fit on the linear plot in Fig.7. This suggests
that the remarkable tendency of cells to be attached to the glass
is probably ascribed to its high zeta potential which may arise from
fixed silicate anions. The pronounced cell attachment to the FN-

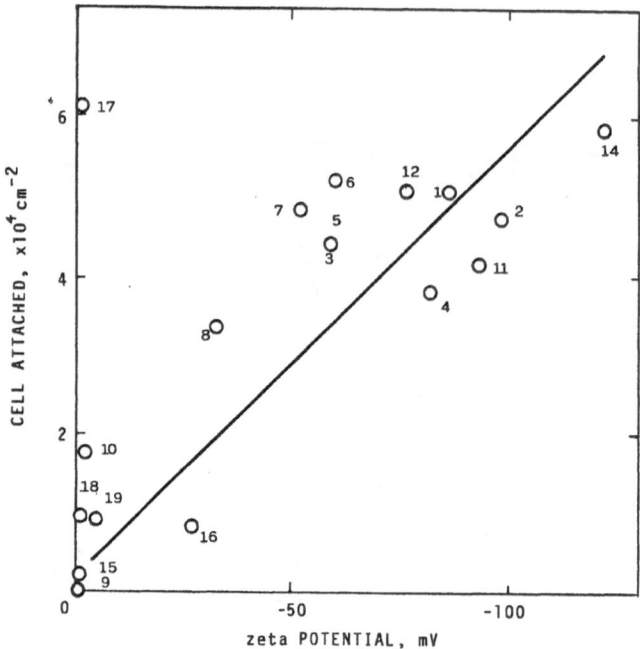

Fig.7. Effect of zeta potential of surfaces on L cell adhesion
        at 37 °C. (60 min incubation using 1.14 x 10^5 L cells/cm^2
        without FCS)
        (See Fig.3 for explanation of film material type code)

grafted PE cannot be explained in terms of water wettability nor zeta
potential.

The above findings that the cell attachment becomes more enhanced
with the increasing negative charge of the substrate surface are
somewhat confusing. The cell surface has generally a negative charge,
which must exhibit a repulsive interaction against a similarly negatively
charged surface. If such an electrostatic interaction plays an important
role in the cell attachment, divalent cations like Ca^{2+}, present in
the culture medium, should contribute significantly to the cell attachment. In fact, the addition of EDTA by 10 mM to the medium resulted
in a reduction of the number of cells attached, as shown in Figure
8. The striking result brought about by the EDTA addition is an
appreciable decrease in the number of cells attached to the glass
and the FN-grafted surface both of which would otherwise provide
a substrate excellent for the cell culture. Gardner and Hanna[19] have
also reported that the cell attachment is reduced by about 50 % in
the presence of EDTA.

DISCUSSION

It is well known that various modes of molecular interaction
are operative when two surfaces come into contact. Among the important
molecular interactions are the van der Waals interaction, the electrostatic interaction, the hydrophobic interaction, the hydrogen bonding,

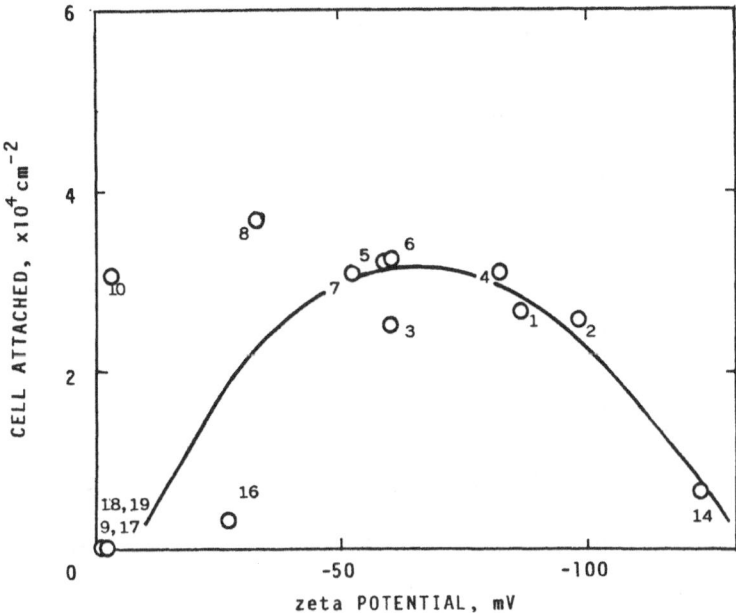

Fig.8. Effect of zeta potential of surfaces on L cell adhesion
in the presence of 10 mM EDTA. (60 min incubation
using 1.14×10^5 L cells/cm^2 without FCS)
(See Fig.3 for explanation of film material type code)

and the charge transfer interaction. In addition, the biospecific
interaction or affinity interaction characteristic to pairs such
as enzyme-substrate, antigen-antibody, acceptor-receptor, should
be mentioned in the biological system. However this interaction
can be, in principle, included in the normal physicochemical interactions as described above.

In the case of biomaterials implanted in body, a quantitative
analysis of the interaction between the material and the living body
is highly complicated, because other substances such as water molecules,
ions, saccharides, proteins are existing between the two interacting
surfaces. Nevertheless, we may be allowed to make an assumption
that other interactions than the van der Waals, the electrostatic
(Coulombic), and the biospecific (affinity) interactions are insignificant in the present case.

In a previous study[13] we evaluated the work of adhesion between
a protein and different polymer surfaces in the surroundings of water
as a function of the interfacial free energy between the polymer
surface and water. The surface free energies of the polymers and
the protein as well as their interfacial free energy against water
were estimated under an assumption that only the van der Waals interaction should be operative in the adhesion. The conclusion attained
in the study was as follows; when the polymer surface is extremely
hydrophilic, adhesion (adsorption) of the protein would not take
place, but with decreasing hydrophilicity or increasing hydrophobicity
of the surface, the amount of protein adsorbed significantly increases
to a maximum and then gradually decreases. This dependence of the

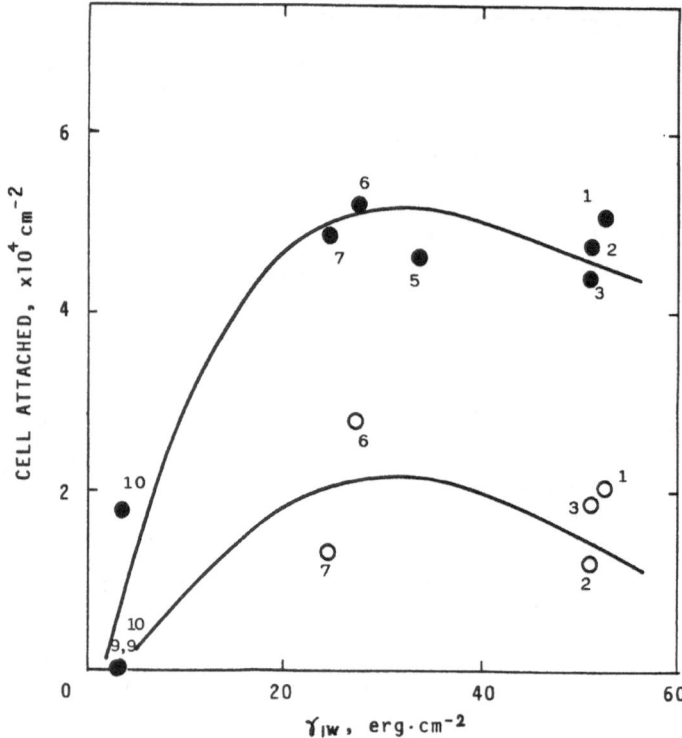

Fig.9. Effect of the interfacial free energy γ_{lW} of surfaces against water on L cell adhesion at 4 °C and 37 °C. (60 min incubation using 1.14×10^5 L cells/cm^2 without FCS) (See Fig.3 for explanation of film material type code) (○): 4 °C, (●): 37 °C

protein adsorption on the hydrophilicity of the polymer surfaces is quite similar to those observed in Figs.3 to 6. Although protein and cell are basically different in surface properties, the difference would not be so important in the discussion on their interaction to a polymer surface. From the data on contact angles the interfacial free energy between the polymer surface and water can be calculated for the films unless they have any grafted chain on the surface. Figure 9 demonstrates the plot of the number of attached L cells against the interfacial free energy. As is seen, no substantial difference is observed in the dependence of cell attachment in Fig.3 and Fig.5, to that shown in Fig.9.

This finding does not mean that the attachment of cells to the foreign surfaces can be explained merely in terms of the van der Waals interaction. It is true that the van der Waals force is very important in the cell adhesion, but one cannot deny the contribution of the Coulombic force to the adhesion as is evident from Figs.7 and 8, where an appreciable influence of the surface potential on the cell adhesion is clearly substantiated. There is not always a correlation between the zeta potential and the hydrophilicity of the substrate surfaces. For instance, cellulose is very hydrophilic and has a very low zeta potential because of nonionic OH groups, while glass is very hydrophilic but has a high zeta potential because

Table I. L cell Adhesion onto Protein-grafted
Polyethylene

Surface	Control, x 10^3	% of control		
		+EDTA	4 °C	+FN
FN-grafted	110	5	0	101
Collagen-grafted	16	0	0	512
BSA-grafted	16	16	0	200

of SiO$^-$ groups. Hydrophobic polymers such as PE and silicone exhibit relatively high zeta potentials, probably because small ions may be strongly adsorbed on the surfaces from the aqueous phase. The reason why PVA gives almost zero zeta potential was discussed in detail in a previous paper[20] and it was concluded that the PVA material contacting with water has water-soluble chains immobilized on the surface, resulting in formation of a diffuse layer with a high water content. This diffuse layer should reduce the zeta potential of PVA to almost zero and prevent protein molecules and cells from the strong interaction with the surface.

It has sometimes been pointed out that very wettable surfaces provide substrates excellent for the cell adhesion[5,6,21]. These include glass, corona-treated polystyrene, and sulfonated polystyrene. As they generally possess negatively charged groups on the surfaces, it follows that their excellent cell attaching ability is probably ascribed to the presence of anionic groups on the surface[22]. It is likely that adhesion of cells having a net negative charge onto a material surface having also a negative charge may be effected by insertion of divalent cations like Ca^{2+} in between[19,23-25], leading to a reduction of high repulsive free energy. This explanation is supported by the EDTA effect on the cell adhesion as is shown in Fig.8. Indeed, when the divalent cations are removed from the incubation medium by the addition of EDTA, the cell adhesion is greatly reduced, especially in the case of glass. The fact of incomplete suppression of the cell attachment by the EDTA addition seems to indicate evidence that the cell adhesion is governed not only by the Coulombic interaction but also by the van der Waals interaction, as discussed above.

Finally, it is necessary to discuss the contribution of the affinity interaction to the cell adhesion. It is quite clear that the biological interaction is not available if the adhesion substrate is composed only of synthetic substances. Therefore, the surfaces that have any concern with the affinity interaction are only those grafted with proteins. As is seen from Figs.3, 4, and 6, the experimental data with respect to the FN-grafted surface are markedly deviated from the others, indicating that this surface has a nature substantially different from the others. This is not an unexpected finding, because it is widely recognized that the FN molecule has binding sites for cells[26,27]. Therefore, the extraordinarily high cell adhesion ability observed for the film surface-grafted with FN is easily understandable[11,28-30]. This expectation that the cell attachment to the FN-grafted surface is due to the biospecific interaction, is also supported by the results shown in Figs.5 and 8. Cells do not attach to the FN-grafted surface if the incubation temperature is lowered to 4 °C

Table II. Summary of Cell Attachment and Contribution of Different Interaction

Surface	Typical examples	Contact angle (deg.)	Cell adhesion	Intensity of interaction with cell		
				van der Waals	Coulombic	Biospecific
very hydrophilic, clear, and nonionic	cellulose	10-40	+	++	+	-
very hydrophilic, clear, and ionic	glass	10-40	+++++	++	++++	-
hydrophilic, diffuse, and nonionic	PVA, PAAm-grafted[a]	20-50	+	+	+	-
hydrophilic-hydrophobic	PMMA, PSt[b]	50-80	++++	++++	+++	-
very hydrophobic	silicone, fluorinated	90-110	+++	++	++++	-
protein-grafted	BSA-	40-50	+	+	+	-
protein-grafted	FN-, collagen-[c]	40-50	+++++	+	+	+++++

a) polyacrylamide, b) polystyrene, c) with free FN
Key; - zero, + very weak, ++ weak, +++ moderate, ++++ strong, +++++ very strong

or Ca^{2+} is removed from the medium, simply because the cells will lose the biological activity to bind the FN molecules under such changed conditions[31,32].

As Figs.3 and 6 indicate, BSA and collagen, grafted to the PE film, seem not to exert a strong biospecific force in the cell attachment, but to render the film surface very hydrophilic, that is, to change the clear surface of PE to the diffuse one. It is interesting to note that the cell attachment to the collagen-grafted PE was greatly enhanced by an addition of soluble FN molecules to the incubation medium. The result is given in Table I, where the other results on the protein-grafted surfaces are tabulated. As is apparent, the effect of FN addition is observed most remarkably for the collagen-grafted surface. The strong affinity of collagen to FN[33] gives an clear explanation for this observation.

Table II summarizes qualitatively the effect of the surface properties on the cell attachment and the contribution of the different molecular interactions to the cell attachment. The classification and the contact angle data given in Table II should be considered as only for comparison. As is clearly demonstrated, it may lead to an erroneous conclusion if the observed results on the cell adhesion are interpreted only in terms of the hydrophilicity or the contact angle of the material surfaces. Further, Table II suggests that the material with a diffuse surface is the most suitable as the biomaterial which exerts the minimum interaction with the contacting living body, while the material grafted with FN may be applicable if a strong adhesion with the tissue is required.

ABSTRACT

Cell adhesion to conventional hydrophilic and hydrophobic films as well as polyethylene(PE) films grafted with synthetic and biological polymers were all studied in a serum-free medium using L cells and HeLa S3 cells. The number of cells attached after one hour of incubation at 37 °C was plotted against the water contact angle of the film. For both of these cell groups, the experimental data obtained, except for the glass and the fibronectin(FN)-grafted PE, fit on a single curve which exhibits a maximum at the contact angle around 70°. On the other hand, the number of cells attached varies to a first approximation linearly with the zeta potential of the surfaces with the exception of the FN-grafted PE. Few cells attached to the almost zero zeta potential surfaces. These results indicate that the cell attachment to the substrates cannot be explained simply in terms of surface wettability nor surface potential. It was therefore concluded that the van der Waals interaction and the Coulombic interaction were concurrently operating between the cells and the synthetic substrates, while the cell attachment to the FN-grafted surface was due mainly to the biospecific interaction.

REFERENCES

1. For instance, "Biomaterials: Interfacial Phenomena and Applications", S.L. Cooper and N.A. Peppas eds., Adv. Chem. Ser. 199, ACS, Washington, D.C. 1982.
2. For instance, A. Ludwicka, B. Jansen, T. Wadström, L.M. Switalski, G. Peters, and G. Pulverer, Attachment ot Staphylococci to Various Synthetic polymers, in "Polymers as Biomaterials", S.W. Shalaby, A.S. Hoffman B.D. Ratner and T.A. Horbett eds., Plenum Publishing Corporation, 1984, p.241.

3. R.J. Good, Theory of the Adhesion of Cells and the Spontaneous Sorting-out of Mixed Cell Aggregates, J. theor. Biol. 37: 413 (1972).

4. D. Gingell, Computed Force and Energy of Membrane Interaction, J. theor. Biol. 30: 121 (1971).

5. S.C. Dexter, Influence of Substratum Critical Surface Tension on Bacterial Adhesion — in situ studies, J. Colloid Int. Sci. 70, No.2: 346 (1979).

6. D.R. Absolom, A.W. Neumann, W. Zingg, and C.J. van Oss, Thermodynamic Studies of Cellular Adhesion, Trans. Am. Soc. Artif. Intern. Organs 25: 152 (1979).

7. A.W. Neumann, O.S. Hum, D.W. Fransis, W. Zingg, and C.J. van Oss, Kinetic and Thermodynamic Aspects of Platelet Adhesion from Suspension to Various Substrates, J. Biomed. Mater. Res. 14: 499 (1980).

8. R. Srinivasan and E. Ruckenstein, Kinetically Caused Saturation in the Deposition of Particles or Cells, J. Colloid Int. Sci. 79, No.2: 390 (1981).

9. E. Ruckenstein and R. Srinivasan, Comments on Cell Adhesion to Biomaterial Surfaces: The Origin of Saturation in Platelet Deposition — Is it Kinetic or Thermodynamic ? J. Biomed. Mater. Res. 16: 169 (1982).

10. H.K. Kleiman, R.J. Klebe, and G.R. Martin, Role of Collagenous Matrices in the Adhesion and Growth of Cells, J. Cell Biol. 88: 473 (1981).

11. F. Grinnell, Cellular Adhesiveness and Extracellular Substrata, Int. Rev. Cytol. 53: 65 (1978).

12. Y. Tamada and Y. Ikada, Dynamic Adsorption and Desorption of Proteins, Markromol. Chem. Suppl. 9: 85 (1985).

13. Y. Ikada, M. Suzuki, and Y. Tamada, Polymer Surfaces Possessing Minimal Interaction with Blood Components, in "Polymers as Biomaterials", S.W. Shalaby, A.S. Hoffman, B.D. Ratner, and T.A. Horbett eds., Plenum Publishing Corporation, 1984, p.135.

14. M. Vuento and A. Vaheri, Purification of Fibronectin from Human Plasma by Affinity Chromatography under Non-Denaturing Conditions, Biochem. J. 183: 331 (1979).

15. R.V. van Wagenen and J.D. Andrade, Flat Plate Streaming Potential Investigations: Hydrodynamics and Electrokinetic Equivalency, J. Colloid Int. Sci. 76, No.2: 305 (1980).

16. R.L. Schnaar, P.H. Weigel, M.S. Kuhlenschmidt, Y.C. Lee, and S. Roseman, Adhesion of Chicken Hepatocytes to Polyacrylamide Gels Derivatized with N-Acetylglucosamine, J. Biol. Chem. 253, No.21: 7940 (1978).

17. T. Matsunaga and Y. Ikada, Dispersive Component of Surface Free Energy of Hydrophilic Polymers, J. Colloid Int. Sci. 84, No.1: 8 (1981).

18. G.M. Kolodny, Effect of Various Inhibitors on Readhesion of Trypsinized Cells in Culture, Exp. Cell Res. 70: 196 (1972).

19. J.L. Gardner and M.H. Hanna, Calcium, Cellular Adhesion and Aggregation Competence in the Cellular Slime Mold Polysphondylium Violaceum, Exp. Cell Res. 137: 169 (1982).

20. Y. Ikada, H. Iwata, F. Horii, T. Matsunaga, M. Taniguchi, M. Suzuki, W. Taki, S. Yamagata, Y. Yonekawa, and H. Handa, Blood Compatibility of Hydrophilic Polymers, J. Biomed. Mater. Res. 15: 697 (1981).

21. F. Grinnell, M. Milan, and P.A. Srere, Attachment of Normal and Transformed Hamster Kidney Cells to Substrata Varying in Chemical Composition, Biochem. Med. 7: 87 (1973).

22. N.G. Maroudas, Adhesion and Spreading of Cells on Charged Surfaces, J. theor. Biol. 49: 417 (1975).

23. C. Rappaport, Monolayer Cultures of Trypsinized Monkey Kidney Cells in Synthetic Medium. Application to Poliovirus Synthesis, Proc. Soc. Exptl. Biol. Med. 91: 464 (1956).

24. L. Weiss, Studies on Cellular Adhesion in Tissue Culture, Exp. Cell Res. 21: 71 (1960).

25. L. Weiss, The Adhesion of Cells, Int. Rev. Cytol. 9: 187 (1960).

26. K.M. Yamada and K. Olden, Fibronectins-Adhesive Glycoproteins of Cell Surface and Blood, Nature 275: 179 (1978).

27. E. Ruoslahti, E.G. Hayman, E. Engvall, W.C. Cothran, and W.T. Buther, Alignment of Biologically Active Domains in the Fibronectin Molecules, J. Biol. Chem. 256, No.14: 7277 (1981).

28. R.C. Hughes, S.D.J. Pena, J. Clark, and R.R. Dourmashkin, Molecular Requirement for the Adhesion and Spreading of Hamster Fibroblasts, Exp. Cell Res. 121: 307 (1979).

29. F. Grinnell and M.K. Feld, Adsorption Characteristics of Plasma Fibronectin in Relationship to Biological Activity, J. Biomed. Mater. Res. 15: 363 (1981).

30. M. Hook, K. Rubin, A. Oldberg, B. Öbrink, and A. Vaheri, Cold-Insoluble Globulin Mediated the Adhesion of Rat Liver Cells to Plastic Petri Dishes, Biochem. Biophys. Res. Commun. 79: 726 (1977).

31. R.J. Klebe, Cell Attachment to Collagen: The Requirement for Energy, J. Cell Phys. 86: 231 (1975).

32. R.J. Klebe, J.R. Hall, P. Posenberger, and W.D. Dickey, Cell Attachment to Collagen: The Ionic Requirements, Exp. Cell Res. 110: 419 (1977).

33. E. Engvall and E. Ruoslahti, Binding of Soluble Form of Fibroblast Surface Protein, Fibronectin, to Collagen, Int. J. Cancer, 20: 1 (1977).

IMMOBILIZATION OF ACID PHOSPHATASE IN POLY-HEMA GELS

M. Cantarella, L. Cantarella, A. Gallifuoco, L. Pezzullo, and F. Alfani

Chemical Engineering Department, University of Naples P.le Tecchio, 80125 Naples, Italy

INTRODUCTION

Natural polymers, such as gelatin, sepharose, carrageenan, cellulose, albumin, are worldwide used either as carrier for biocatalyst immobilization or as stabilizing agent for preventing thermal deactivation. However, these materials possess poor mechanical properties, low temperature tollerance and reduced resistance to microbial attack and, consequently a large industrial utilization is unexpected.

Therefore, many methods have been proposed hitherto to replace natural with synthetic polymers which could represent a valid solution to the previously cited bottlenecks.

Enzymes and cells have been immobilized in polymeric matrices of acrylamide gels (1,2), in membrane of polyvinyl alcohol (3), in polyurethane matrices (4), and other synthetic carriers. A very exhaustive review on immobilized microbial cells was recently presented (5).

So far, there has been a great interest in the immobilization of enzymes and cells in hydrogels prepared using acrilic monomers. Gels were prepared (6) by heterogeneous suspension copolymerization of acrylate and methacrylate monomers containing the hydroxyl group and the diacrylates and dimethacrylates in the presence of inert solvents, the concentration of which controls the internal structure of the gel, namely, pore size distribution, specific surface and the number of active groups. In this method, such macroporous carriers were activated with bromocyanogen for binding biologically active compounds.

A modification of the previous immobilization technique was proposed (7). Glycidylmethacrylate was graft polymerized on different polysaccharide supports in the presence of H_2O_2 and Fe^{+2}. Horseradish peroxidase and glucose oxidase were immobilized by direct reaction on sepharose copolymers with a synthetic polymer percentage ranging from 25 to 50%. These immobilized enzyme systems should present interesting properties since the matrix consists of a hydrophilic skeleton and of hydrophobic regions, the synthetic polymer, to which the enzyme is covalently bound. However, the coupling efficiency was in general low, 0.8 - 1.5%, and the presence of limitations to diffusion of substrates and products indicates that significant conformational changes and mobility reduction of the enzymatic macromolecules are introduced during immobilization

Further on, direct binding through oxirane groups was shown to be relatively slow and to bring about an activity loss at low enzyme concentration (8). A more efficient immobilization of penicillin amidase on bead-form macroporous carriers of glycidyl methacrylate and ethylene dimethacrylate copolymers was achieved after chemical modification of the amino matrix with glutaraldehyde.

The advantages of radiation induced polymerization for immobilization of enzyme are discussed in (9). The desired proposal of this immobilization method is to trap a considerable part of the enzyme on the porous surface of carrier polymers, thus improving the disadvantages of methods previously described. In a very preliminar investigation (10) glucoamylase, invertase and β-galactosidase were immobilized by using 2-hydroxyethyl acrylate and dimethylacrylamide under γ-ray irradiation. Enzyme leakage from the gel was a major problem to eliminate, since increased irradiation improves enzyme retention but the activity of the gels is very low. Only in the case of dimethylacrylamide gels, leakage could be eliminated by irradiation of 2.0 Mrad and the gel possesses very high activity. Pursuing the optimization of this immobilization method, it was observed (9) that enzyme radiation damage could be sufficiently retarded at low temperature, between -24°C and -78°C, and that the porosity of the polymerized composite can be varied by cooling rate and temperature. The enzyme is either distributed on the polymer surface or partially isolated in the pores and the polymer matrix. Therefore, enzyme activity recovery varies with carrier porosity which depends on monomer concentration. Some extent of enzyme leakage was still observed where the composites had a large porosity. Finally, the effect of polymer matrix on enzyme activity and thermal stability of thermolysin immobilized by radiation polymerization in polymer of 2-hydroxyethylmethacrylate, 2-hydroxyethyl acetate, trimethyrolpropane trimethacrylate and methoxytetraethyleneglycol was recently studied (11). The body of experimental results indicates that the optimum monomer concentration varies with the size of immobilized enzyme. As the dimension of the protein molecule decreases, the enzyme activity at high monomer concentrations (70-90%, v/v) increases owing to the appearance of the entrapped enzymes. Further on, the preparation of the immobilized enzymes using relatively high monomer concentrations was found to improve thermal stability, although the enzyme activity was slightly lower.

A third method for enzyme immobilization in hydrogels was originally proposed in (12). The enzyme, glucose oxidase, was entrapped in 2-hydroxyethyl methacrylate gel plus diethyleneglycoldimethacrylate as crosslinking agent. The reaction was carried out at room temperature for 14 hours and the initiator, di(sec-butyl)peroxydicarbonate, was used as 1% solution in methanol.

The microenvironment of the hydrogels affects the reactivity of the enzymes and the preparation of immobilized enzymes with desired properties can be achieved controlling the hydrogel structure. Polyvinylpyrrolidone (PVP) was added to the polymerization solution for varying the structure and the properties of the gel matrix. However, in such gels of poly-HEMA and PVP the enzyme is probably not homogeneously distributed and the gels appear to be heterogeneous and probably have a microporous stucture. Therefore activity is reduced for larger particle size because of diffusional effects and for very small particles because of enzyme elution in the grinding operation. Nevertheless, this method of enzyme entrapment in hydrogels is promising especially because the biocatalyst is protected from bacterial degradation since bacteria and their proteolytic enzymes cannot enter the gel matrix. Recently, this method was studied for the immobilization of yeast cells entrapped in poly-HEMA plus ethylenedimathacrylate as crosslinking agent (13). Enzyme immobilization in the same carrier with acid phosphatase as model is discussed hereafter.

The purpose of the present study was to propose a modified immobilization method in which polymerization temperature is reduced, reaction time is shortened by UV radiation and the porosity of polymer matrix is changed by the addition of hydrophilic low molecular weight polymers in order to reach a satisfactory recovery of enzyme activity in the gel and an improved thermal stability of the biocatalyst.

EXPERIMENTAL

Materials

The experiments were carried out with acid phosphatase (EC 3.1.3.2) from potato and its artificial substrate, p-nitrophenyl phosphate (sodium salt), supplied from Boehringer Biochemia (Italy). The monomer, 2-hydroxyethylmethacrylate (HEMA) was from Rohm Gmbh (Germany), the crosslinking agent, ethylendimethacrylate (EDMA), was from BDH (England) while the initiators of the polymerization reaction, H_2O_2 and $FeSO_4$, were from Baker (Holland) and BDH (England) respectively. Polyethylene glycol (PEG) and polyvinylpyrrolidone (PVP) were from Merck (F.R.G.) and Ega-CHEMIE (F.R.G) respectively.

All other chemicals were pure grade reagents from usual sources.

Preparation of immobilized enzyme

Poly-2-hydroxyethylmethacrylate (Poly-HEMA), with entrapped acid phosphatase, was prepared by free radical polymerization of 2-hydroxyethylmethacrylate using ethylenedimethacrylate as crosslinking agent and H_2O_2 plus Fe^{+2} as initiators. The immobilization technique adopted in this study is suitable for all mixtures of HEMA and water in which the volume percentage of organic species ranges from 30 to 100%.

In a standard preparation, 6 ml of water are added to 6 ml of HEMA and 60 mg of EDMA and traces of oxygen, which is an inhibitor of polymerization reaction, are eliminated from the solution by bubbling N_2 at 4 °C for 10 minutes. Afterwards, aqueous solutions of enzyme, H_2O_2 and $FeSO_4$ are rapidly and consecutively added to give a final concentration of 1.2 mg/ml, 5 mM and 1.25 mM respectively and a total volume of 20 ml. The solution is maintained in N_2 atmosphere for 5 more minutes at 4 °C and stirred. When low molecular weight polymers, PVP, PEG, are used to modify the porosity of the carriers, a concentrated solution is prepared in advance and the addition occurs before those of initiators and enzyme.

Finally, the mixture is introduced in the volume between two flat pyrex plates perfectly sealed with rubber and silicone and allowed to react at 4 °C. The time required for the polymerization slightly varies between two preparations and mainly depends on reaction temperature and the presence of UV radiation. At the above temperature and with UV source at 365 nm, immobilization of acid phosphatase in poly-HEMA is accomplished in approximately one hour. Then, gel is removed from the plates, rinsed several times, dried under vacuum at refrigerator temperature, and ground with a mortar and pestle. The gel particles are sieved and the fraction smaller than 30 mesh was used for kinetic studies.

Assay of acid phosphatase activity

The enzyme assay was carried out colorimetrically measuring the adsorption at 405 nm after alkalinization of the sample with an equal volume of 1 M NaOH solution and determining the p-nitrophenol liberated at 30 °C during 10 min of incubation, in 2 ml of reaction volume and 2 mM of p-nitrophenyl phosphate in 50 mM sodium citrate/citric acid buffer, pH 5.6. A molecular adsorption coefficient of 18.500 litre per mol and cm at 405 nm was used.

Activity recovery, A_R, was defined as a ratio of immobilized enzyme activity to that of native enzyme of the same quantity as used for immobilization.

RESULTS AND DISCUSSION

Effect of polymerization chemicals on native enzyme activity

Enzyme damage, caused by the chemicals present in the polymerization mixture, can be prevented by limiting their concentrations and the temperature at which the immobilization is carried out. The percentage of ethylendimethacrylate is very low, the highest level at the beginning of polymerization is 0.3 wt%, and it was proved that for prolonged time enzyme activity is unaffected by its presence. Similarly, polyvinylpyrrolidone and polyethylenglycol, in the range of explored percentages from 1 to 5 wt%, do not interfere with acid phophatase activity.

On the contrary, it was observed that HEMA and the initiators can denature the native enzyme and their effect on the activity was investigated to find the suitable conditions for immobilization.

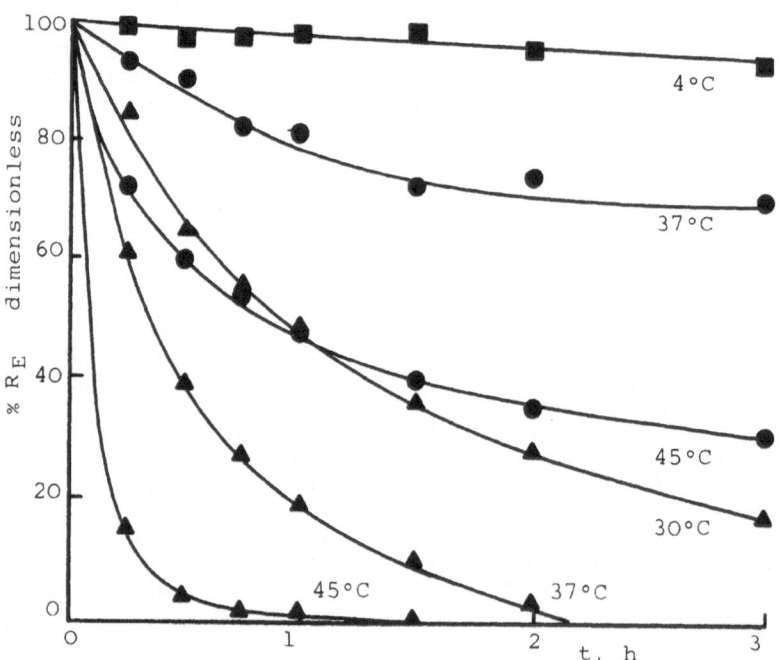

Fig. 1 - Residual activity of native acid phosphatase versus time at different HEMA concentrations and temperature.
Purified : ▲ - 30 wt%, ● - 3 wt%. Commercial : ■ - 30 wt%

An aqueous solution of native enzyme of the same concentration as used for immobilization, 1.2 mg/ml, was stored at different temperatures varying the concentration either of HEMA or of initiators. At regular time intervals, the residual activity of the enzyme, R_E, was assayed and monitored. The relative activity of the native enzyme, defined as the ratio of the activity of chemically contaminated enzyme to that of the same amount of uncontaminated enzyme, was calculated and plotted in Fig.s 1 and 2 versus time as function of chemicals concentration and temperature.

The rate of acid phosphatase deactivation is very fast at 45 °C and 37 °C whatever HEMA concentration was and becomes almost zero independently of monomer concentration at temperature below 10 °C. At intermediate values of temperature the rate of enzyme deactivation depends on the weight percent of HEMA in the incubation volume. The results of Fig. 1 refer to tests performed with distillated HEMA, purged of hydroquinone traces, and of hydroquinonemonomethylether, 40 ppm and 220 ppm, which are present in commercial HEMA and are used as stabilizers. However, commercial HEMA poorly damages the enzyme at 4°C, since after two hours, 95.3% of initial biocatalyst activity is safe at 30 wt% of monomer. Therefore, it is obvious that damage of organic species can be remarkably reduced by carrying out the enzyme immobilization at low temperature. On the other hand, according to Fig. 2, mixtures of $FeSO_4$ and H_2O_2, which have been preferred to other initiators since their activity is scarcely dependent on temperature, even at 4 °C can induce significant decay of catalyst activity if their concentration is increased. Low concentrations of these chemicals are not suitable because of lower polymerization yields after 4-5 hours, and the preferred conditions were 4 °C, 1.25 mM and 5 mM concentration of $FeSO_4$ and H_2O_2 respectively, which, during 60-90 minutes of immobilization length gives rise to 4.2% of acid phosphatase deactivation at most.

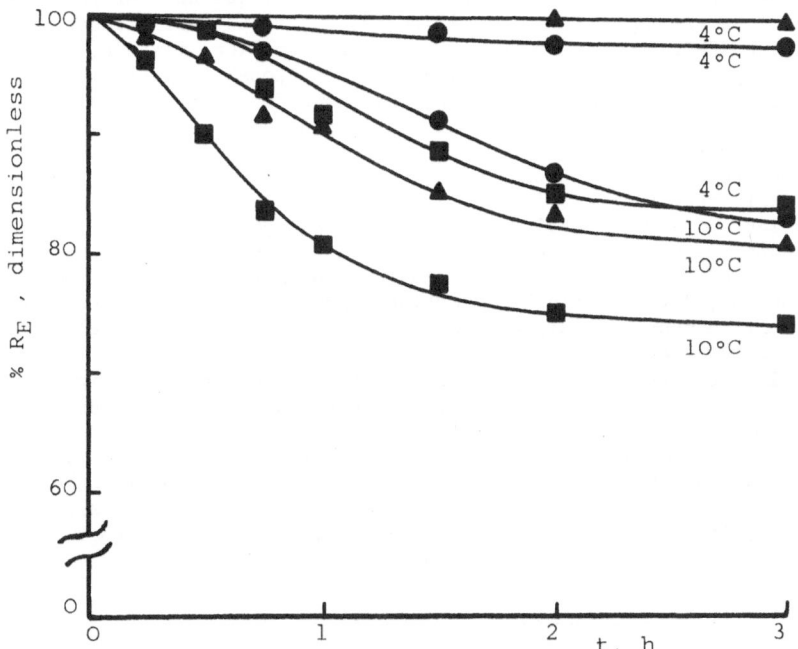

Fig.2 - Residual activity of native acid phosphatase vs. time at various initiator concentrations and temperature: ■ -2.5 $FeSO_4$ and 10 mM H_2O_2, ● -1.25 $FeSO_4$ and 5 mM H_2O_2, ▲ -0.25 $FeSO_4$ and 1.25 mM H_2O_2

Optimum condition of the immobilization reaction

Once the optimum temperature and concentration of initiators were determined in such a way as to minimize acid phosphatase deactivation, different immobilization procedures were tested in order to reach high activity recovery and to shorten the time needed to produce a well formed gel.

At first, the concentration of crosslinking agent (EDMA) was varied from 1 to 5% by weight of the organic phase but neither activity recovery, A_R, nor polimerization length, t_p, showed a clear trend. The values were always comparable each other, since t_p ranges between 100 min and 150 min, and A_R varies from 4.81 to 10%. In a second set of preparations, a saturating concentration of the specific substrate for acid phosphatase (p-nitrophenylphosphate) was added into the polymerization mixture, however in this condition the enzyme was not enough stabilized by its substrate and the activity recovery and time of immobilization were in the range quoted above.

On the other hand, some advantages were observed using a UV source during the second phase of reaction when the mixture is allowed to react in a pyrex mould. The film of gel assumes a firm rubbery aspect in approximately 50 minutes, and the immobilized enzyme activity is close to 14% of the native enzyme one.

However, the observed low values of activity recovery compared with the expected enzyme deactivation which can be caused by polymerization chemicals, suggested the conclusion that most of the biocatalyst is entrapped in the gel but is segregated and cannot react with the substrate.

Different inert molecules were used to increase the characteristic dimensions of gel pores and to facilitate the mass transfer of reactant and product in the carrier.

The addition of NaCl (10^{-5} mM) did not improve the activity recovery and caused a small increase of time required for gel formation (1h). In a new series of immobilization tests synthetic linear polymers, PVP and PEG, were introduced in the reaction mixture, and their percentage was varied between 1 and 5 % of total weight. The molecular weight of PVP was 10.000 daltons, while two different molecular weights, 6.000 and 1.500 daltons, of PEG were tried. Linear synthetic polymers, such as PVP and PEG, were satisfactory used (14) to retard enzyme thermal deactivation and in polymerization mixtures to modify the physical properties of film and membranes. Two benefits were expected by carrying out the immobilization of acid phosphatase in the presence of these polymers. At the end of preparation, during the rinsing of gel, PVP and PEG should be released and leave a much more porous matrix and further on, during polymer formation, should protect the enzyme towards microenvironmental modifications especially caused by local hot spots and attack of chemicals.

The results partially confirmed the aims of the investigation since the activity recovery of the enzyme was further on increased up to 27.5%

Under the different conditions explored, A_R varied from 21.1 to 27.5% depending on the polymer and its concentration in the mixture. This latter part of the investigation is still in progress. Nowadays, the results indicate that a larger percentage of polymer is needed where low molecular weight polymers are used as additive.

Kinetic properties of immobilized acid phosphatase

In Table 1 values of the Michaelis constant, K_m, and the activation energy, E_a, for both native and immobilized acid phosphatase are reported. The enzymatic reaction obeys a Michaelis-Menten rate equation

Table 1 : Kinetic parameters of native and immobilized acid phosphatase

	A_R, %	K_m , mM	E_a, cal/mol
Native enzyme		0.36	9695
Poly-HEMA	10.0	0.38	8200
Poly-HEMA gel with 3% PVP	19.2	0.48	7809
Poly-HEMA gel with 3% PEG 6000	23.4	0.42	8620
Poly-HEMA gel with 5% PEG 1500	26.0	0.41	7266

and the temperature dependence in the range from 25 to 45 °C follows the Arrhenius law.

For the immobilized enzyme K_m is slightly larger than the corresponding value for the free enzyme. This result could be determined either by a lower specificity of enzyme in the gel towards substrate or more probably, by the presence of diffusional resistance to substrate transport in the carrier. According to this second hypothesis reactant concentration in the gel is smaller than the one in the bulk and the calculated values of Michaelis constant are apparent. Moreover, since in the presence of reaction mechanisms partially controlled by diffusional limitation the activation energy of the process becomes smaller and this finding was observed in the present investigation, the conclusion was reached that p-nitrophenyl phosphate does not freely diffuse in the gel. This result is valid either for the unmodified gel or for the gels prepared using PVP and PEG as additives. However, since the differences of K_m and E_a between native and immobilized enzyme are small, the catalytic reaction is the prevalent rate controlling step.

Enzyme thermal stability

The rate of thermal inactivation of both native and immobilized enzyme was investigated in a membrane reactor according to the method reported in (15). The reactor was a stirred and thermostated Amicon cell (mod.52) equipped with ultrafiltration membranes YM 10, the molecular weight cut-off of which (10.000 daltons) ensure total retention of free acid phosphatase. The experiments were carried out with 0.5 mg of native acid phosphatase and with 100 mg of gel equivalent to 0.37 mg of enzyme and the enzyme activity was measured by using a 2 mM PNPP concentration in 50 mM sodium citrate/citric acid buffer, pH 5.6

The rate of p-nitrophenol formation, r, is reported in the semilogarithmic plot of Fig. 3 versus process time, t, and show an apparent increase during the first 10-15 hours. This initial transient phase is due to product accumulation in the reactor and can be predicted by an analysis of the response to an addition of the generated product. The apparent maximum of reaction rate occurs at a process time, the value of which is directly proportional to reactor volume and inversely

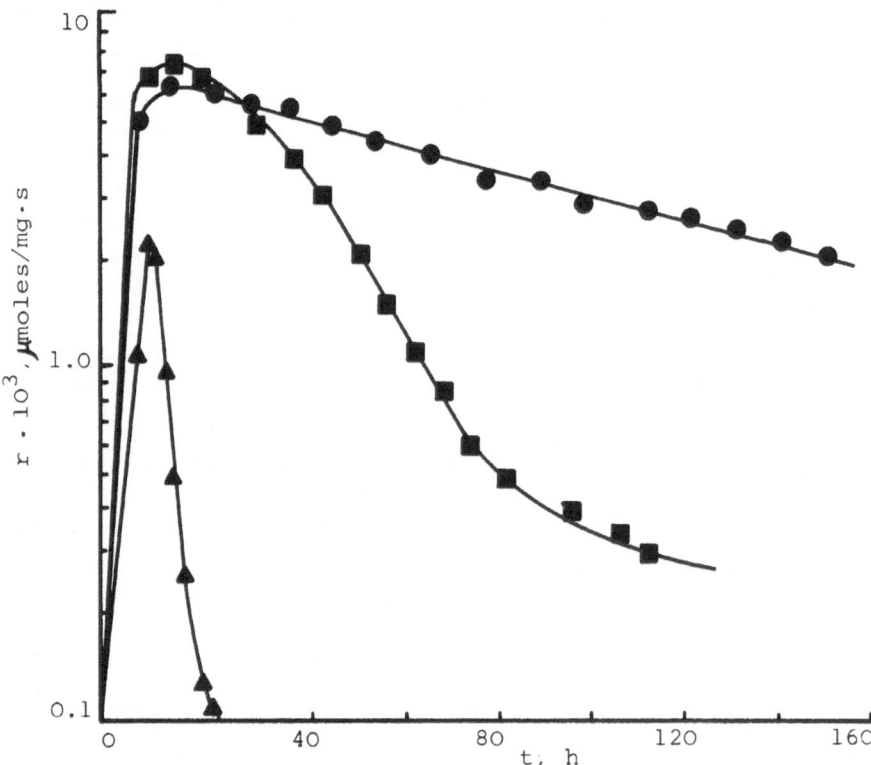

Fig. 3 - Thermal inactivation of acid phosphatase. ▲ - native enzyme at 45 °C. Immobilized enzyme in poly-HEMA plus 5 wt% of PEG 1500, ● - 30 °C, ■ - 45 °C

proportional to flow rate. From this time onwards, any observed decrease in the reaction rate is caused only by a reduction in the concentration of active enzyme and therefore the rate of thermal deactivation of the enzyme can be easily increased using this portion of the curve.

This method becomes unusable when the rate of inactivation is very fast and the enzyme half life, defined as the time required to reach 50% of catalyst deactivation, is approximately equal or shorter than the time needed for the steady-state situation to be attained.

The results plotted in Fig. 3 show that the immobilization in gel has largely improved enzyme thermal stability and that, of course, the rate of thermal inactivation is an increasing function of the temperature. In the case of a limited activity decay, such as the situation observed at 30 °C, the linear behaviour in the semilogarithmic plot indicates that deactivation obeys a first order mechanism. At higher temperature, 45 °C, the data show different rates of deactivation with process time, and this phenomenon, which is commonly observed (16,17), was shown (18) to be determined for acid phosphatase by the transition from the native structure to the totally inactive one throughout a series mechanism of first order irreversible reactions.

Finally, the apparent lower reaction rate exhibited by native acid phosphatase with respect to the gel at 45 °C depends on the large extent of enzyme deactivation occured during the initial transient phase of the system.

ACKNOWLEDGEMENT

This study was supported by a grant for scientific research of the Ministry of Education and in part by research contract CEE-GB-075-I-(TT) in the framework of the Biomolecular Engineering Programme.

LIST OF SYMBOLS

A_R = activity recovery, dimensionless
E_a = activation energy, cal/mol
K_m = Michaelis constant, mM
r = reaction rate, μmol/mg sec
R_E = relative activity, dimensionless
t = process time, h
t_p = polimerization time, min

REFERENCES

1. Morikawa Y., Karube I., Suzuki S., 1980, Continuous production of Bacitracin by immobilized living whole cells of Bacillus sp., Biotechnol. Bioeng., 22, 1015-1023

2. Freeman A., Haronowitz A., 1981, Immobilization of microbial cells in crosslinked, prepolymerized, linear polyacrylamide gels: antibiotic production of immobilized Streptomyces clavuligerms cells, Biotechnol. Bioeng., 23, 2747-2759

3. Imai K., Shiomi T., Sato K., Fujishima A., 1983, Preparation of immobilized invertase using poly(vinyl Alcohol) membrane, Biotechnol. Bioeng., 25, 613-617

4. Klein J., Kluge M., 1981, Immobilization of microbial cells in polyurethane matrices, Biotechnol lett., 3, 65-70

5. Fukui S., Tanaka A., 1982, Immobilized microbial cells, Ann. Rev. Microbiol., 36, 65-70

6. Turkova J., Hubslkova O., Krivakova M., Coupek J., 1973, Affinity chromatography on hydroxyalkyl methacrylate gels. I. Preparations of immobilized chymotrypsin and its use in the isolation of proteolytic inhibitors, Biochem. Biophys. Acta, 322, 1-8

7. D'Angiuro L., Cremonesi P., Cantafi R., Mazzola G., Vecchio G., 1980, Enzyme immobilization on polyglycidylmethacrylate graft-copolymer of different polysaccharides, Angew. Makromol. Chem., 91, 161-178

8. Drobnik J., Sandek V., Svec F., Kalal J., Voytisek V., Barta M., 1979, Enzyme immobilization techniques on poly(glycidyl methacrylate-co-ethylene dimethacrylate) carrier with Pennicillin Amidase as model, Biotechnol. Bioeng., 21, 1317-32

9. Kaetsu I., Kumakura M., Yoshida M., 1979, Enzyme immobilization by radiation induced polymerization of 2-hydroxyethylmethacrylate at low temperatures, Biotechnol. Bioeng., 21, 847-861

10. Maeda H., Suzuki H., Yamauchi A., Sakimac A., 1975, Preparation of immobilized enzymes from acrylic monomers under γ-ray irradiation, Biotechnol. Bioeng., 17, 119-128

11. Kumakura M., Kaetsu I., Kobayashi T., 1984, Properties of thermolysin immobilized in polymer matrix by radiation polymerization, Enzyme Microb. Technol., 6, 23-26

12. Hinsberg I., Kapoulas A., Korus R., O'Driscoll K., 1972, Gel entrapment of enzymes: Kinetic studies of immobilized glucose oxidase, Biotechnol. Bioeng., 16, 159-168

13. Cantarella M., Migliaresi C., Tafuri M.G., Alfani F., 1984, Immobilization of yeast cells in hydroxyethylmethacrylate gels, <u>Applied Microbiol & Biotechnol.</u>, <u>20</u>, 233-238
14. Alfani F., Cantarella M., Cirielli G., Scardi V., 1984, The use of synthetic polymers for preventing enzyme thermal inactivation, <u>Biotechnol. Lett.</u>, <u>6</u>, 345-350
15. Greco G., Albanesi D., Cantarella M., Gianfreda L., Palescandolo R., Scardi V., 1979, Enzyme inactivation and stabilization studies in ultrafiltration reactor, <u>Eur. J. Appl. Microbiol. Biotechnol.</u>, <u>8</u>, 249-261
16. Alfani F., Albanesi D., Cantarella M., Scardi V., 1982, Effects of temperature and shear on the activity of acid phosphatase in a tubular membrane reactor, <u>Enzyme Microb. Technol.</u>, <u>4</u>, 181-184
17. Bouin J.C., Hultin H.O., 1982, Stabilization of glucose oxidase by immobilization/modification as a function of pH, <u>Biotechnol. Bioeng.</u> <u>24</u>, 1225-1231
18. Gianfreda L., Marrucci G., Grizzuti N., Greco G., 1984, Acid phosphatase deactivation by a series mechanism, <u>Biotechnol. Bioeng.</u>, <u>26</u>, 518-527

INTERACTIONS BETWEEN DERIVATIVES OF INSOLUBLE POLYSTYRENE AND HUMAN ANTIBODIES TO FACTOR VIII:C

N. Belattar, D. Gulino, J. Jozefonvicz and *Y. Sultan

Laboratoire de Rescherches sur les Macromolécules-Université Paris-Nord, Avenue JB. Clément-93430 Villetaneuse, France
*Laboratoire d'Hémostase, Hôspital Cochin, Unité Inserm 152
27, rue du Faubourg St-Jacques, 75674 Paris, France

SUMMARY

In order to obtain completely synthetic adsorbents mimicking the interaction FVIII:C-AntiVIII:C, crosslinked polystyrene was modified by various amino acids or by some of their derivatives. The syntheses of the resins were achieved by a two step process:crosslinked polystyrene was first chlorosulfonated and subsequently amino acids were attached onto the polymer.Then, the "in vitro" removal of Anti VIII:C from hemophiliac's immunoglobulins G (IgG) was tested by measuring simultaneous adsorptions of either IgG or Anti VIII:C onto the polymer surfaces. Among the different resins, some of them relatively possess specificity towards Anti III:C as they can adsorb 60% of Anti VIII:C and only 16% of IgG from the starting material. Another ones unspecifically adsorb Anti VIII:C as well as the overall IgG.

INTRODUCTION

Human antibodies that neutralize factor VIII procoagulant activity (F VIII:C, antihemophilic factor) are detected in 5% to 10% of polytransfused patients with hemophilia A [1,2] and as autoantibodies [3] that spontaneously developp in previously healthy individuals. These antibodies (Anti VIII:C) cause serious difficulties for affected patients as they inactivate either FVIII:C produced by patients with autoantibodies or any transfused FVIII preparations. So, management of haemorrages in these patients is difficult and may be obtained by intensive plasma exchanges immediately followed by FVIII infusions in order to restore normal haemostasis [4]. So, four liters of replacement fluid which can be plasma protein fraction (PPF) or fresh frozen plasma (FFP) may be usually daily injected for as long as the clinical state dictates. Problems due to plasma exchange procedure such as availability of sufficient blood products of the appropriate groups, hepatitis or LAV/HTLV3 infections [5] and drastic decrease of some proteins with biological activity [6], can be resolved using a technique based on specific immunoadsorption. For instance, Nilsson et al.[7] have reported extracorporeal adsorption on a protein A-Sepharose in a patient with hemophilia B and high titer antibodies specific for Factor IX (AntiFIX). Indeed, since protein A reacts with the Fc part of IgG[8], protein A bound to Sepharose can be used as an immunosorbent for isolating the overall IgG and consequently for AntiFIX removal. The plasma separated intermittently in a cell centrifuge passed through a column of protein A-Sepharose, then the treated plasma was retransfused. Unfortunately, the Anti-FIX, as well as the total immunoglobulin content, decreased to one-fifth

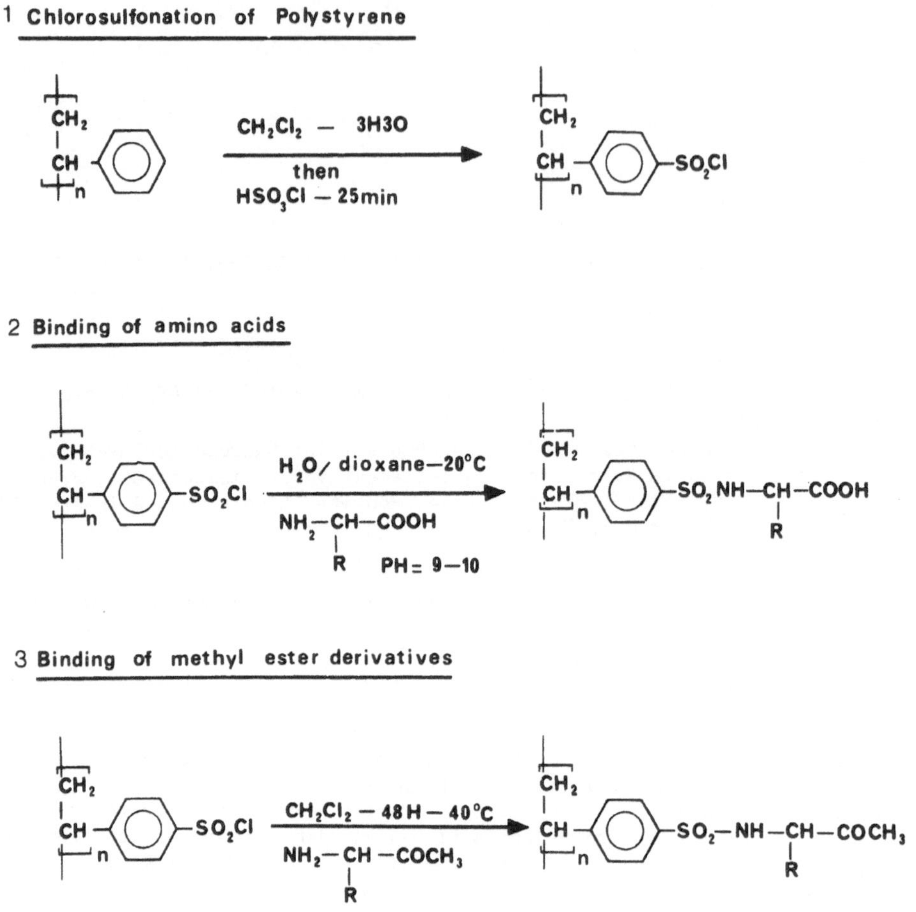
1 Chlorosulfonation of Polystyrene
CH2Cl2 — 3H30
then
HSO3Cl — 25min
SO2Cl
2 Binding of amino acids
H2O/ dioxane—20°C
NH2—CH—COOH
R PH= 9—10
SO2Cl
SO2 NH—CH—COOH
R
3 Binding of methyl ester derivatives
CH2Cl2 — 48 H — 40°C
NH2— CH —COCH3
R
SO2Cl
SO2— NH —CH—COCH3
R

Fig. 1 : Attachement of different amino acid derivatives onto crosslinked polystyrene.

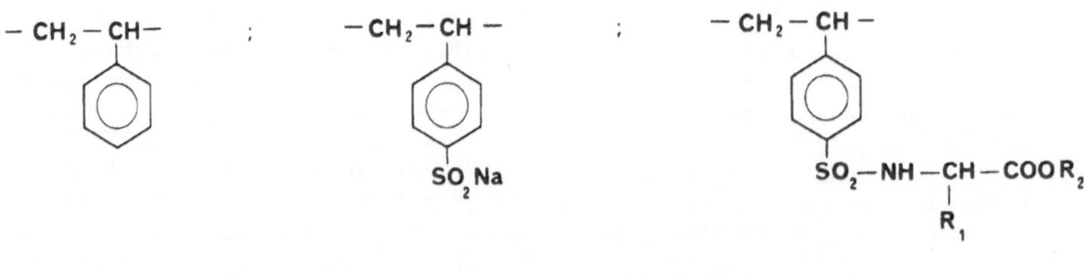
— CH2 — CH—
— CH2 — CH —
SO2 Na
— CH2 — CH —
SO2—NH—CH—COOR2
R1

Fig 2 : Structure of the modified polystyrene.

of the original values. Then, the remaining antibodies were neutralized by infusion of factor IX concentrate. Contrary to conventional plasma exchanges, no sign of hemolysis, complement activation or activation of the coagulation system seemed to be induced by this substitution therapy. Alike, Nilsson and al.[9] improved the treatment of hemophiliac B with inhibitor by binding purified factor IX concentrate on a Sepharose gel. This technique allows the specific removal of Anti IX, eliminating unspecific adsorption of others plasma proteins, as previously described. In connection with this, an another procedure for removing high titer Anti VIII:C by extracorporeal circulation might be realized over immobilized FVIII:C on a chromatographic gel. Rapid loss of antigenic properties for FVIII:C makes this method inapplicable to patients with hemophilia A. We hypothesized that FVIII:C might possess a minimum active sequence composed of amino acids, such as arginyl residue, able to bind Anti VIII:C. Consequently, completly synthetic resins with suitable chemical substituents mimicking this minimum sequence might interact with Anti VIII:C. Based upon this hypothesis, crosslinked polystyrene was substituted by various amino acids or their derivatives. Experimental data demonstrated that the resins induced an "in vitro" removal of Anti VIII:C from hemophiliac immunoglobulins G.

MATERIALS AND METHODS

Preparation and characterization of the resins.

The interaction between FVIII:C an Anti VIII:C involving and unknown amino acid sequence, different L amino acids were linked to polystyrene matrix (Table 1).

Table 1 : Characterization of the studied resins
* Capacity : content of amino acid sulfamide group (determined by acidic titration and elemental analysis).

linked amino acids		capacity* meq/g
0	(SO$_3$NA)	0
L Alanine	(Ala)	2.7
L Phenylalanine	(Ø Ala)	1.1
L glutamic acid	(Glu)	0.8
L arginine	(Arg)	1.5
nitro L arginine	(Nitro Arg)	1.1
L arginine methyl-ester	(AOM)	0.6
nitro L arginine methyl ester	(NAOM)	1.9
L methionine	(Meth)	1.4
L Threonine	(Threo)	1.0
L Proline	(Pro)	1.7
L hydroxyproline	(OHPro)	2.2

In order to determine the most suitable chemical groups, some amino acid derivatives were also fixed. For example, to prove the effect of free guanidin group upon the interaction with Anti VIII:C, polystyrene was substituted either with arginine (free guanidin group) or with nitroarginine (guanidin blocked with nitro group). Likewise, to evidence the influence of carboxyl groups, arginine or nitroarginine were replaced by their respective methyl esters as substituents on the polystyrene matrix.
Linkage of amino acid derivatives was achieved by a two step process[10,11,12], (Fig.1) except for lysine. Firstly, crosslinked polystyrene beads (40-70 μm) were chlorosulfonated by chlorosulfonic acid in dichloromethane. Secondly, attachment of different amino acids to chlorosulfonated polystyrene was performed in a mixture of water and dioxane at pH=9-10. On the other hand, coupling of ester derivatives was realized in dichloromethane during 48 hours at 40°C in order to avoid the hydrolysis of the

ester group. The synthesis of the resin substituted by L-lysine NH2-CH(COOH)-(CH2) 4-NH2 required a three step process. Indeed, in order to mask the ε-amino group of lysine, Nε-ButylOxyCarbonyl-L-lysine (SIGMA, USA) was fixed onto the chlorosulfonated polystyrene. Then, the polymer was hydrolysed with a 4N sodium hydroxyde solution to remove the protecting group. The resulting polymer had exclusive α-amino groups linked to the matrix ($|$-⟨O⟩-SO2-NH-CH-(COOH)-(CH2)4-NH2).

This serie of steps resulted in resins partially substituted with sulfonic acid sodium salt groups and amino acid sulfamide sodium salt groups (Fig. 2).

In order to eliminate completely any impurity that might interact with blood proteins, all samples were successively washed with 1.5 M sodium chloride and 1 M sodium citrate solutions. Then, samples were suspended and equilibrated in Michaelis buffer, filtered, washed several times with water and dried under vacuum. The content of the amino acid derivatives linked to polystyrene was determined by acidic titration after hydrolysis of the modified resins if necessary. It was also estimated from the elemental analysis of nitrogen. The number of sulfonated groups was determined from the chlorosulfonyl content as previously described [12]. After crushing, the average particle diameters of the swollen resins, determined using a quantitative TAS miscroscope, varied from 5μm up to 10 μm.

Biological assays

* Plasma samples

Plasmas, prepared from blood of normal subjects and from one VIII:C deficient patient with an Anti VIII:C alloantibody, were collected into 3.8% trisodium citrate (1 vol. anticoagulant for 9 vol. of blood). A pool of plasma from 15 healthy subjects was used as reference normal plasma for VIII:C and Anti VIII:C assays. All plasma samples and aliquots of the reference pool were stored at a temperature of -70°c.

* Isolation of immunoglobulin fraction

Immunoglobulins G were isolated from one hemophiliac's plasma with an Anti VIII:C titer of 640 Bethesda U/ml [13]. Defibrinated plasma was dialysed against phosphate buffer (0.005M pH=6.5) overnight at 4°c. After centrifugation, the supernatant was passed through a column of DEAE cellulose, equilibrated with the same buffer. Fractions containing IgG were pooled, concentrated and then extensively dialysed against NaCl 0.15 M. The final preparation obtained from Hemophiliac's plasma (IgGh) contained 50 Bethesda units Anti VIII:C/mg IgG.

* IgG concentrations

They were measured by radial immunodiffusion method (RID method) using I.C.L. plates (I.C.L. Scientific, Fountain Valley, California). The "low level" and "ultra low level" kits were both used according to the IgG concentrations to be tested.

* Factor VIII:C assays

Factor VIII procoagulant activity (VIII:C) was measured by partial thromboplastin time (PTT) using Factor VIII deficient human plasma as substrate [14] (Diagnostica Stago, France). Pooled normal plasma, prepared as previously described, served as standard (1U/ml) for measurement of VIII:C.

* Factor VIII inhibitor measurements

The conditions for Anti Factor VIII assay (Anti VIII:C) were those suggested by the Bethesda Conference[13] In each assay, 0.1 ml of a dilution of the Anti VIII:C plasma or test material (IgG) were incubated with 0.1ml of pooled normal plasma for two hours at 37°c. The factor VIII procoagulant activity of this mixture was then compared to that control tube in which Michaelis buffer was incubated with the normal plasma. One unity of Anti VIII:C activity is defined as that which inactivates 50% of the procoagulant activity of the control sample during the two hour incubation. Thus, the antibody titer, expressed in Bethesda units, was obtained by determining the reciprocal of that dilution of test plasma or IgG preparation that inactivated 0.5U of VIII:C during the two hour incubation.

* Adsorption procedure

Adsorption measures were realized by incubating IgGh (50µl) at various concentrations either with a given suspension of resin (50 µl;2-10mg/ml) or with Michaelis buffer. After 30 mn incubation time, the mixture containing IgGh were centrifugated. The supernatants were used to simultaneously determine the initial and the residual concentrations of IgGh and Anti VIII:C activity. The assays performed in the presence of buffer allowed the determination of initial concentrations of IgGh or Anti VIII:C before adsorption. By replacing the buffer with the polymer suspension, the amount of residual IgGh and Anti VIII:C in the supernatants can be estimated. Consequentely, the level of adsorbed IgGh or Anti VIII:C could be deduced from the difference between the initial and the residual concentrations.

a) Adsorption isotherm of IgGh

On a RID plate, two standards with known amounts of immunoglobulins G (0.15 3.00 mg/ml) were run concurrently with four supernatants containing IgGh. For maximum reproductibility, a precision micropipette was used for dispensing 5 µl of supernatant in each well. To avoid distorted results, samples corresponding to IgGh concentrations before and after adsorption were delivered in two adjacent wells. The accuracy of this RID method was verified by testing different IgG titers obtained by successive dilutions.

b) Adsorption isotherm of Anti VIII:C

Quantitative determination of Anti VIII:C was made using a standard curve which correlated VIII:C assay with the concentrations of Anti VIII:C (Fig. 3). The concentrations of Anti VIII:C were varied from 0 up to 2 Bethesda U/ml by diluting the previously fractionnated IgG. In fact, to construct the standard curve, the various dilutions of IgGh with Anti VIII:C were firstly incubated with normal plasma during 2 hours at 37°c. Secondly after a 1/10 dilution, 0.1 ml of these mixtures were incubated with 0.1 ml of hemophilic plasma reagent and 0.1 ml of activated cephalin during ten minutes. Then, the coagulation times (CT) were measured after an addition of 0.1 ml CaCl2 solution (0.025M). A prolongation of CT was observed as soon as the Anti VIII:C titer increased. (Fig.3). Practically, after suitable dilutions, CT were consecutively performed on supernatants corresponding to Anti VIII:C, firstly, before adsorption and, secondly, after adsorption. Compared to blank assays before adsorption, CT were shortened for supernatants in contact with the resin when Anti VIII:C adsorption occurred. The residual amounts of Anti VIII:C were determined using the standard curve and the diluting factors.

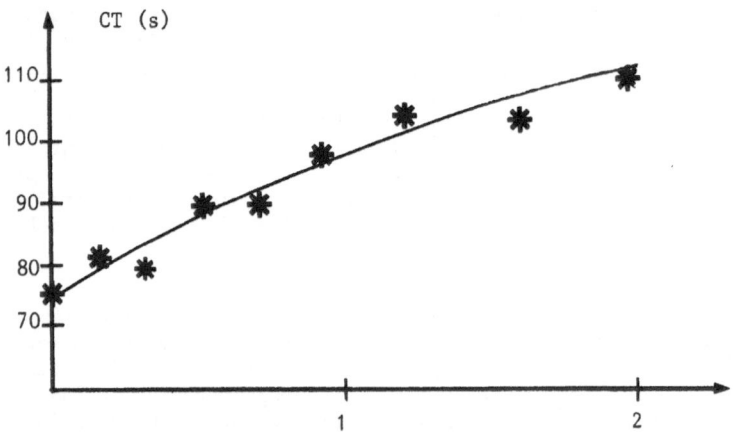

Fig 3 : Standard curve.

Anti VIII:C (U/ml)

RESULTS

Biological assays were performed on crushed resins and consequently, the specific surface varied between the different polymers. So, direct comparisons of the Anti VIII:C isotherms obtained with the various resins had to be avoided. To remedy this inconvenient, it was absolutely necessary to both determine the Anti VIII:C and IgGh isotherms, for the same polymer. The overall IgG were supposed unspecifically adsorbed onto the polymers.

Preliminary assays were essentially performed to determine the range of the resin concentration which had to be used for the both constructions of Anti VIII:C and IgGh isotherms. Indeed, too high concentrations of resins entailed the whole adsorption of antibodies without reaching equilibrium between adsorbed and free proteins. For an upper concentration of IgGh (about 1.6 mg/ml), the suitable concentration of the polymer was ranging from 2 up to 10 mg/ml to accurately determine the isotherms.

Furthermore, the most interesting resin might adsorb the highest amount of Anti VIII:C and minimalize the non specific sorption of the overall IgGh. Besides, direct comparison between the two types of isotherms allowed an evaluation of the selectivity of the different resins towards Anti VIII:C. In fact, selectivity (S) was expressed by the relation :$S=(Pi2/Pi1)-1$, where Pi1 and Pi2 were respectively the slopes drawn in one point of either IgGh or Anti VIII:C isotherms corresponding to the same initial IgGhi concentration. Practically, S was estimated for IgGhi=0(S1) and for IgGhi=IgGhmax (S2) i.e. when saturation occurred. (Fig.4). The knowledge of the specific activity of IgGh (50 Bethesda Units Anti VIII:C/mg IgG) allowed the use of a double scale for the X and Y axis drawing in the isotherm constructions and thus, permitted the direct comparison of both isotherms.

When a resin was unspecific, superposition of the two types of isotherms was observed and the resulting S value was zero. On the contrary, when the adsorption of Anti VIII:C was promoted in comparison to those of IgGh, the Anti VIII:C isotherms became upper than the IgGh one and the S value was superior to zero.

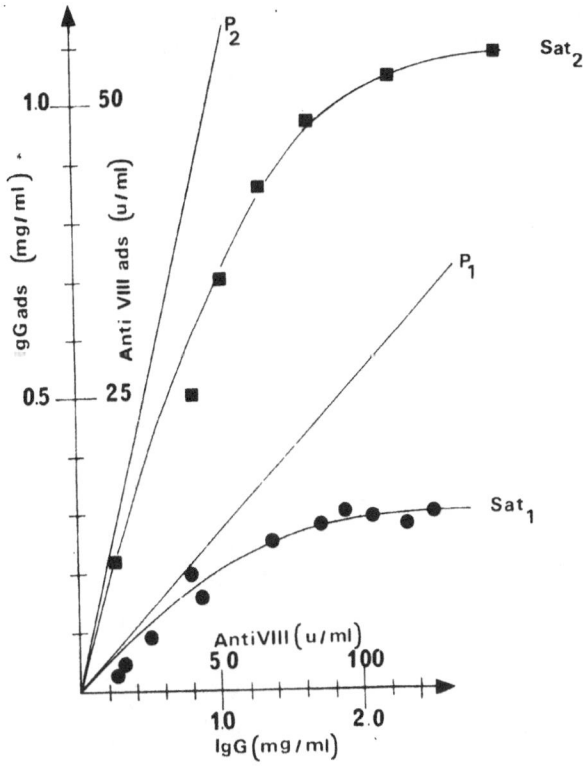

Fig 4 : Determination of the selectivity

Isotherms were performed on resins substituted either by arginine or nitroarginine as also by their respective methyl ester derivatives.

RESINS	PA	PNA	PAOM	PNAOM
R_1	H	H	CH_3	CH_3
R_2	H	NO_2	H	NO_2

Blocking of guanidin group induced a drastic decrease on the selectivity as shown for nitroarginine and nitroarginene methyl ester. Replacement of carboxylic groups by methyl ester groups slightly increased the selectivity (Fig. 5).

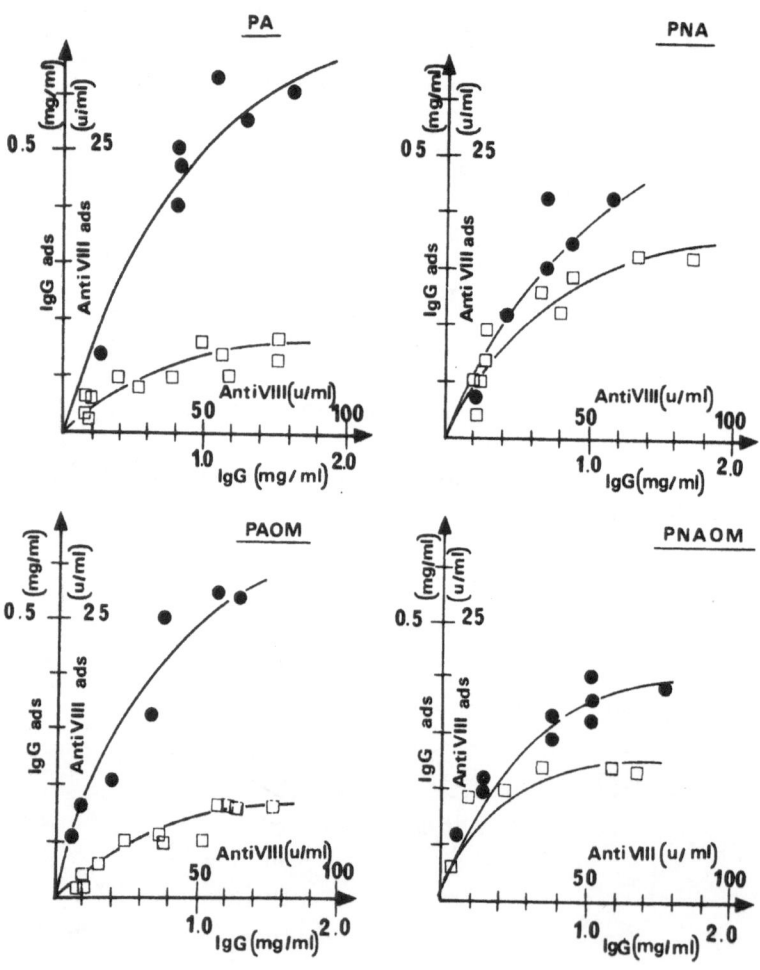

Fig 5 : Adsorption isothermes

●   Anti VIIIc

□   IgG$_h$

Table 2 : Selectivity of the resins

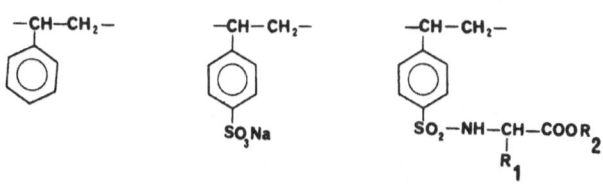

BOUND AMINO ACIDS	R_1	R_2	S_1	S_2
	0	0	2.2	1.7
ALA	$-CH_3$	H	2.2	2.1
$\emptyset$ALA	$-CH_2-\bigcirc$	H	2.0	
GLU	$-(CH_2)_2-COOH$	H	3.3	3.2
ARG	$-(CH_2)_3-NH-C\begin{smallmatrix}NH\\NH_2\end{smallmatrix}$	H	2.1	
NITROARG	$-(CH_2)_3-NH-C\begin{smallmatrix}NH\\NH-NO_2\end{smallmatrix}$	H	0.1	0.3
ARGININE METHYL ESTER	$-(CH_2)_3-NH-C\begin{smallmatrix}NH\\NH_2\end{smallmatrix}$	CH_3	3.2	
NITROARGININE METHYL ESTER	$-(CH_2)_3-NH-C\begin{smallmatrix}NH\\NH-NO_2\end{smallmatrix}$	CH_3	0.7	0.7
METH	$-(CH_2)_2-S-CH_3$	H	0.9	1.0
THREO	$-CH\begin{smallmatrix}CH_3\\OH\end{smallmatrix}$	H	2.7	2.1
PRO	$\begin{smallmatrix}NH-\\CH_2\,CH_2\,CH_2\\NH-\end{smallmatrix}$	H	1.2	1.3

Values for selectivity obtained with the different resins were summed up in table 2. Comparison of S for resins substituted by alanine and phenylalanine demonstrated no improvement due to the binding of phenyl groups. In fact, the best substituents for the resins were : glutamic acid, hydroxyproline or arginyl methyl ester. At biological pH=7.35, these groups bore negative or positive charges. So, electrostatic interactions might induce the specificity of the resins. In order to ensure this hypothesis, different polymers grafted with substituents without charge such as glutamic acid dimethyl ester will be also studied.

Attemps to improve selectivity will be carried out by modifying the structure of the resins. For example, it will be interesting to obtain resins substituted with different levels of the same amino acid derivative. Indeed, in order to minimalize non specific sorption of undesirable proteins onto these synthetic sorbents and to exploit as much as possible the affinity of the ligand, lower content of amino acid derivatives attached onto the matrix should be used [15]. Furthermore, introduction of spacer arm between the polystyrene matrix and the amino acid substituent should enhance the accessibility to the proteins. Consequently, by placing the ligand at the end of the extension arm such that it protrudes into the buffer, there should be a substantial increase in the strength of the interaction between Anti VIII:C and the coupled ligand.

The best resins will have to be tested in plasma where other proteins such as albumin, fibrinogen will concurrently complete with IgG_h in the mechanism of adsorption. If selectivity in plasma towards Anti VIII:C is retained, these specific sorbents will be packed in columns and assays for Anti VIIIc removal, performed under conditions mimicking the blood flow, will be undertaken. The last step of this study will be the use of the synthetic sorbents to remove, by extracorporeal circulation, Anti VIII:C from hemophiliac'plasmas according to the method described by Nilsson et al.

REFERENCES

1. H.S. STRAUSS, Acquired circulating anticoagulants in hemophilia A, New Engl J Med. 281; 866 (1969)

2. R.BIGGS, Jaundice and antibodies directed against factor VIII and IX in patients treated for haemophilia or Christimas disease in the United King dom, Br. J. Haematol. 26; 313 (1974)

3. S.S. SHAPIRO,M. HULTIN, Acquired inhibitors to the blood coagulation factors, Semin Thromb. Hemostas. 1; 336 (1975).

4. G.W.SLOCOMBE; A.C. NEWLAND; M.P. COLVIN and B.T. COLVIN, The role of intensive plasma exchange in the prevention and management of haemorrhage in patients with inhibitors to Factor VIII, Br. J. Haematol.47; 577 (1981)

5. J.L. MARX, Health officials seek ways to halt AIDS. Science ; 219; 271 (1983).

6. Y. SULTAN; A. BUSSEL; P. MAISONNEUVE ; M. POUPENEY; X. SITTY and P. GAJDOS, Potential danger of thrombosis after plasma exchange in the treatment of patients with immune disease. Transfusion; Sept-Oct (1979).

7. I.M. NILSSON; S. JONSSON; S.B.SUNDQVIST; A. AHLBERG; S.E. BERGENTZ, A procedure for removing high titer antibodies by extracorporeal Protein A - Sepharose adsorption in hemaphilia : Substitution therapy and surgery in a patient with hemophilia B adn antibodies, Blood, 58; 1; 6, 38 (1981).

8. H. HJELM; K. HJELM and J. SJOQUIST, Protein A from staphylococcus aureus. Its isolation by affinity chromatography and its use as an immunosorbent for isolation of immunoglobulins. FEBS. Letters 28; 1; 73 (1972).

9. C. FREIBURGHAUS; S.B. SUNDQUIST; H. SANDBERG and I.M. NILSSON, Elimination of specific antibodies in whole blood in a continuous extracorporeal system, Thromb Haemost. 50; 1; 208 (1983)

10. C. FOUGNOT; J. JOZEFONVICZ; M. SAMAMA; L. BARA, New heparin-like insoluble materials : Part I, Ann. Biomed. Engineer; 7; 429 (1979).

11. S. GOLDSTEIN; A. GULKO; G. SCHMUCKLER, Synthetic aspects of selective ion exchanger, Isr J. Chem. 89. 3 (1972).

12. M.A. PETIT and J. JOZEFONVICZ, Synthesis of copper II complexes of asymmetric resins prepared by attachement of α amino acid to cross-linked polystyrene. J. Appl. Polym. Sci. 2589 (1977).

13. C.K. KASPER; L.M. ALEDORT; R.B. COUNTS; J.R. EDSON; J. FRATANTONI; D. GREEN; J.W. HAMPTON: M.W. HILGARTNER ; J. LAZERSON; P.H. LEVINE; C.W. Mc MILLAN; J.G. POOL; S.S. SHAPIRON; N.R. SHULMAN; J. VAN EYS, A More uniforme measurement of FACTOR VIII inhibitors. Thromb. Diath. Haemorrh 34; 869 (1975).

14. R.T. BRECKEN RIDGE; O.D. RATNOFF, Studies on the nature of the circulating anticoagulant directed against antihemophilic factor : with notes on an assay for antihemophilic factor blood 20; 137 (1262).

15. J. TURKOVA, Specific sorbents for high performance liquid affinity chromatography and large scale isolation of proteinases. Affinity Chromatography and related techniques. T.C. J. Gribnau, J. Visser and R.J.F. Nivard (Edit) 1982. Elsevier Scientific Publishing.

SYNTHETIC POLYMER-COLLAGEN COMPOSITES FOR BIOMEDICAL USE

Miroslav Štol, Miroslav Tolar and Milan Adam

Research Institute for Rheumatic Diseases
Institute of Physiology of Czechoslovac Academy
of Sciences, Prague, Czechoslovakia

INTRODUCTION

Chemical purity and inertness of classical implantable synthetic materials warrants their non-toxicity and non-antigenicity in respect to the recipient´s organism (Bruck, 1973). However, chemically inert plastics are rather poor substrates for attachment of animal cells. If implanted into the organism, they are surrounded by a fibrous capsule.

Practically all animal cells need attachment to a solid surface for their life. Their normal metabolism, growth and expression of differentiated characters are anchorage-dependent (Grinell, 1978).

One of the ways, how the adhesivity of cells to substrate can be enhanced, is combination of artificial and biological materials.

A group of synthetic hydrogels was developed by Wichterle and Lím (1960). The poly(2-hydroxyethyl methacrylate)-polyHEMA is frequently used as a prosthetic material. It is translucent, non-toxic, non-antigenic, fairly hydrated, permeable for ions and small molecules, very resistant to enzymatic digestion. However, animal cells do not attach, grow and differentiate satisfactorily on its surface (Folkman and Moscona, 1978; Štol et al., 1985).

We have shown that copolymers of polyHEMA with methacrylic acid were able to support growth of transformed cell lines in vitro (Tolar and Štol, 1968; Tolar et al., 1969). Civerchia-Perez et al. (1980) incorporated soluble collagen into the polyHEMA hydrogel structure by mixing dissolved components before the free-radical crosslinking polymerization. The growth of fibroblasts and smooth muscle cells (Toselli et al., 1984) was greatly enhanced, although the content of collagen in the composites was rather low (maximally 3-4 %, w/v). Elastin incorporated in a similar way (Toselli et al., 1983) improved adhesion and growth of endothelial cells.

Several years ago, a new technique was suggested by us (Štol et al., 1980; Štol et al., 1985), which enabled incorporation of any amount of acid soluble collagen (ASC) or insoluble collagen (ISC) into the composite with polyHEMA, ranging from 0 to 100 % (w/w).

The biological properties of the composites were tested in vitro (Štol et al., 1985) and in vivo (Štol et al., in preparation). Skeletal myogenesis was chosen for testing in vitro, because formation of myotubes from myoblasts is dependent on presence of collagen (Hauschka and Konigsberg, 1966). The same composites were implanted into a popliteal region of rats and their interaction with surrounding tissues was followed. The aim of this study was to specify quantitatively the combinations of polyHEMA and collagen suitable for the in vitro and in vivo use.

MATERIALS AND METHODS

Preparation of liquid mixture of polyHEMA and collagen

Uncrosslinked polyHEMA was prepared separately by free-radical solution polymerization of HEMA monomer (SPOFA, Prague) and it was then thoroughly purified (Wichterle, 1971).

The insoluble fibrillar collagen was prepared according to Steven (1976). The acid soluble collagen was prepared following the procedure described by Bazin and Delaunay (1976). In both cases, bovine hides were used as a source of collagen.

The procedure of mixing polyHEMA and collagen in any desired ratio consisted of several steps:

(a) Preparation of 10 % (w/v) polyHEMA stock solution in 33 % (v/v) aqueous acetic acid.

(b) Preparation of 2 % (w/v) stock dispersion of insoluble collagen fibrils or stock solution of acid-soluble collagen in 1 % (v/v) aqueous acetic acid.

(c) Mixing of the two polymeric components in a desired ratio. Before mixing, the content of acetic acid in the collagen dispersion or solution was raised to 33 % (v/v) to avoid precipitation of polyHEMA from the mixture.

More detailed information on materials and methods used can be found in Štol et al. (1985).

Coating of cultivation surfaces

The resulting liquid mixture was used to form a thin layer on the cultivation surface of plastic Petri dishes 60 mm in diameter (Koh-I-Noor, Czechoslovakia) under sterile conditions. The solvent was evaporated from the layers at room temperature. Then the layers were rinsed with sterile double-distilled water and dried out. The layers were rehydrated by adding some cultivation medium several hours prior to seeding of the dissociated muscle cells. Sterility of the dishes can be maintained with help of a germicide lamp.

Chick skeletal muscle cell cultures

A previously described technique of explantation was used (see Tolar et al., 1983). Limb muscles from 10-day-old chicken embryos were mechanically dissociated and $2 . 10^5$ cells were plated per dish. The feeding medium consisted of Eagle´s minimal essential medium (Eagle, 1955; ÚSOL Prague) supplemented with 10 % fetal calf serum (Veterinary School, Brno) and 5 % chicken embryonal extract (Paul, 1975) and was changed twice a week. The cultures were maintained at 37 °C at humidified atmosphere containing 5 % carbon dioxide in air.

The density of myotubes formed in the cultures was assessed using a network of perpendicular and horizontal lines 1 mm apart (Normatex, Holland), which was fixed to the underside of the floor of each plastic Petri dish. Irrespectively of their orientation, living myotubes crossing a line 1 mm long were counted in each square. Twenty counts were made per each dish and a mean value was calculated. At least three dishes were evaluated for each type of a polyHEMA-collagen composite layer. The consecutive counts on the 6th and the 8th day of cultivation were made in the same frames of the network, so that they could be compared. The differences between experimental and control data were estimated using Student´s t-test.

Solid implants into a popliteal region of rats

Solid discs (1 mm thick and 10 mm in diameter) were prepared from the liquid mixture of the polyHEMA-collagen composites, as described above. The only exception was addition of 0.25 % (v/v) glutaraldehyde (Merck) for crosslinking of the two polymeric components. Plastic dishes made from hard polyvinylchloride were used as non-adherent molding containers for preparation of the solid composite plates. Calculated amount of the liquid mixture of the corresponding composite was poured into the container and solvent was then evaporated in a dust-free box at room temperature. The discs were then cut out from the solid composite plates.

The solid discs were sterilized by irradiation with a dose 2.5 Mrad using linear high-frequency electron accelerator (LINAC, Tesla Prague). They were implanted into a popliteal region of Wistar rats (3 months old). Following a planned time period, they were taken out and examined histologically and electron microscopically. Scanning electron microscope JEOL JSM-35 (under 25 kV) was used.

RESULTS

PolyHEMA-collagen composites containing 90, 80, .., 10 % (w/w) of polyHEMA and 10, 20, .., 90 % (w/w) of fibrillar or soluble collagen were prepared as layers for in vitro testing and as discs containing fibrillar collagen for testing in vivo. Pure polyHEMA and pure collagen were used as controls. For the sake of simplicity, only the percentage of collagen will be indicated further on.

Several characteristics of the composites were followed:

Compatibility of polyHEMA and collagen in the composites

Although the fibrillar collagen and polyHEMA are well miscible in all proportions when dissolved in a common solvent, removal of the solvent leads to instability of the system. This feature is less expressed, if low concentrations of fibrillar collagen are used (less than 30 %, w/w). Here, the polyHEMA forms a continuous phase. However, if more than 40 % (w/w) of fibrillar collagen is present, the synthetic polymer tends to form discrete globular particles.

No such difficulties occurred, when soluble collagen (ASC) was used in the composites.

In vitro myogenesis on the composite layers

No myotubes were formed on the pure polyHEMA surface, although the mononuclear cells were able to become attached to and grow on this surface to a minor extent. Even the composites containing as much as 40 % of fibrillar collagen proved inadequate as substrates for the formation of myotubes. An appreciable number of myotubes developed on the composite layer containing 50 % of fibrillar collagen. Significantly increased myotube densities were noted, when composites containing 60 - 90 % of collagen were used. However, the density of myotubes, which occurred on a pure collagen surface, was not achieved with any type of the composites. In both 6 and 8-day-old muscle cultures, the presence of collagen in the composite cultivation substrates had a favourable effect on myotube formation. (See figures 1 and 2.) Normally, the density of myotubes developing from myoblasts increases in the cultures for 4-5 days, remains stable for a further 1-2 days and decreases due to the degeneration of myotubes in the absence of motor innervation (see Tolar et al., 1983). Therefore, the total number of myotubes per mm was lower in the 8-day-old cultures.

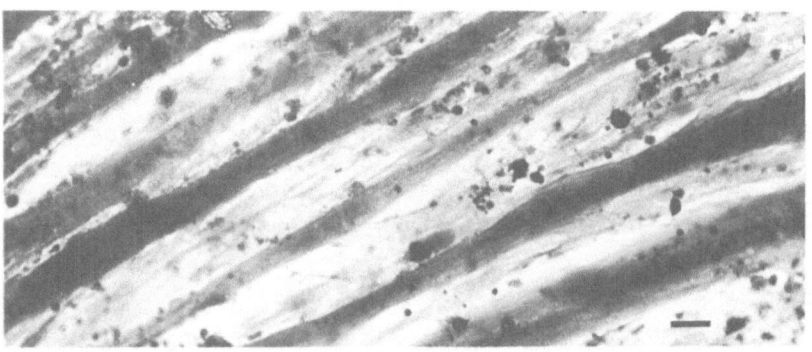

Fig. 1. Growth and differentiation of dissociated chicken embryo muscle cells seeded on a layer formed by a composite containing 70 % (w/w) of fibrillar collagen. The 6-day-old culture was stained by haematoxylin-eosin. (The bar indicates 10 μm.)

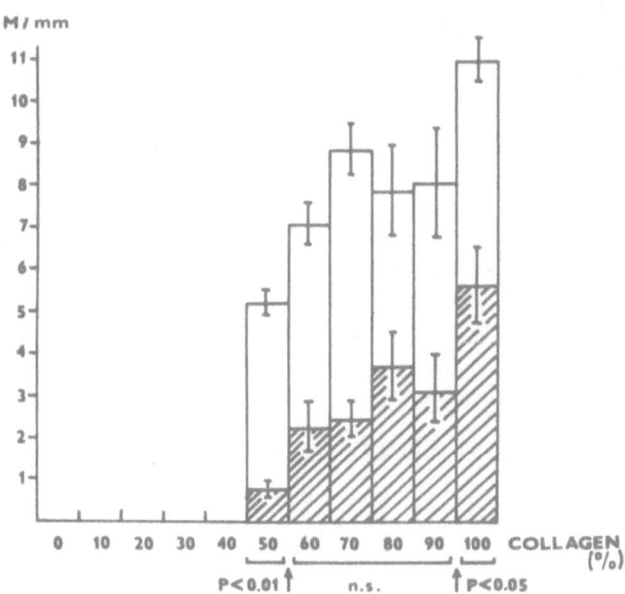

Fig. 2. Densities of myotubes (M/mm) ascertained in 6-day-old
(empty columns) and 8-day-old (hatched columns)
muscle cell cultures maintained on artificial layers
consisting of polyHEMA and fibrillar collagen in the
concentrations ranging from 0 to 100 % (w/w). The myotubes were formed only on layers containing 50 % and
more of fibrillar collagen.

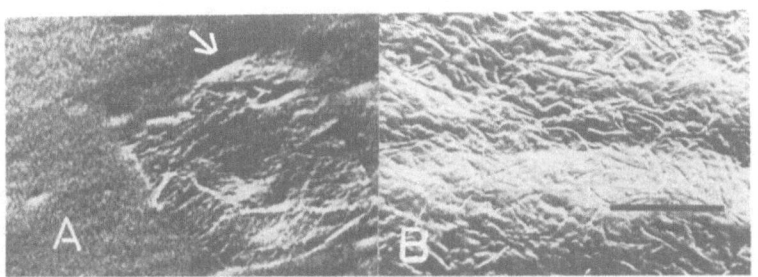

Fig. 3. Scanning electron micrographs showing structure of
the cultivation surfaces (bar indicates 10 /um).
(A) The composite containing 40 % (w/w) fibrillar
collagen; the fibrils are covered by a homogenous
layer of polyHEMA. They can be exposed mechanically
(arrow).
(B) The composite containing 60 % (w/w) fibrillar
collagen; a rich net of collagen fibrils is freely
accesible to the cultivated cells.

As seen in the scanning electron microscope, the surface of polyHEMA alone and of composites containing 10-40 % (w/w) fibrillar collagen was quite smooth. It was formed only by polyHEMA (Fig. 3A). The collagen fibrils embedded in polyHEMA remained uncovered on the surface of the composites containing more than 50 % (w/w) fibrillar collagen (Fig. 3B). It seems comprehensible that only exposed collagen fibrils were able to support myogenesis.

If similar composites of soluble collagen with polyHEMA were tested, all the concentrations used supported formation of myotubes. No significant differences in myotube densities were observed. It is very probable that soluble collagen proteins can form with polyHEMA more uniform composite structure and that collagen macromolecules are exposed on the very fine surface layer of the composite.

Alterations of the composite discs implanted in vivo

Solid discs with the same range of fibrillar collagen content, as described above, were implanted into a popliteal region of rats. They were examined histologically 3, 6 and 12 months after the implantation, and in some cases also by scanning electron microscope. Here, the results obtained by Štol, Holuša, Cífková and Tyráčková (Štol et al., in preparation) will be briefly described.

(a) Pure polyHEMA   The implant remained firm and undisturbed for the whole time period. It became surrounded by a thick fibrous capsule. Celular reaction consisted in presence of foreign-body multinuclear cells, some polymorphonuclear leukocytes and eosinophils.

(b) 10 - 20 % collagen   Superficial layer of the polyHEMA matrix was exposed, so that the surface of the explant became uneven. The implant remained firm for the whole time period. A fibrous capsule was thin and closely attached to the implant. Cellular reaction consisted of foreign-body multinuclear cells and some polymorphonuclear leukocytes.

(c) (30) - 40 - 50 % collagen   The implants were ivaded to an increasing depth by fibrous tissue containing capillaries. The polyHEMA matrix was exposed in the form of globules of different size. The central part of the implant was edematous and deformed. A cellular reaction was low and a fibrous capsule was quite thin, often missing.

(d) (60) - 70 - 90 % collagen   The implant was totally disintegrated 12 months after the operation. Globular particles of polyHEMA dissappeared from the site of implantation.

(e) Pure collagen   The implants were fully resorbed within three months after the operation.

All the methods and techniques used in this particular study, and the results obtained by the authors cited above, will be published elsewhere.

DISCUSSION

It has been confirmed by the results of this study that the ability to anchor the animal cells and support their growth and differentiation is greatly enhanced, if composites of poly-HEMA and collagen are used as cultivation substrates, instead of pure polyHEMA. It was found independently by us (Štol et al., 1980; Štol et al., 1985) and another group of the authors (Civerchia-Perez et al., 1980; Toselli et al., 1983, 1984).

The technological procedure described by us is preferentially suitable for formation of thin layers or coatings of solid supports. It can be applied on surfaces of cultivation vessels, on knitted blood vessel prostheses or prostheses of various tube-like organs, where the thin layer of a composite would prevent leakage of liquids from inside and would support growth of cells on their inner and outer surfaces.

The technique for preparation of protein-polyHEMA hydrogels worked out by Civerchia-Perez et al. (1980) and Toselli et al. (1983) is based on crosslinking copolymerization of HEMA monomer with small amount of ethylene dimethacrylate in presence of a soluble protein. As an inert solvent is used ethylene glycol. Inherent to its nature, maximally 4 % (w/v) of collagen content may be reached by this method. Further, impurities such as free monomer, unlinked oligomers, rests of catalyst and solvent remain in the copolymers (Brynda et al., 1985; Cífková et al., 1985) and have to be removed additionally.

These disadvantages are circumvened by our method. The purified uncrosslinked polyHEMA and collagen are mixed together in a common solvent in any ratio ranging from 0 to 100 %. The mixture solidifies, when the solvent is evaporated.

The in vitro formation of myotubes and also scanning electron microscopic examination of layers prepared for cultivation of cells showed that heavier collagen fibrils sedimented during drying out of the solvent, so that the surface layer consisted of pure polyHEMA in the composites containing upto 40 % (w/w) of fibrillar collagen. The content of the fibrillar collagen had to be increased to 50 % and more, in order to become exposed and biologically effective.

In contrary, even the lowest amount (1 %, w/w) of soluble collagen (ASC) tested was similarly efficient in promotion of myotube formation as a layer of pure soluble collagen.

It is suggested, therefore, to use in the in vitro experiments soluble rather than insoluble or particulate components in the composites. Possibility to incorporate any quantity of collagen (or another or several other proteins) into the composite layer makes our technique especially well suited for studies on mechanism of attachment of various cell types.

In vivo, the composites triggered a local cellular reaction and were attacked by macrophages. Relatively high content of fibrillar collagen in the composites (30 % and more) seemed to be unsuitable for the in vivo use. In such cases, the implants were disintegrated and globular particles of polyHEMA were liberated and phagocytosed. This fact has several undesirable consequences:

(1) Such an implant can fulfill its mechanical role only for a limited time period and rather soon collapses.

(2) Phagocytosed polyHEMA particles are resistant to enzymatic hydrolysis and persist in the organism (Bílý, 1977; Větvička et al., 1983; Fornůsek et al., 1985).

(3) Higher content of fibrillar collagen increases probability of calcification (Cífková et al., in preparation).

Thus, the low (less than 20 %) concentrations of fibrillar collagen are to be preferred for the in vivo use. In this way, the recipient´s fibrous capsule is thinner than in the case of the pure polyHEMA implant. The core of the implant remains undisturbed and its mechanical properties (including reinforcement by collagen fibrils) are maintained.

It can be suggested that combination of both soluble and insoluble collagens would further improve attachment of cells to the surface of the implant, as it did in the experiments in vitro. It is also possible to aply multiple coatings differing in quality and quantity of the admixed collagen types on various kinds of the implants or their parts.

It can be concluded that using our method of preparation any defined content of soluble or insoluble colagen proteins can be incorporated into composites with polyHEMA. It widens the range of possible combinations that can be favourably utilized for the in vitro and in vivo studies and practical applications.

ACKNOWLEDGEMENTS

We wish to express our gratitude to Professor Claudio Migliaresi and to Professor Luigi Nicolais, Department of Materials and Production Engineering, University of Naples, for their interest in our work and kind invitation to publish our results in this book.

REFERENCES

Bazin, S., Delaunay, A., 1976, Preparation of acid and citrate soluble collagen, in: "The Methodology of Connective Tissue Research", D A. Hall, ed., Joynson-Bruvvers, Oxford.
Bílý, B., 1977, Gel-collagen sponge implantation in spongious bone of the pig´s mandible 3 years after the implantation, Acta Facult. Med. Univ. Brunensis, 55:53.
Bruck, S. D., 1973, Aspects of three types of hydrogels for biomedical applications, J. Biomed. Mater. Res., 7:387.
Brynda, E., Štol, M., Chytrý, V., Cífková, I., 1985, Washing-out of impurities from poly(2-hydroxyethyl methacrylate) hydrogels, J. Biomed. Mater. Res., in press.

Cífková, I., Brynda, E., Mandys, V., Štol, M., 1985, Diffusion
 of impurities from the implanted poly(2-hydroxyethyl
 methacrylate) and its influence on acute inflammation
 development, J. Biomed. Mater. Res., in press.
Civerchia-Perez, L., Faris, B., LaPointe, G., Beldekas, J.,
 Leibowitz, H., Franzblau, C., 1980, Use of collagen-
 hydroxyethyl-methacrylate hydrogel for cell growth,
 Proc. Natl. Acad. Sci. USA, 77:2064.
Eagle, H., 1955, The specific amino acid requirements of a mammalian cell (strain L) in tissue culture, J. Biol. Chem.,
 214:839.
Folkman, J., Moscona, A., 1978, Role of cell shape in growth
 control, Nature, 273:345.
Fornůsek, L., Větvička, V., Zídková, J., Kopeček, J., 1985,
 Hydrophilic polymeric microspheres: their use in immu-
 nologic methods, Makromol. Chem. Suppl., 9:125.
Grinnell, F., 1978, Cellular adhesiveness and extracellular
 substrata, Int. Rev. Cytol., 53:65.
Hauschka, S. D., Konigsberg, I. R., 1966, The influence of
 collagen on the development of muscle clones,
 Proc. Natl. Acad. Sci. USA, 55:119.
Paul, J., 1975, "Cell and Tissue Culture", Churchill Living-
 stone, Edinburgh-London-New York.
Steven, F. S., 1976, Preparation of macromolecular collagens,
 in: "The Methodology of Connective Tissue Research",
 D. A. Hall, ed., Joynson-Bruvvers, Oxford.
Štol, M., Tolar, M., Adam, M., Čefelín, P., Kálal, J., 1980,
 Composite polymeric materials for biological and medi-
 cal use, Czechoslovak Patent Application, PV 5125-80.
Štol, M., Tolar, M., Adam, M., Poly(2-hydroxyethyl methacry-
 late)-collagen composites which promote muscle cell
 differentiation in vitro, Biomaterials, 6:193.
Tolar, M., Štol, M., 1968, The use of poly(glycol methacrylate)
 -PGMA- for cultivation of cells in vitro (in Czech),
 Čs. Fyziol., 33:613.
Tolar, M., Štol, M., Kliment, K., 1969, Study on surgical
 sewing material coated with Hydron, J. Biomed. Mater.
 Res., 3:305.
Tolar, M., Michl, J., Dlouhá, H., Teisinger, J., 1983, Defined
 medium supplemented with growth-promoting alpha-globulin
 supports the formation, differentiation and innervation
 of myotubes in cell cultures from chicken embryo,
 Mol. Physiol., 3:151.
Toselli, P., Faris, B., Oliver, P., Wedel, N., Franzblau, C.,
 1983, Preservation of ultrastructure of cells cultured
 on protein-hydroxyethylmethacrylate hydrogels,
 J. Ultrastr. Res., 83:220.
Toselli, P., Faris, B., Oliver, P., Franzblau, C., 1984, Ultrastructural studies of attachment site formation in aortic smooth muscle cells cultured on collagen-hydroxy-
 ethylmethacrylate hydrogels, J. Ultrastr. Res., 86:252.
Větvička, V., Fornůsek, L., Kopeček, J., Přikrylová, D., 1983,
 Phagocytosis of 2-hydroxyethylmethycrylate copolymer
 particles by different types of macrophages, Folia Bio-
 logica (Prague), 29:424.
Wichterle, O., 1971, Hydrogels, in: "Encyclopedia of Polymer
 Science and Technology", Vol. 15, J. Wiley, New York.
Wichterle, O., Lím, D., 1960, Hydrophilic gels for biological
 use, Nature, 185:117.

DIRECT ASSAY OF PHAGOLYSOSOMAL HYDROLASE BY

FLUOROGENIC SUBSTRATE-BINDING MICROSPHERES

T. Uchida*, K. Suzuki**, S. Hosaka*, and T. Fujikura**

* Basic Research Laboratories, Toray Industries Inc.
  1111 Tebiro, Kamakura, 248 Japan
** Dept. of Pathology, Radiation Effects Research
   Foundation   5,2 Hijiyama Park, Minami, Hiroshima
   730 Japan

ABSTRACT

    A new method of directly measuring phagolysosomal hydrolase has been
developed. 4-methylumbelliferyl-β-D-glucuronide(4MUGL)-binding microspheres and decanoyl fluorescein(DF)-binding microspheres were prepared.
The microspheres were phagocytosed by human peripheral neutrophils, then
within phagolysosomes immobilized 4MUGL and immobilized DF were hydrolyzed by β-glucuronidase and lipase respectively. These reactions were
confirmed by the facts that 4MU, which is fluorescent, was released from
4MUGL-binding microspheres and DF binding microspheres turned fluorescent, when the microspheres were phagocytosed. Each enzyme activity
within phalysosomes was examined by measuring the increase of fluorescence intensity without cell rupture.

INTRODUCTION

    Phagocytes enclose foreign substances (microorganisms, lipids,
wastes, etc.) into phagolysosomes, where to kill and digest them with
highly reactive oxygen and lysosomal hydrolase. Among many kinds of
hydrolase in lysosomes, β-glucuronidase is one of the representative
enzymes of lysosomes. Accordingly, the measurement of the β-glucuronidase
activity within phagolysosomes would be useful for examining the digestive
activity of phagocytes. Esterase is also an important enzyme within
phagolysosomes, because the enzyme scavenges lipid on the arterial wall.
Therefore, the measurement of the lipase activity within phagolysosomes
could be useful for the study of the prevention and diagnosis of atherosclerosis.
    There have been, however, some difficulties in direct assay of enzymes
released into the phagolysosomes of the cells, while the activities of
phagocytic enzymes released into the medium by exocytosis have been measured
in vitro system (1~8). We have recently developed a new method, which
makes possible directly to measure highly reactive oxygen within phagosomes,
using luminol-binding microspheres (9). In this paper, we report a new
method of directly measuring the activity of hydrolase (β-glucuronidase
(10) and lipase (11), within phagolysosomes using fluorogenic substrate-
binding microspheres.

MATERIALS AND METHODS

Preparation of 4MUGL-binding microspheres

Carboxylated hydrophilic microspheres (2µm in diameter) were prepared in the following method (9) : glycidyl methacrylate, 2-hydroxy-
ethyl methacrylate, methacrylic acid and triethyleneglycol dimethacrylate
were mixed at a molar ratio 65:20:10:5. Twenty four grams of the monomer
mixture were dissolved in 76g of ethylpropionate, and 0.13g of 2,2'-azobis
(2,4-dimethyl-4-methoxyvaleronitril) were then added to the resultant
solution. Polymerization was carried out at 40°C for 3 hr. The precipitated particles were aminated by ammonia and hydrolyzed by dilute sulfuric
acid.
    One ml of an aqueous solution of $NaIO_4$ (2mg/ml) and 1 ml of dimethyl-
sulfoxide solution containing 4MUGL (2mg) were mixed and stirred at 25°C
for one hour. After adjusting pH of the reaction solution to 85 with NaOH,
an equal volume of the aminated microsphere suspension (1%) was added to
the solution and allowed to react for two hours at 25°C.

Preparation of DF-binding microspheres

    Preparation method of the microspheres was similar to that of 4MUGL-
binding microspheres : 2-hydroxyethyl methacrylate, methyl methacrylate,
glycidyl methacrylate, methacrylic acid and triethyleneglycol dimethacrylate were mixed at a molar ratio of 40:35:10:10:5. Polymerization and
amination were carried out in the same way as above-mentioned, to prepare
hydrophilic microspheres with amino groups (1.5 µm in diameter). 14 mg
of dacanoyl chloride, 10 mg of dichloro-triazinyl fluorescein and 100 µl of
triethylamine were dissolved in 900 µl of dimethylformamide. The mixture
was stirred at 25°C for 6 hr and triazinyl-decanoyl fluorescein was produced. Five ml of the suspension of aminated microspheres (1%) was mixed
with 62.5 µl of dichloro-triazinyl decanoyl fluorescein solution and the
pH of the mixture was adjusted to 8.0. The suspension was stirred at 40°C
for 2 hr and decanoyl fluorescein (DF)-binding microspheres (1 %) was obtained. One ml of DF-microspheres (1 %) was mixed with 100 µl of fresh
human serum and incubated at 37°C for 30 min. Thus opsonized DF-microspheres were obtained.

Separation of neutrophils

    Mononuclear cells were removed from normal human peripheral blood by
Ficoll-Hypaque centrifugation. Erythrocytes were lysed with ammonium
sulfate solution, and neutrophils were resuspended in PBS.

Assay of phagolysosomal β-glucuronidase

    100 µl of neutrophil suspension (4 x 10^6 /ml) and 100 µl of micro-
sphere suspension (2 x 10^8 /ml) were mixed and incubated at 37°C. After
incubation for various periods, the mixture was centrifuged at 250 g for
3 min. The supernatant was used to determine the amount of extracellular
4MU in the medium. The cell pellet was solubilized by 50 µl of a mixture
of 1 % Triton X - 100 and ethylene glycol monomethyl ether (1:1), and used
to determine the amount of intracellular 4MU. The fluorescence intensities of the samples were measured at 450 nm by use of exciting light of
365 nm.

Assay of phagolysosomal lipase

Ten μl DF-binding microsphere suspension (5.6x10^6/ml) were added to 100 μl of neutrophil suspension (4x10^6/ml), and the mixture was incubated at 37°C for various periods. The fluorescence intensity of the suspension was measured at 515nm by use of exiting light of 490nm.

RESULTS AND DISCUSSION

When 4MUGL-binding microspheres were incubated with neutrophils, they were phagocytosed by neutrophils without opsonization in the same way as opsonized zymosans. As phagocytosis proceeded, the amount of the 4MU released from the microspheres increased, and the ratio of extracellular 4MU to intracellular one was constantly 3:1 (Fig.1). Fig.2 shows that 4MU detected in the medium was not produced there but within the phagolysosomes of neutrophils. These results show that we can measure the activity of lysosomal enzymes within phagolysosomes during phagocytosis by measuring the amount of 4MU in extracellular fluid. We presumed the model for this reaction in Fig.3. This method makes possible the measurement of the activity of lysosomal enzyme secretion into phagosomes in intact neutrophils during phagocytosis. By this method, lysosomal enzyme secretion following phagocytosis was measured, that is, only phagocytic activity or only lysosomal enzyme release can not be separately measured. This is called active phagocytosis. This active phagocytosis may express effective phagocytosis in vivo as well as bactericidal and cytotoxic activity of neutrophils.

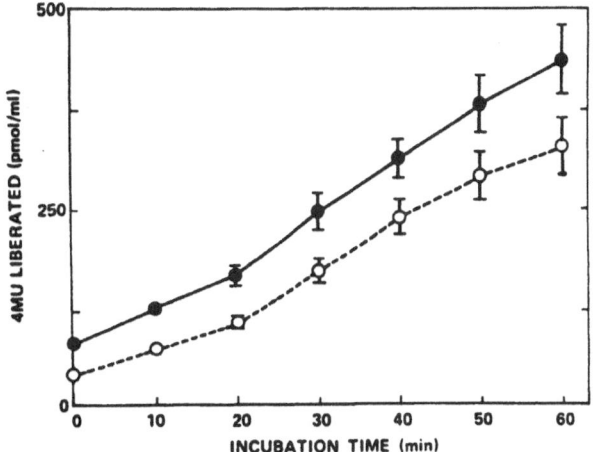

Fig.1   Time course of 4MU liberation from phagocytosed microspheres. Neutrophils were incubated with 4MUGL-binding microspheres for 0,10,20,30,40,50 and 60 min. Total liberated 4MU (●-●) and extracellular 4MU (O-O) were measured by the fluorescence as described in MATERIALS & METHODS. The mean values of triplicate experiments are shown.

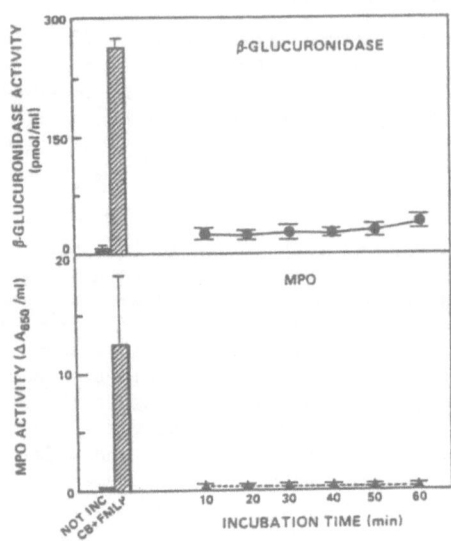

Fig.2   Time course of leakage of β-glucuronidase and myeloperoxidase.
         Neutrophils were incubated with 4MUGL-binding microspheres for
         0,10,20,30,40,50 and 60 min. Maximal extracellular release of
         both enzymes from neutrophils was obtained with stimulation of
         cytochalasin B and fMet-Leu-Phe as a control experiment (shaded
         bar, not incubated; hatched bar, incubated for 60 min). Leakage
         of β-glucuronidase and myeloperoxidase with phagocytosis into
         extracellular fluid were determined by the procedures described
         in MATERIALS & METHODS.

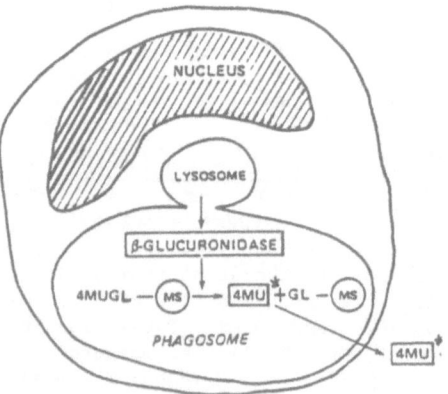

Fig.3   Scheme for liberation of fluorescent 4MU by lysosomal β-glucuroni-
         dase from 4MUGL-binding microspheres.

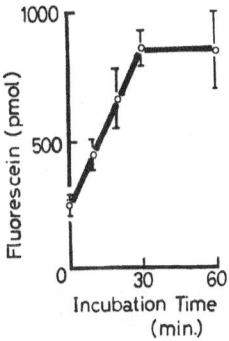

Fig.4　Time course of the hydrolysis of microsphere-bound decanoyl fluorescein by esterase in neutrophils. The assay method is described
in MATERIALS & METHODS. The mean values of triplicate experiments
are shown.

　　　　Opsonized DF-binding microspheres were phagocytosed by neutrophils in
the same way as 4MUGL-binding microspheres, and the amount of the produced
fluorescein increased with time (Fig.4). From these results, we presumed

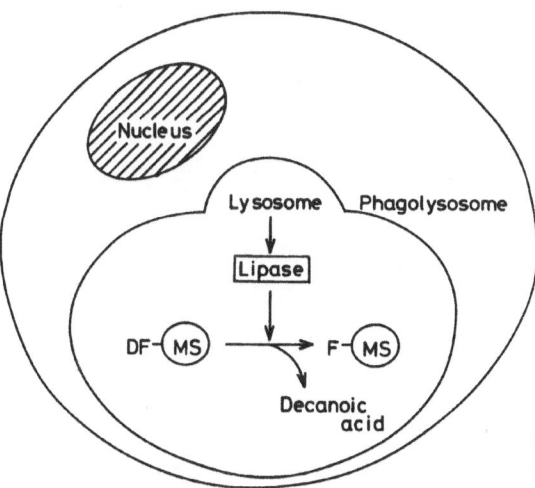

Fig.5　Scheme for hydrolysis of DF bound to the microspheres by lipase
within a phagolysosome.

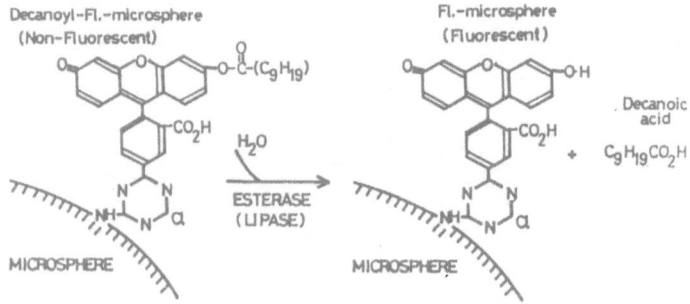

Fig.6   Reaction scheme of the conversion of non-fluorescent microspheres
          (substrate) into fluorescent ones caused by the action of esterase
          (especially lipase).

that the immobilized DF was hydrolysed by lipase within phagolysosomes,
releasing a substituted fluorescein as shown in Fig.5 and 6.
     In order to confirm the reaction scheme above proposed, the flow cyto-
metric analyses of the incubation mixture of DF-binding microspheres and
neutrophils were carried out. As can be seen in Fig.7a and Fig.7b, neither
DF-binding microspheres nor neutrophils emitted fluorescence per se. DF-
binding microspheres were distinguished from neutrophils by the light
scattering intensity. After the mixture of DF-binding microspheres and
neutrophils was incubated at 37°C for 60 min, the fluorescence intensity
of the higher light scattering region increased. On the other hand, the
fluorescence intensity of the lower light scattering intensity region

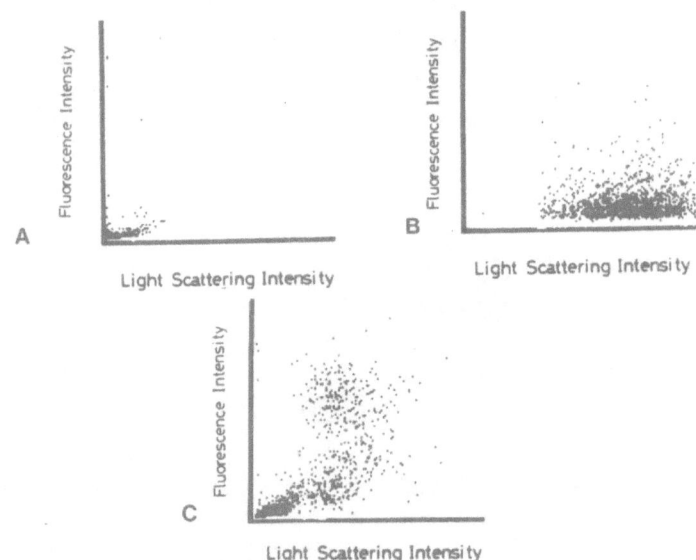

Fig.7   Flow cytometric analysis of DF-binding microspheres, neutrophils
          and the mixture of them.
          (A) DF-binding microspheres
          (B) Neutrophils
          (C) The mixture of DF-binding microspheres and neutrophils after
              incubation at 37°C for 60 min.

154

remained slight (Fig.7c). These resuls indicate that only microspheres within the cells became fluorescent. In other words, decanoyl fluorescein bound to microspheres was hydrolyzed only within the cells and not outside the cells. Thus, phagolysosomal esterase was assayed by use of DF-binding microspheres without the rupture of cells. Since decanoyl fluorescein is an ester of a fatty acid with long chain, it is efficiently hydrolyzed by lipase (12,13).

We hope that the new method of phagolysosomal esterase assay presented here would be useful for the study of the role of phagocytes in the prevention of atherosclerosis.

CONCLUSION

We have devised a new method to measure the digestive activity of phagocytes (Fig.8). The method utilizing fluorogenic substrate- binding microspheres (2μm) makes possible the direct measurement of hydrolase (β-glucuronidase and lipase) within phagolysosomes.

ACKNOWLEDGEMENTS

We thank Ms.Miura,K.(TORAY) for her assistance at the preparation for DF-binding microspheres, Dr.Sakatani T. and Dr.Sasagawa,S. (RADIATION EFFECTS RESEARCH FOUNDATION) for their assistance at the assay of β-glucuronidase, Mr.Soga,H., Ms.Shinoda,A. and Ms.Kurima,T. (Showa Denko Inc.) for their technical assistance at the flow cytometric measurement.

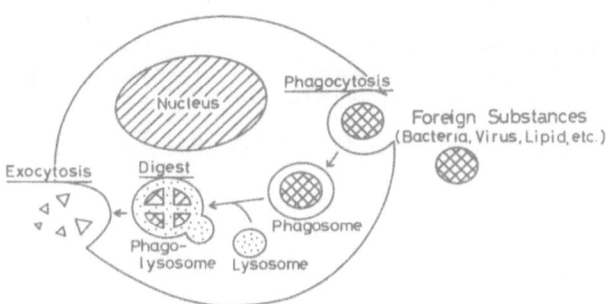

Fig.8   Scheme for digestive function of phagocytes.

REFERENCES

1. Weissmann,G.,Zurier,R.B.,Spieler,P.J. and Goldstein,I.M. (1971)
   Mechanism of lysosomal enzyme release from leukocytes exposed to immune
   complexes and other particles. Exp.Med. Part 2, 134, 149-164s.

2. Bainton,D.F. (1973) Sequential degranulation of the two types of polymorphonuclear leukocyte granules during phagocytosis of microorganisms.
   J.Cell.Biol. 58,249-264.

3. Allen,R.C.,Mills,E.L.,McNitt,T.R. and Quie,P.G. (1981) Role of myeloperoxidase and bacterial metabolism in chemiluminescence of granulocytes
   from pacients with chronic granulomatous disease. J.Infect.Dis. 144,
   344-348.

4. Goldstein,I.,Hoffstein,S.,Gallin,J. and Weissmann,G. (1973) Mechanism
   of lysosomal enzyme release from human leukocytes: Microtuble assembly
   and membrane fusion induced by a component of complement. Proc.Natl.
   Acad. Sci. 70,2916-2920.

5. Amano,D.,Kagosaka,Y.,Usui,T.,Yamamoto,S. and Hayaishi,O. (1975)
   Inhibitory effects of superoxide dismutases and various other proteins
   on the nitroblue tetrazolium reduction by phagocytizing guinea pig polymorphonuclear leukocytes. Biochem.Biophys.Res.Commun. 66, 272-279.

6. Reiss,M. and Roos,D. (1978) Differences in oxygen metabolism of phago-
   cytosing monocytes and neutrophils. J.Clin.Invest. 61, 480-488.

7. Suzuki,K.,Swenson,C.,Sasagawa,S.,Sakatani,T.,Watanabe,M.,Kobayashi,M.
   and Fujikura,T. (1983) Age related decline in lysosomal enzyme release
   from polymorphonuclear leukocytes after N-formyl-methionyl-leucyl-
   phenylalanine stimulation. Exp.Hematol. 11,1005-1013.

8. Wright,S.D. and Silverstein,S.C. (1983) Receptor for C3b and C3bi
   promote phagocytosis but not the release of toxic oxygen from human
   phagocytes. J.Exp.Med. 158,2016-2023.

9. Uchida,T.,Kanno,T. and Hosaka,S. (1985) Direct measurement of phago-
   somal reactive oxygen by luminol-binding microspheres. J.Immunol.
   Methods. 77, 55-61.

10. Suzuki,K.,Uchida,T.,Sakatani,T.,Sasagawa,S.,Hosaka,S. and Fujikura,T.
    Active phagocytosis of polymorphonuclear leukocytes by fluorescence
    liberation from phagocytosed microspheres. submitted for publication.

11. Uchida,T.,Hosaka,S. and Miura,K. (1985) Direct measurement of phago-
    lysosomal esterase activity. Biochem.Biophys.Res.Commun. 127,584-589.

12. Guilbert,G.G. and Kramer,D.N. (1964) Fluorometric Determination of
    lipase,acylase,alpha- and gamma- chymotrypsin and inhibitors of these
    enzymes. Anal.Chem. 36,409-412.

13. Fleisher,M. and Schwartz,M.K. (1971) An automated,fluorometric proce-
    dure for determing serum lipase. Clin.Chem. 17,417-422.

THE SMALLER DIAMETER VASCULAR GRAFT - A BIOMATERIALS CHALLENGE

Allan S. Hoffman, Buddy D. Ratner, Andy M. Garfinkle,
Larry O. Reynolds, Thomas A. Horbett and *Steven R. Hanson

University of Washington, Center for Bioengineering and
Chemical Engineering Seattle, WA
*Scripps Clinic and Research Foundation, La Jolla, CA

ABSTRACT

The surface composition of a biomaterial can have an important influence on biologic responses. The surface composition of a synthetic vascular graft may be modified in a number of ways. In this paper we report on a surface treatment using a gas discharge which deposits a thin coating onto the graft surface, significantly changing its surface chemistry, but without measurable change in porosity, compliance or surface topography. Treatments with tetrafluoroethylene (TFE) gas yield dramatic improvements in both thrombo-and emboli-resistance of the graft, based on *in vitro* measurements and *ex vivo* tests using a baboon.

Keywords: Small Diameter vascular graft; gas or plasma discharge treatments; fluoropolymer coatings; biomaterial surface modification.

INTRODUCTION

Improvement of the small diameter vascular prostheses represents a major research challenge today. Despite the large number of available products, the graft of choice remains the saphenous vein[1-4]. This situation is a reflection of the limited understanding regarding the key factors which influence the patency of small diameter grafts. Clearly, graft material properties, host-related variables and the surgical technique are all important factors. It has been very difficult to independently investigate and separate the effects of each of the key variables. Thus, there has been much speculation and debate regarding the relative importance of each of these factors in a given usage situation.

The important material properties which can influence protein and cell interactions at the biomaterial-biologic interface are listed in Table I. It is probable that surface composition and topography most strongly influence the composition and organization of the initial adsorbed protein layer. It is this layer which mediates subsequent cellular events at that interface. Thus, a great deal of effort has gone into surface modifications and characterization of polymeric biomaterials.

TABLE I. IMPORTANT BIOMATERIAL PROPERTIES IN AIR OR AQUEOUS SOLUTIONS

SURFACE

1) TOPOGRAPHY	(ROUGHNESS, POROSITY, DEFECTS)
2) "COMPOSITION"	("POLARITY", IONIC CHARGE, OXIDATION, CONTAMINATION, H_2O)
3) MORPHOLOGY	(DOMAINS, CRYSTALLITES)
4) MECHANICS	(MOVEMENTS-MOLECULAR, OR MACROSCOPIC)

BULK

"EXCHANGES"	(ABSORPTION, DESORPTION, BIODEGRADATION)

Polymer surfaces are commonly modified by physical deposition of other compounds (e.g., surfactants or polymers), by direct chemical modification of the polymer surface (e.g. oxidation, hydrolysis, sulfonation, etc.) or by chemical bonding of a different polymer (graft copolymerization) or "polymer-like" composition (gas discharge polymerization). Figure 1 illustrates the different ways which can be used to generate free radicals on a surface for subsequent chemical bonding of a coating composition.

Radio frequency glow discharge (RFGD) treatment has been used to modify biomaterial surface compositions without changing bulk material properties[7-9]. In plasma or glow discharge polymerization, an organic compound (monomer) in the gas phase is introduced into a vacuum system containing the substrate material to be treated. The monomer gas is then subjected to an electric discharge, typically a capacitively or inductively coupled radio frequency discharge. Active species are generated in the gas due to absorption of this excitation energy.

Two types of reactions between a gas plasma and a polymer surface can be distinguished. These are net etching or deposition reactions. The plasma may react directly with the polymer, e.g., to oxidize

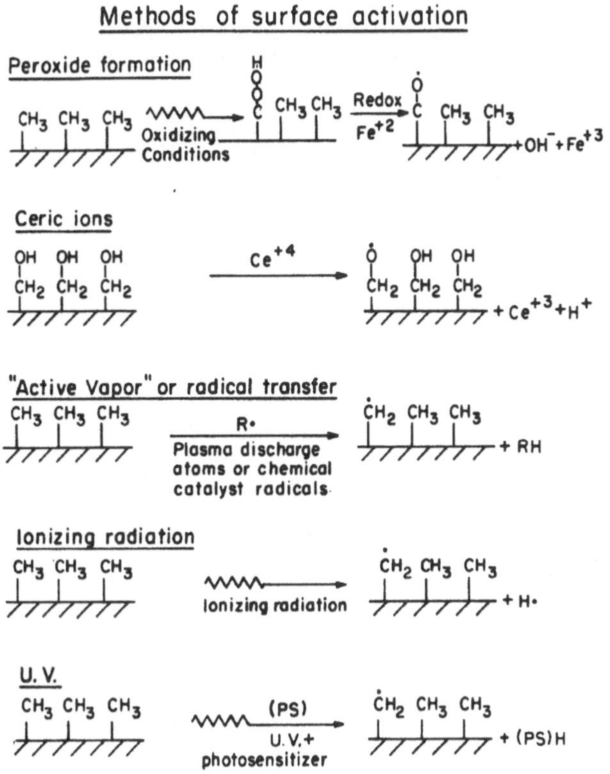

Methods of surface activation

Fig. 1. Examples of techniques and reactions for generating radicals on surfaces. (Note: The precise nature of the radical intermediates formed has not been elucidated in some cases. Representation in this figure show schematically radical species which might be formed.)[6]

it and to ablate surface atoms in an etching discharge, or the plasma may create polymer radicals by chain transfer reactions and these radicals can then initiate other reactions with the gases, such as graft polymerization. Short or long chain oligomers or polymers may also form in the gas phase and deposit directly on the surface. The predominant reaction will depend on the compositions of the gases and the polymer substrate, and also on the discharge conditions.

The gaseous compounds useful for polymer deposition may be common gases and saturated or aromatic organic compounds as well as typical vinyl type monomers. The coatings are usually very thin - perhaps only a monomolecular layer - and they are presumed in some cases to be highly crosslinked. One would expect these coatings to remain well-adhered to the substrate polymer under normal conditions. The plasma discharge process only affects the polymer surface, and, conversely, the surface must "see" the plasma to be affected. Polymer films generated in this manner have been shown to be ultra-thin, tightly bound and pinhole free[10].

The structure and composition of the plasma modified surface depends on the plasma conditions (e.g., gas pressure and flow rate, continuous or pulsed discharge, energy input, etc). The compositional makeup may be broad and contain a variety of chemical groups. In addition, the coatings may not always be deposited uniformly on the substrate surface since the plasma characteristics can vary significantly over very short distances within the reactor. Thus, it is important to select reactor conditions which lead to the deposition of uniform films of the desired composition.

Electron spectroscopy for chemical analysis (ESCA) provides a valuable means of analyzing the surface composition of these films, where extensive monomer rearrangement in the deposited polymer is very common[11]. ESCA is particularly useful in the structural analysis of fluorocarbon films. Surface compositional information not readily obtained by other techniques can be acquired by analyzing the $C1_s$ level binding energy shifts resulting from carbons which are covalently bound to fluorine atoms.

159

TABLE II. VARIATION OF SURFACE COMPOSITION
WITH TIME EXPOSED TO TFE GLOW DISCHARGE

RFGD EXPOSURE TIME (MIN.)	C/F
0	∞
1	0.98
5	0.92
20	0.86
30	0.72
40	0.72
PTFE (THEOR.)	0.50

SUBSTRATE: MEDIUM POROSITY DACRON KNIT

RFGD CONDITIONS: 15 WATTS, 0.2 TORR

TABLE III. COMPOSITION OF RFGD FLUOROPOLYMER
COATINGS ON THE LUMEN OF HIGH POROSITY DACRON
KNIT VASCULAR GRAFTS

ESCA SAMPLE LOCATION (CM)*	C/F
+ 3	0.68
-3	0.71
PTFE (THEOR.)	0.50

*0 = CENTER OF SAMPLE

RFGD CONDITIONS: 15 WATTS
0.2 TORR, 30 MINUTES

160

TABLE IV. HOMOGENEITY OF FLUOROPOLYMER COATING
ON A MEDIUM POROSITY DACRON KNIT VASCULAR GRAFT

ESCA SAMPLE LOCATION (CM)*		C/F
LUMEN	+ 1	0.70
	-1	0.67
OUTER SIDE	+ 1	0.70
	-1	0.71
PTFE (THEORETICAL)		0.50

* 0 = CENTER OF GRAFT

RFGD CONDITIONS: 15 WATTS,
0.2 TORR, 30 MINUTES

TABLE V. THE EFFECT OF TFE GLOW DISCHARGE
TREATMENT ON DACRON GRAFT POROSITY

DACRON GRAFT	UNTREATED CONTROL	TFE TREATED TREATED
WOVEN	1.8 ± 0.2	1.8 ± 0.1
MEDIUM POROSITY KNIT	19.2 ± 1.0	20.9 ± 2.1
HIGH POROSITY KNIT	28.7 ± 2.1	28.5 ± 1.3

UNITS = CC OF WATER/CM^2-MIN x 10^{-2}

Glow Discharge Reaction Vessel

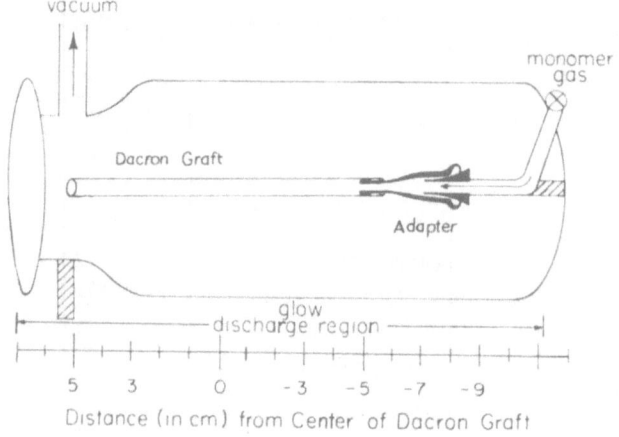

Fig. 3. Schematic of plasma discharge reaction vessel.

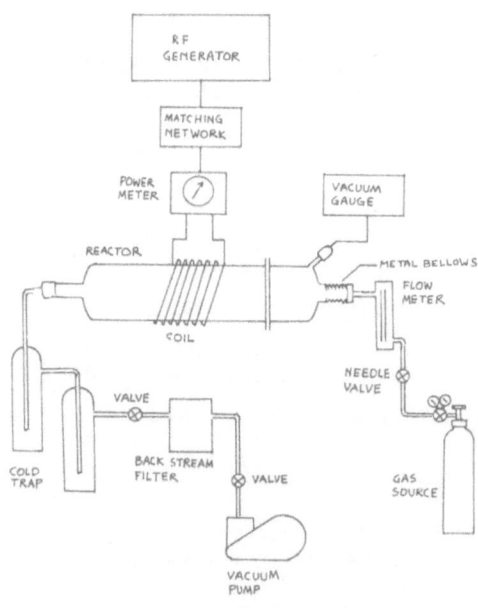

Fig. 2. Schematic of plasma discharge system.

previously reported for bovine fibrinogen. Plasma fibrinogen concentrations were determined by measuring the extinction of a redissolved clot induced by thrombin. This procedure is essentially as described by Ratnoff and Menzie[27] except that the clot was redissolved in a sodium dodecyl sulfate solution and the absorbance measured at 280 nm. The buffer used in all instances was 0.01 M phosphate, 0.12 M NaCl. 0.02% sodium azide, pH 7.4 ("CPBSz").

The adsorption experiments were done at 37°C using a flow system run at 100 ml/min. The woven Dacron vascular graft and TFE-treated woven Dacron vascular graft samples were 1 cm. long and 0.04 cm. in diameter. They were placed inside special tubular glass holders tapered at the downstream end to prevent loss of the sample but still allow free, uniform access of the flowing plasma to all parts of the grafts. Uniformity of access and rinsing was tested using a colored dye. For each adsorption time period, three samples in series were inserted into the flow system, degassed 37°C CPBSz was pumped through the system to eliminate air bubbles. Plasma was then passed through for the desired time, after which buffer was pumped through again for 1 min, and the samples removed and dip rinsed several times in each of three separate buffer solutions. The samples were then placed in a counting tube containing buffer and the retained radioactivity was measured on a γ-counter (TM Analytic Model 1185). Separate samples were run for each time point shown; thus, flow was not interrupted to remove samples. The next day, the samples were transferred to fresh counting tubes containing fresh buffer and the retained radioactivity remeasured. The latter, "overnight soak" data were used in the results presented. The retained radioactivity was converted to amount adsorbed, $\mu g/cm^2$, by dividing the retained radioactivity (in cpm) by the specific activity of the plasma fibrinogen pool (in cpm/μg) and the total (inside & outside) planar equivalent area of the samples. The true area of these woven samples is undoubtedly much higher than the area used. Thus, although the adsorption figures measured (Fig. 4) appear much higher than observed previously for flat film samples, this is a reflection of the lack of true area measurement.

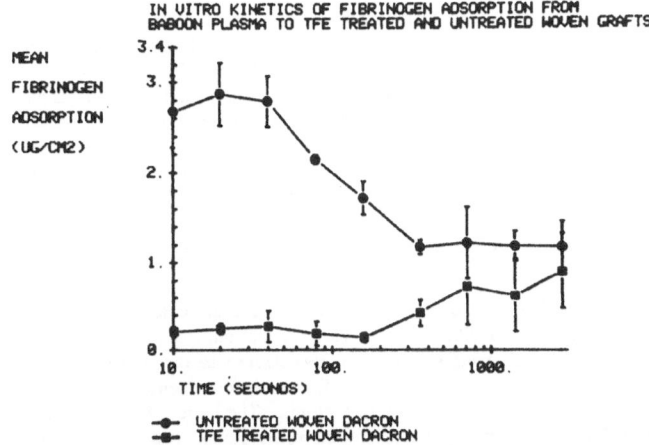

Fig. 4. Kinetics of *in vitro* fibrinogen adsorption from baboon plasma to TFE treated and untreated woven grafts.

D. The Baboon Model

The thrombotic response to variations in graft surface composition was investigated using the *ex vivo* baboon femoral shunt model[13,18]. Twenty healthy male baboons (*papio cynocephalus*), each supporting an external femoral arterial-venous shunt, were studied at the Regional Primate Center at the University of Washington. Animals weighting between 9-15 kg were supported in restraining chairs during the experiments. All were free from drugs and were maintained on a regular diet. This model has yielded measurements which have been shown to be unaffected by the surgical shunting procedure, independent of the implanted Silastic A-V shunt and reproducible between the different experimental animals[13-19].

E. Preparation of Graft Materials

Cleaned, untreated or recently glow discharge TFE treated Dacron prosthetic arterial grafts were rendered impervious to blood leakage by a surrounding layer of heat-shrink Teflon. All segments studied were approximately 10 cm in length and were kept rigid by this surrounding layer. Two 20 cm-long segments of Silastic tubing (1/8 in. ID, Dow Corning) were forced over the heat-shrink covered graft at either end to facilitate connection into the A-V shunt system. Graft materials were filled with sterile Ringer's citrated dextrose solution (RCD) and the bubbles were removed by ultrasonic treatment prior to use. The graft segment filled with RCD was connected with 3/4 inch long Teflon adaptors to the two 35 cm Silastic segments of the implanted A-V shunt. The graft materials were assigned randomly to the various animals for the study.

F. Blood Flow Measurements

Blood flow through the shunt-vascular graft system was measured using a Doppler ultrasonic flowmeter (L & M Electronics, Daly City, CA). The transducer probe was a clip-on variety with a fixed crystal designed to fit snugly on the Silastic shunt itself. Flow measurements were made by attaching the probe to the downstream Silastic shunt and recording for one minute the mean system blood flow. Blood flow, which rarely varied more than 15% over the one minute measurement interval, was then averaged to yield a single value. The probe was moved at regular time intervals between the various animals being simultaneously studied. This type of probe set-up thus permitted repetitive and reproducible flow measurements.

Blood flow was always checked in the Silastic shunt before graft placement. A minimum blood flow of 100 ml/min was required before starting an experiment. Flow rates lower than this indicated shunts which were likely to occlude. Blood flow rates before and after graft placement and before and after sampling were regularly compared to check for problems resulting from handling. While large changes in blood flow occurred rarely, changes which typically led to occlusion were more gradual and showed a continued downward trend over time periods greater than one hour. The range of initial shunt blood flow rates in the different animals which were studied was 140-320 ml/min. with a mean of 285 ml/min.

G. *In Vitro* Microembolization Studies

A laser light scattering system has been developed for the detection of microemboli in whole blood flowing under real time conditions[21-23]. The size and number of these aggregates was detected using a computer-based optical sizing system. The detectable size range in this study is from 80μm to 800μm diameter particles. Provisions were made for continuous recirculation of fresh whole baboon blood within the flow system at a velocity of 100 ml/min. The barrel of a 20 cc polypropylene syringe served to dampen the pulsatile flow generated by the roller pump. It also served as a reservoir where entrapped air bubbles could be removed.

Five centimeter lengths of graft material were prepared using the procedure from the *ex vivo* shunt study. The system was first filled with a pH balanced isoton solution to pre-wet the material surfaces. Fresh whole baboon blood stabilized with ACD was introduced into the syringe barrel. This was pumped through the graft system, displacing all of the isoton. This procedure insured that there were no material-air interfaces. Each material was tested a minimum of five times and each time blood was drawn from a different baboon donor. Embolization measurements were made from 1 ml ID cuvettes for a period of 1 minute every other minute for the 90-minute time period of a typical experiment.

164

RESULTS AND DISCUSSION

A. RFGD Treatment and Surface Characterization

ESCA was used to evaluate the degree of substrate coverage and the chemical structure of the deposited fluoropolymer films. The ESCA spectra of the TFE treated surfaces are shown in Fig 5. It is clear that the spectrum of the TFE treated surface is very different from that of either Teflon PTFE or Dacron. There is a significant proportion of carbons in the polymer surface which are bound to fewer than two fluorine atoms, indicating that the RFGD fluoropolymer coating is probably highly crosslinked. A significant concentration of CF_3 groups was also detected in the surfaces. The presence of these low energy groups was reflected by the critical surface tension values of about 13.0 dynes/cm which were measured on these surfaces. These are lower than the γ_c values of about 18.5 dynes/cm[28] reported for PTFE Teflon.

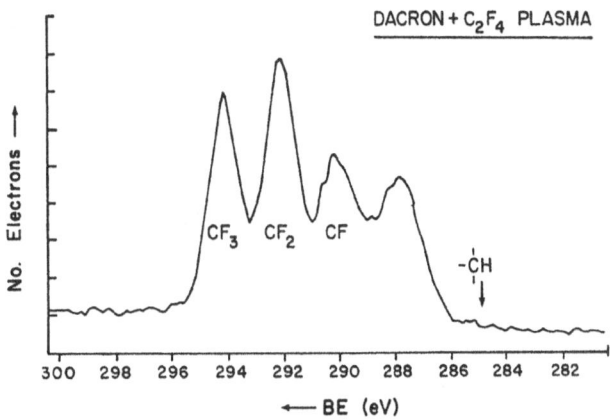

Fig. 5. High resolution $C1_s$ ESCA spectrum for a TFE treated Dacron surface.

B. The Baboon Model: An *Ex Vivo* Patency Study

The thrombotic response to changes in graft surface composition was investigated in the baboon model for the different porosity grafts. Graft patency was markedly improved for TFE treated woven and knit Dacron grafts when compared with the untreated grafts, for all time periods studied. These results are shown in Figures 6 and 7. Woven grafts that were glow discharge treated with argon alone occluded more rapidly than the untreated Dacron grafts. This result suggests that the specific gas treatment chosen, and not the glow discharge process *per se*, is responsible for the improved patency. Although the high porosity, untreated Dacron knit grafts occluded more rapidly than the untreated woven Dacron, after one week the patency results were equally poor. Vascular graft surface chemistry thus appears to be a more important determinant of short-term patency than either porosity or surface texture.

C. *In Vitro* Microembolization Studies

In addition to thrombotic occlusion, and important further consideration is the tendency of these treated surfaces to produce thromboemboli. Table VI shows results of *in vitro* embolization studies using fresh whole baboon blood. These laser scattering studies indicate that all TFE treated graft surfaces produce significantly fewer emboli than the untreated control graft surfaces. Thus, the improved baboon shunt patency of the TFE treated Dacron graft materials is probably not due to increased embolization. It should be noted that the rate of emboli production on Goretex graft surfaces is also markedly reduced by the TFE glow discharge treatment.

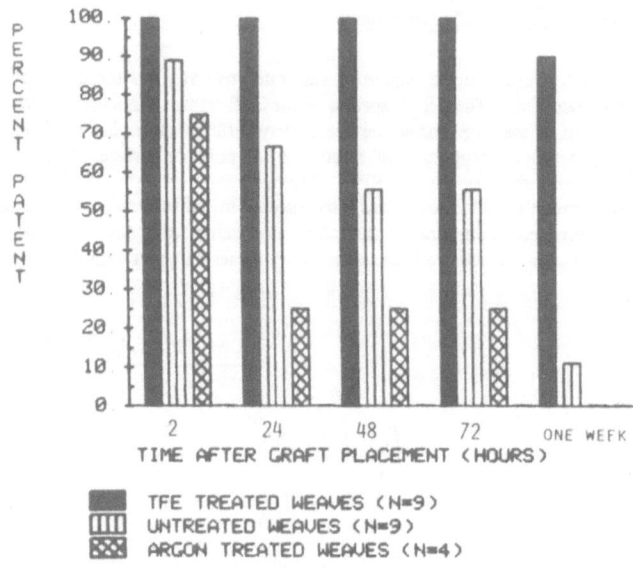

Fig. 6. Graft patency for small diameter, low porosity Dacron weave grafts before and after argon and TFE glow discharge treatments.[14]

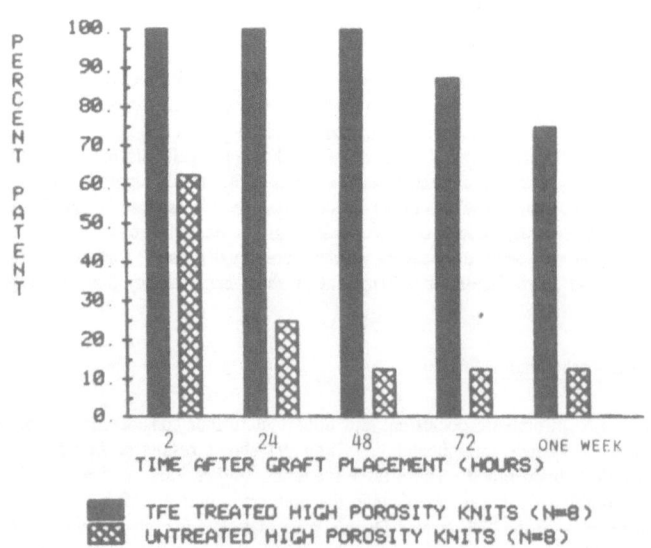

Fig. 7. Graft patency for small diameter, high porosity Dacron knit grafts before and after TFE glow discharge treatment.[14]

TABLE VI. RELATIVE *IN VITRO* EMBOLIZATION RATES OF SILASTIC TUBING AND TFE TREATED AND UNTREATED VASCULAR GRAFT MATERIALS

	GORETEX	DACRON KNIT	DACRON WEAVE	SILASTIC TUBING
UNTREATED CONTROL	8.56 ± 4.39	8.68 ± 0.34	8.32 ± 1.28	1.0 ± .27
TFE TREATMENT	1.55 ± 0.32	2.69 ± 0.78	3.72 ± 0.89	---

UNITS = (PLATELETS/DAY-CM^2) x 10^8

SIZE RANGE = 80-800 μ

We have also analyzed the real-time production of emboli volume during the initial 90 minute period and these data are compared in Figures 8-10 for the three vascular grafts before and after treatment. Dramatic reductions in emboli production due to the TFE treatment can be noted. These data support the conclusion that the *in vitro* embolization process from graft surfaces is strongly dependent on chemical composition and is much less influenced by surface texture.

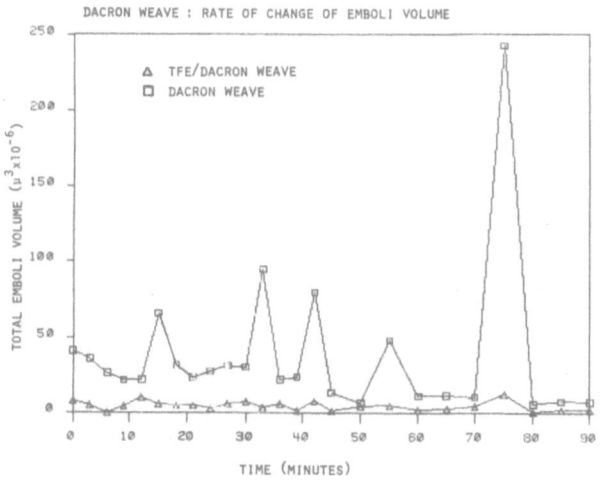

Fig. 8 Rate of change of total emboli volume as a function of time for TFE treated and untreated Dacron weaves.

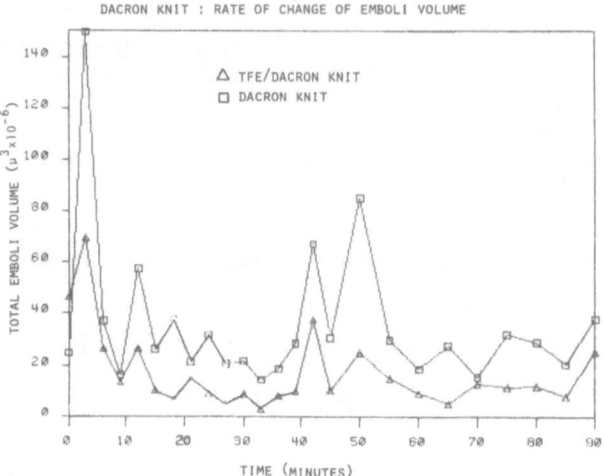

Fig. 9. Rate of change of total emboli volume as a function of time for TFE treated and untreated Dacron knits.

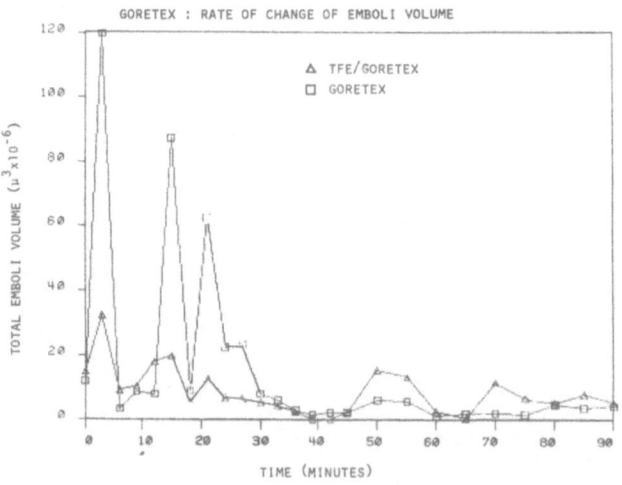

Fig. 10. Rate of change of total emboli volume as a function of time for TFE treated and untreated Goretex.

168

D. Doppler Blood Flow and Fibrinogen Adsorption Studies

Graft blood flow rates in the baboon shunt were measured using a Doppler flow meter. The values were normalized for each experiment to the initial graft blood flow rate using the relationship:

$$\frac{blood\ flow\ rate\ at\ time\ t}{initial\ blood\ flow\ rate} \times 100 = relative\ blood\ flow\ rate$$

For all similar materials, the relative blood flow measurements were averaged across each sampling time period to yield a mean relative graft blood flow value ($\pm$ 1.0 S.D.). The changes in mean relative graft blood flow with time after graft placement are shown in Figures 11 and 12.

Blood flow rates in the acute contact time period, 0-2 hours, proved to be a good indicator of subsequent graft patency. In this time period, flow rates in the untreated grafts decreased more rapidly with time and to a greater extent than was observed for the TFE treated grafts.

These results suggest that the cross-sectional area available for blood flow is reduced by luminal thrombus buildup which is strongly dependent on graft surface chemical composition. Ultimately, this selective thrombus accumulation is reflected in the improved patency exhibited by the TFE treated grafts. The *in vitro* fibrinogen adsorption kinetics (Fig 4) strongly support the premise that the initial protein layer deposited during the first seconds to minutes has an important influence on such thrombus buildup. Thus, the overall thrombogenic and thromboembolic processes are strongly dependent on graft surface composition.

Woven grafts which were glow discharge treated only with argon exhibited an even more rapid decrease in blood flow rate over the initial two hours than the untreated Dacron weaves. ESCA data (Table VII) show that the C/O ratio decreases slightly after argon treatment, suggesting that the argon contained traces of oxygen which oxidized the surface somewhat. These results again support the conclusion that graft surface chemistry plays a key role in thrombus formation within small diameter grafts.

Surface texture had less influence than surface chemistry on both early blood flow changes and graft patency. For the untreated Dacron grafts, blood flow decreased most rapidly in the very textured, high porosity velour knits. However, this poor response was also similar to that of the argon glow discharge treated weaves with smoother surfaces. Finally, patency at one week did not correlate with graft surface texture.

Our results contrast with those of two other groups. Yasuda and Gaziki[29] state that materials having "high surface oxygen contents" after a TFE gas discharge treatment (presumably also containing O_2 in the reactor) exhibit greater "blood" compatibility than those surfaces with essentially carbon and fluorine (and only minor amounts of oxygen) in their ESCA spectrum. This is in contrast to our data on the argon vs. TFE treatment of the Dacron weaves. Also, our TFE treated surfaces contained insignificant oxygen contents. They based their conclusion on *in vitro* whole blood coagulation tests performed in small cells or tubes; they did not use flowing, fresh primate blood and this could account for the different conclusions reached.

In another publication, Didisheim et al.[24] have recently stated that texture is the important variable and surface chemistry is not important in blood-surface interactions. This conclusion was based on platelet attachment to TFE plasma discharge-treated smooth-walled Silastic tubing and woven Dacron material. A canine experimental model was used. This conclusion is not supported by our studies in the baboon model. Differences between the two animal models may account for these contrasting results.

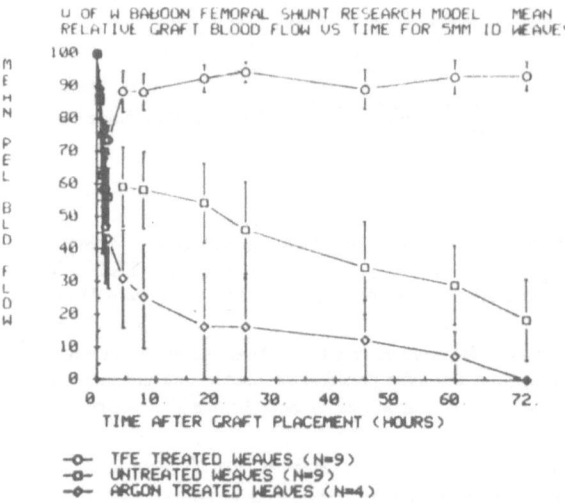

Fig. 11. Variation of mean relative graft blood flow with time for small diameter, low porosity Dacron weave grafts before and after argon and TFE glow discharge treatments.[14]

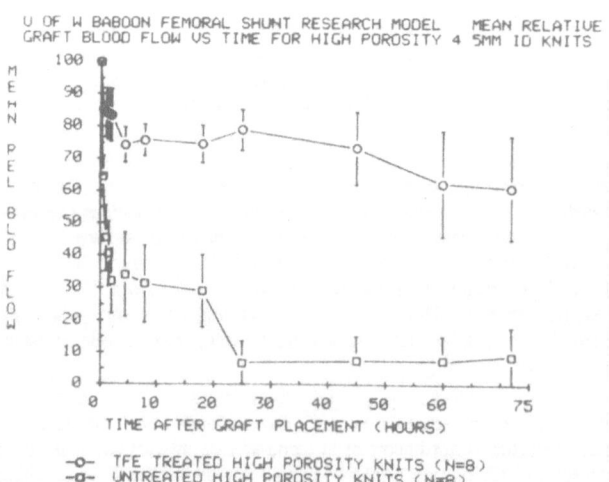

Fig. 12. Variation of mean relative graft blood flow with time for small diameter, high porosity Dacron knit grafts before and after TFE glow discharge treatment.[14]

170

TABLE VII. EFFECT OF AN RFGD ARGON TREATMENT
ON DACRON SURFACE COMPOSITION

ESCA SAMPLE LOCATION (CM*)		C/O
CLEANED, UNTREATED PET		2.67
PET (THEOR.)		2.50
ARGON TREATED PET	-3	2.36
	+1	2.38

* 0 = CENTER OF SAMPLE

RFGD CONDITIONS: 15 WATTS,
0.2 TORR, 5 MINUTES

General Discussion

In the acute contact time period, the rapid changes in blood flow observed for all grafts reflect the development of a thrombus layer on the graft surface. This process appears to be highly dependent on graft surface chemical composition. With increasing time, the flow rates in the untreated grafts continued to decrease, while the TFE treated grafts maintained relatively high flow rates. These results are shown in Figures 4 and 6. The cross-sectional area available for blood flow is probably being progressively reduced by thrombus buildup in the untreated grafts, as seen by the blood flow decrease with time. The character of the initial thrombus layer thus appears to have a strong influence on subsequent thrombus accumulation and related graft patency.

As noted above, greater rates of embolization were observed *in vitro* for untreated control grafts, when compared with TFE treated grafts. This increase in *in vitro* embolus production could be due to differences in the adhesiveness or uniformity of the deposited thrombus layer. In addition, embolization could re-expose underlying material and therefore delay the formation of a stable, potentially less reactive thrombus layer.

CONCLUSIONS

The RFGD treatment of small diameter Dacron vascular grafts with TFE monomer results in a dramatic improvement in patency when studied in the *ex vivo* femoral shunt baboon model. Blood flow rate proved to be a good early indicator of subsequent graft patency. *In vitro* embolization rates were found to markedly decrease after TFE treatment of graft materials. This result indicates that improved *ex vivo* patency was probably not due to increased embolization. This decrease in the rate of embolization also occurred with Goretex grafts, indicating that the TFE glow discharge treatment results in a different and potentially more biocompatible fluoropolymer surface even when compared with PTFE Teflon.

We are investigating the mechanism for the improved patency and decreased embolization exhibited by the TFE glow discharge treated grafts. We are testing the hypotheses that the mechanism involves modified initial protein adsorption, resulting in differences in the adherent thrombus layer and its subsequent reorganization. Further studies are planned to investigate the longer term *in vivo* performance of all of the grafts as implants in the baboon.

ACKNOWLEDGEMENT

The authors would like to acknowledge the support of the NIH, NHLBI, Device and Technology Branch, Grant HL-22163-01 to 05. We would like to thank Cheryl Kruesel for typing the manuscript. We also wish to acknowledge the use of the facilities of the Regional Primate Center at the University of Washington, Dr. Orville Smith, Director.

REFERENCES

1. Dardik, H., ed., "Graft Materials in Vascular Surgery," Year Book Med. Publ., Chicago, (1978).

2. Mortensen, J.D., "Safety and Performance of Currently Available Vascular Prostheses," *ASAIO Journal* 4: 125, (1981).

3. Sawyer, P.N., Kaplitt, M.J., eds., "Vascular Grafts," Appleton-Century-Crofts, New York, (1978).

4. Vascular Prostheses Workshop, sponsored by Devices and Technology Branch, NHLBI, Bethesda, MD. May 14, 1981: NIH Publ. No. 82-1215, (1981).

5. Hoffman, A.S., "Synthetic Polymer Biomaterials - a Review," in "IUPAC - Macromolecules" (eds. H. Benoit and P. Rempp), Pergamon Press, London, p. 321 (1982).

6. Hoffman, A.S., "Ionizing Radiation and Gas Plasma Discharge Treatments for Preparation of Novel Polymeric Biomaterials," *Adv. in Polymer Sci,* 57, 141-157, (1984).

7. Bell, A.T, Hollahan, J.R., eds., "Techniques and Application of Plasma Chemistry," John Wiley and Sons, New York, (1974).

8. Yasuda, H., Morosoff, N., "Plasma Polymerization of Tetrafluoroethylene. II. Capacitive Radio Frequency Discharge," *J. Appl. Polym. Sci.* 23: 1003, (1979).

9. O'Kane, D.F., Rice, D.W., "Preparation and Characterization of Glow Discharge Fluoropolymer-Type Polymers," *J. Macromol. Sci.-Chem.* A10(3): 567, (1976).

10. Hollahan, J.R., Wydeven, T., Johnson, C.C., "Combination Moisture Resistant and Anti-Reflection Plasma Polymerized Thin Films for Optical Coatings," *Appl. Opt.* 13: 1844, (1974).

11. Clark, C.T., Feast, W.J., Kilcast, D., Musgrave, W.K.R., "Applications of ESCA to Polymer Chemistry. III. Structure and Bonding in Homopolymers of Ethylene and the Fluoroethylenes and Determination of the Compositions of Fluoropolymers," *J. Polym. Sci., Polym. Chem. Ed.* 11: 389, (1973).

12. Harker, L.A., Slichter, S.J., Sauvage, L.R., "Platelet Consumption by Arterial Prostheses: The Effects of Endothelialization and Pharmacologic Inhibition of Platelet Function," *Ann. Surg.* 186: 594, (1977).

13. Hanson, S.R., Harker, L.A., Ratner, B.D., Hoffman, A.S., "*In Vivo* Evaluation of Artificial Surfaces with a Non-human Primate Model of Arterial Thrombosis," *J. Lab. Clin. Med.* 95: 289, (1980).

14. Garfinkle, A.M., Hoffman, A.S., Ratner, B.D., Hanson, S.R., "Improved Patency in Small Diameter Dacron Vascular Grafts after a Tetrafluoroethylene Glow Discharge Treatment," *Trans. Soc. for Biomatls., Vol. VII,* p. 337, Washington DC, April 1984 and Trans. ASAIO, XXX, 432-439, (1984).

15. Hoffman, A.S., Garfinkle, A.M., Ratner, B.D., "Surface Modification of Small Diameter Dacron Vascular Grafts after a Tetrafluoroethylene Glow Discharge Treatment," *Trans. Soc. for Biomatls., Vol. VII,* p. 186, Washington DC, (April 1984).

16. Hoffman, A.S., Ratner, B.D., Garfinkle, A.M., Hanson, S.R., "Plasma Gas Discharge Treatment for Improving the Biocompatibility of Biomaterials," Patent pending.

17. Callow, A.D., Ladig, C.B., O'Donnell, T.F., Gembarowicz, R., Keough, E., Ranberg-Laskaris, K., Valeri, R., "Platelet-Arterial Synthetic Graft Interaction and its Modification," *Arch. Surg.* 117: 1447, (1982).

18. Harker, L.A., Hanson, S.R., "Experimental Arterial Thromboembolism in Baboons," *J. Clin. Invest.* 64: 559, (1979).

19. Hanson, S.R., Harker, L.A., Ratner, B.D., Hoffman, A.S., "Evaluation of Arterial Surfaces Using Baboon Arteriovenous Shunt Model," in "Biomaterials, 1980" (eds. G. Winter, D. Gibbons, H. Plenk), John Wiley and Sons, London, pp 519-528, (1982).

20. Hoffman, A.S., Ratner, B.D., Horbett, T.A., Reynolds, L.O., Cho, C.S., Harker, L.A, Hanson, S.R., "Unusual Biological Interactions at Biomaterial Interfaces: Influence of Molecular Surface Character," *Artif. Org.,* in press.

21. Reynolds, L.O., Johnson, C., Ishimaru, A., "Diffuse Reflectance from a Finite Blood Medium -- Applications to the Modeling of Fiberoptic Catheters," *J. Appl. Optics* 15: 2059, (1976).

22. Reynolds, L.O., Simon, T., "Size Distribution Measurements of Microaggregates in Stored Whole Blood," *Transfusion* 20: 669, (1980).

23. Reynolds, L.O., Sanchez, R., "Radiative Transport Solution for Microparticulate Sizing," *Proc. 36th Ann. Conf. on Eng. in Medicine and Biol.* 25: 66, (1983).

24. Wagner, C.D., "Sensitivity of Detection of the Elements by Photoelectron Spectrometry," *Anal. Chem.* 44: 1050, (1972).

25. Weathersby, P.K., Horbett, T.A., Hoffman, A.S., "Solution Stability of Bovine Fibrinogen," *Thrombosis Research,* 10, 245-252, (1977).

26. Horbett, T.A., "Adsorption of Proteins from Plasma to a Series of Hydrophilic-Hydrophobic Copolymers. II. Compositional Analysis with the Prelabeled Protein Technique," *J. Bio. Mat. Res.* 15, 673-695, (1981).

27. Ratnoff, O.D., Menzie, C., "A New Method for the Determination of Fibrinogen in Small Samples of Plasma," *J. Lab. Clin. Med.* 37, 316, (1950).

28. Zisman, W.A., "Influence of Constitution on Adhesion," *Ind. and Eng. Chem.* 55: 19, (1969).

29. Yasuda, H., Guziki, M., "Biomedical Applications of Plasma Polymerization and Plasma Treatment of Polymer Surfaces," *Biomaterials,* 3, 68-77, (1982).

30. Didisheim, P., Tirrell, M.V., Lyons, C.S., Stropp, J.Q., Dewanjee, M.K. "Relative Role of Surface Chemistry and Surface Texture in Blood-Material Interactions," *Trans. Am. Soc. Artif. Intern. Organs* 29: 169, (1983).

SELECTION OF MATERIALS FOR ARTIFICIAL ORGANS

Pierre M. Galletti

Artificial Organ Laboratory
Brown University
Providence, RI 02912

The first artificial organs were conceived for short-term application, often under emergency conditions, and in diseases which otherwise would lead to death within hours or days. Thus the creativity of the pioneers was directed primarily toward device design. In the fabrication of artificial organs, engineers simply took advantage of materials which had been developed for non-medical purposes, and with them were able to build the first artificial kidneys or vascular grafts, cardiac valves, hip prostheses, etc. This heroic image of artificial organ technology persists in the mind of the public because the media tends to focus on the "life saving" aspects of new forms of treatment. The current reality, however, is quite different: pooling together external prostheses and implants, the largest volume of application of artificial organs is not encountered in life-threatening situations. It relates to devices designed to improve functional ability fitness, comfort, convenience, or esthetics (Table 1).

Table 1

ORGAN REPLACEMENT	
Medical Indication	Value sought
Life-threatening situations	Survival
Elective restoration of structure or function	Rehabilitation Relief Freedom Attractiveness
Preventive maintanance of wearable organs	Prophylaxis Insurance

Table 2

U.S. ANNUAL PRODUCTION OF ARTIFICIAL ORGANS
(ca. 1984)

CARDIOVASCULAR PROSTHESES

Blood oxygenators	310,000
Vascular prostheses	150,000
Cardiac pacemakers	140,000
Cardiac valves	70,000

NEUROSURGICAL PROSTHESES

Venrticular shunts	22,000
Stimulators	5,000

ORTHOPEDIC PROSTHESES

Hip prostheses	120,000
Knee prostheses	65,000
Elbow, wrist and finger prostheses	50,000

PLASTIC SURGERY PROSTHESES

Breast prostheses	105,000
Facial reconstruction prostheses	12,000
Penile prostheses	6,000

DENTAL PROSTHESES 20,000

SENSORY PROSTHESES

Middle ear prostheses	50,000
Implantable lenses	280,000
Other visual prostheses	40,000

METABOLIC PROSTHESES

Implantable minipumps	4,000
Artificial kidneys	6,500,000

Most of these artificial organs are intended for long-term or permanent use, and most will be implanted in the body, rather than merely connected to it (Table 2). Thus, stability of material properties and appropriate matching of implants with live structures are becoming paramount issues. Tissue-material interactions dominate the long-term and effectiveness of implants. Indeed, biomaterials science has become the basic science of organ transplantation.

This view is translated in a growing appreciation of the time element in the evaluation of biomaterials. Nowadays, the relevant time scale is measured in months or years, rather than hours and days. As a result, a new aspect of tissue-material interactions has to be considered: most components of living structures (cells, cell organelles and specific organic molecules) have cycle durations or turnover rates which are shorter than the expected service period of an implant. Thus the investigator has to be concerned not solely with the initial phenomena in tissue-material interactions, but also with the continuing, hopefully stable and quasi-permanent aspects of these interactions, that is to say the effects of biomaterials on the fine aspects of structure and functions of live tissues.

In this new perspective on artificial organs, the relevant life science is cell biology. My purpose is to illustrate, by way of examples, some long-term aspects of implants in a functionally relevant biological environment, drawing your attention to the manner in which living tissues react to a material, rather than the complementary phenomenon, the adaptation or decay of a material exposed to a warm, humid and corrosive milieu, which has been the focus of attention for biomaterials science in the past few days.

I intend to show that we are currently on the threshold of a new era in the history of artificial organs, the era of custom-designed materials for specific biological purpose. Several advances contribute to usher in this revolutionary approach: progress in our capability to synthetize new polymers, ceramics and alloys; generalized availability of analytic equipment to characterize the surface and bulk properties of new materials; and a growing appreciation of the molecular and cellular factors (e.g., cell surface receptors, attachment factors, cell junctions, intracellular contractile proteins, etc.) which control the biological side of tissue-material interactions.

Selection of materials in relation to function to be served

The variety of functions to be served by biomaterials is best illustrated if one considers the broad field of cardiovascular and pulmonary devices: there is a need for structural materials to build the scaffold and insure the mechanical stability of non-moving parts; for duct materials to serve as new passageways for biological fluids; for flexing materials to answer the need for cyclical deformation of natural organs which rely on contractile proteins; for exchange materials to mediate the transport of gases, liquids or solutes; finally, for transducer materials to generate the signals which eventually permit regulation of function through appropriate feedback control (Table 3).

Selection of materials in relation to interfacing potential

Another look at materials can be taken if one focusses on the interactions expected between a synthetic substance deliberately

Table 3

CARDIOPULMONARY IMPLANTS

STRUCTURAL/SCAFFOLD MATERIALS	Pacemaker casings
	Inf. vena cava filters
	Periarterial scaffolds
	Valve rings and struts
	Septal and atrial patches
DUCT MATERIALS	Chronic catheters
	External a-v shunts
	Vascular prostheses
	Bypass grafts
	Hydrocephalus shunts
	Airway prostheses
FLEXING MATERIALS	Cardiac valves
	Ventricular assist devices
	Artificial hearts
	Myocardial patches
EXCHANGE MATERIALS	Gas exchange devices
	Solute exchange devices
	Water exchange devices
	Bioartificial organs
TRANSDUCER MATERIALS	Low frequency sensors
	Ultrasonic sensors
	Piezoelectric activators

---

introduced in the body, and the living structures to which it will be exposed.

The most investigated chapter in the catalogue of desirable interactions is that of non-thrombogenic materials (or anti-thrombogenic materials as they were called in the days of initial enthusiasm). This subject has been treated at great length in earlier presentations, and I will merely note two uncommon facts of thrombosis prevention in the cardiovascular system: the glistening appearance of a chronic pulmonary-artery to pulmonary vein shunt (Fig. 1) in the sheep (high flow, low pressure, non-anticoagulated environment), and a scanning electron micrograph (Fig. 2) of the luminal surface of a ULTI carbon-coated dacron patch on the inferior vena cava after 8 weeks (another high flow, low pressure, non-anticoagulated environment).

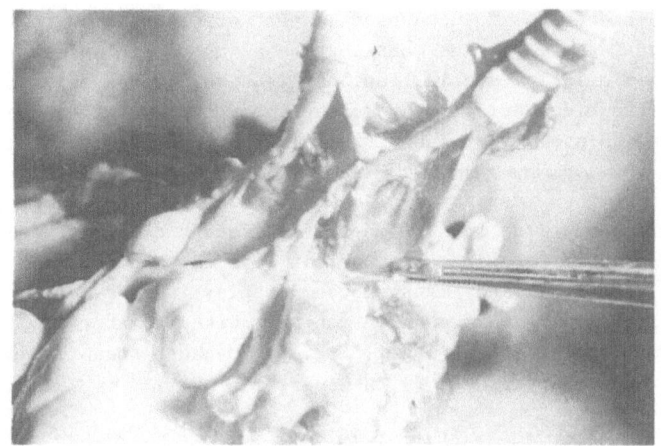

Figure 1.

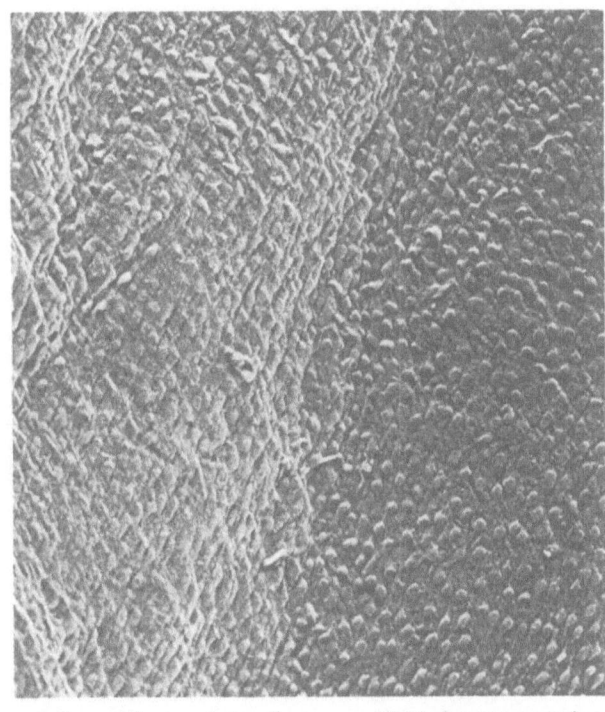

vein wall   interface   ULTI-dacron patch

Figure 2.

A class of growing interest is that of cell anchoring materials.
Cell attachment to a condensed layer of complex proteins is a
generalized feature of the basic unit of most biological structures
(Fig. 3). Similarly, stable implants require the adhesion of living
tissue to the biomaterials they contact, a result which can be achieved
either deliberately or unwittingly, but still without a full
understanding of the interactions involved. A number of "attachment
factors" have been identified (Table 4). Some are non-specific, meaning
that they are of value for a umber of different cell types; others are
specific for a particular cell type, or even for a particular segment of
the cell membrane, and might even display unique "lock and key"
characteristics. Tissue growth in contact with biomaterials is easy to
achieve with cancer cells (e.g., Fig. 4), because transformed cell are
not anchor dependent (Table 5). However, it can be achieved with

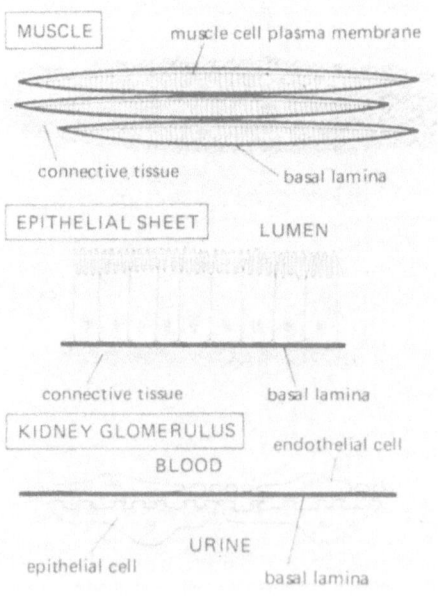

Figure 3.

Table 4

ATTACHMENT FACTORS

NON-SPECIFIC	SPECIFIC
I. Collagens (Type I, III, or IV)	I. Crude separation of basement membrane materials(biomatrix)
II. Laminin	II. Lectins
III. Fibronectin	III. Basement membrane attachment receptors (glycosaminoglycans)
	IV. Cell surface attachment receptors

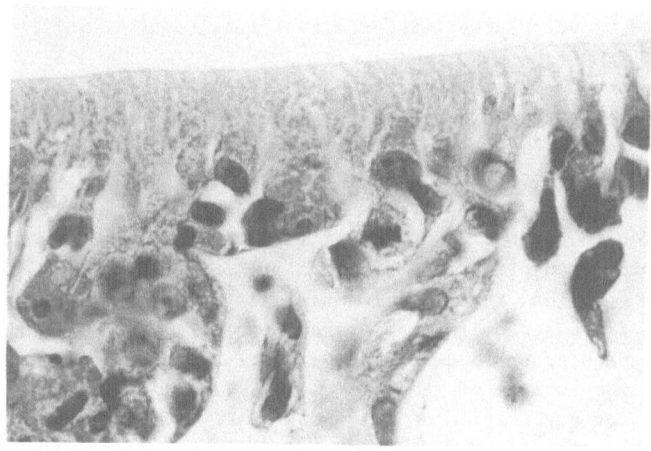

Figure 4.

Table 5

CHARACTERISTICS OF MAMMALIAN CELLS IN T.C. CONDITIONS

NON TRANSFORMED CELLS	TRANSFORMED CELLS
Anchorage dependent	Anchorage independent
Density inhibition of growth	Cell density does not affect growth
Cell contact inhibition of locomotion	No contact inhibition of locomotion
Non tumorogenic when grafted	Tumorogenic when grafted to a susceptible host
Usually available as primary cultures or as cell lines with restricted number of passages	Usually available as continuous cell lines, either by spontaneous, chemical or viral transformation
Respond to hormonal or signal regulations	Usually do not respond to hormonal regulations
Difficult to grow in quantities	Multiply easily in "in vitro" devices
Diploid	Anaploid

normal tissues, as demonstrated with an intrasplenic (Fig. 5) and an intraperitoneal implant (Fig. 6). Used as bioartificial organs, such implants have remained functionally active for several months, demonstrating that a stable, intimate interaction between material and live tissue is very much in the realm of feasibility.

Bioresorbable materials represent another approach to constructive interactions between tissues and synthetic polymers. The concept of bioresorbable structures providing protection, guidance, or simply a matrix for tissue organization is currently blossoming in many areas of artificial organ design: nerve guide channels to facilitate axon regeneration in peripheral and cranial nerves; membrane immunoseparation for endocrine transplants; orthopedic implants such as rods, plates and screws to provide transient mechanical support during spontaneous healing of fractures; moldable implants to fill voids in bones and soft tissues; meshes or braided structures to repair joint capsules, tendons, and even the walls of body cavities. I will illustrate here a cardiovascular application of bioresorbable materials, namely a fully bioresorbable arterial graft, which functions thanks to the progressive reorganization of a vessel wall structures in the slowly disappearing

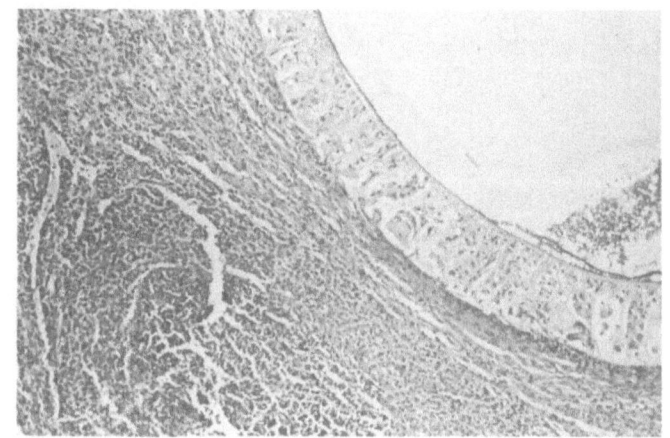

Figure 5.

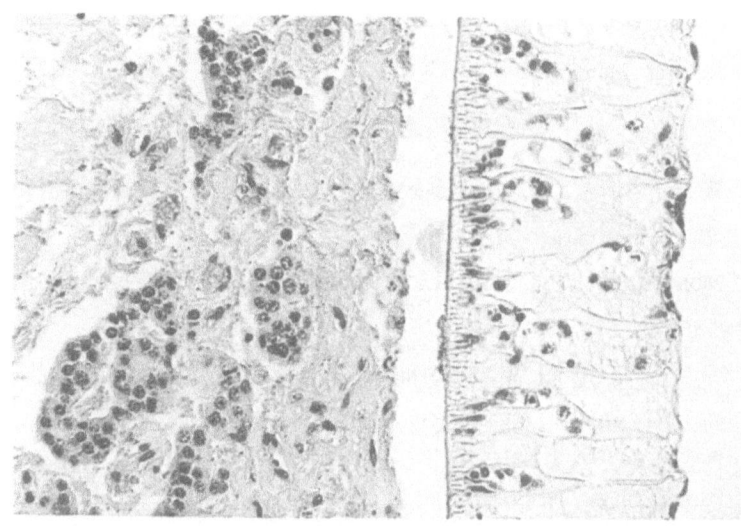

Figure 6.

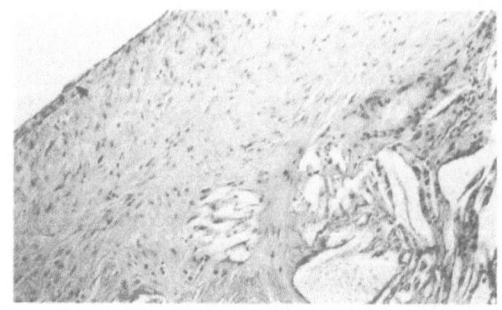

Figure 7.

scaffold provided by the polymer fabric. The repair process (Fig. 7) eventually yields a newly formed vessel wall of appropriate structural strength and hemocompatibility.

CONCLUSIONS

The initial approach to implant compatibility with living tissues focussed on inertness and low reactivity with a view to minimizing the unavoidable "foreign body reaction". Experience has taught that a positive approach, promoting beneficial tissue interactions (Table 6),

Table 6

TISSUE VS. IMPLANT STRATEGIES

DEFEND IMPLANT BY

        low reactivity
        passivation of surfaces
        manipulation of surface characteristics

PROMOTE BENEFICIAL INTERACTIONS BY

        coatings favoring cell attachment
        seeding of live cells
        tissue ingrowth in porous structures
        bioresorbable scaffolds

---

must be taken for long-term implants. The challenge in biomaterial development is no longer limited to a search for stable, low reactivity materials, capable of withstanding the warm, humid and corrosive environment of the mammalian body. New strategies (Table 7) are needed to stabilize the interactions between tissue and implants, taking into account the changing nature of both.

Table 7

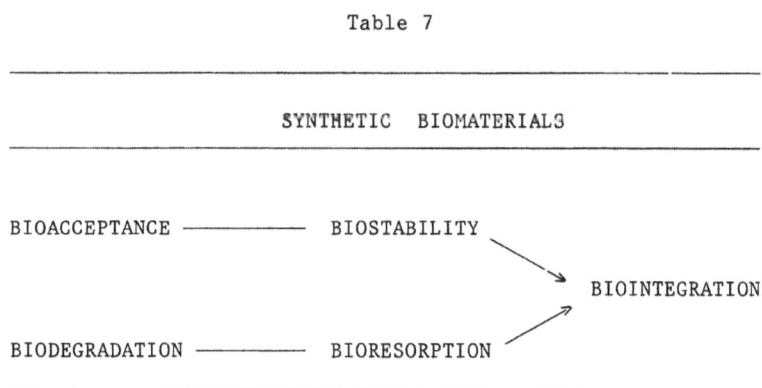

DURABILITY PREDICTION AND MEASUREMENT FOR POLYMERIC BIOMATERIALS

J. L. Kardos, K. P. Gadkaree*, and A. P. Bhate**

Materials Research Laboratory and
Department of Chemical Engineering
Washington University
St. Louis, MO 63130

INTRODUCTION

Synthetic polymer biomaterials are often asked to perform under conditions of large cyclic deformations for long lifetimes. Typical of such applications are Left Ventricular Assist (LVA) pump bladders, heart valve components, and vascular grafts. In designing with these materials and eventually qualifying them for clinical usage, it is necessary to be able to predict the fatigue lifetimes accurately and reliably. To be sure, these materials must be biologically compatible; but even the most perfectly biocompatible material will not be qualified for structural use in humans unless its mechanical longevity can be proven from an accurate data base.

The nature of the fatigue problem is complicated by the statistical spread of cycles-to-failure data at any one stress level. If fatigue life is to be reliably predicted, the statistical nature of the failure must be examined. Curiously, there has not been much effort to do this for polymers (1,2), although fatigue laws involving two-parameter Weibull distributions have often been employed in describing fatigue of metals and reinforced plastics (3).

Usually, the problem of life prediction is treated in a deterministic way by experimentally determining a flaw growth law and then integrating it to a limit of the number of cycles in the lifetime. This has been the approach adopted in the studies on elastomers (4-7) and most investigations of plastics (2) and reinforced plastics (8). This does not help in explaining the scatter observed in the lifetime of specimens which look exactly the same and are subjected to the same conditions.

It is now an accepted fact that fatigue failure initiates at pre-existing flaws. These flaws grow due to fatigue loading and catastrophic failure takes place as the critical size is reached. Thus, knowing the distribution of initial flaw sizes and the fatigue law for flaw propagation, the distribution of flaw sizes after n cycles can be determined and the probability of failure calculated.

---

* Corning Glass Works, Corning, NY 14831
**Tamko, Inc., Joplin, MO 64801

PREDICTIVE FORMAT

Uniaxial Deformation

The calculational format for predicting fatigue lifetimes under cyclic uniaxial deformation was developed recently by Gadkaree and Kardos (9). Their approach is shown schematically in Figure 1. A solid material with perfect molecular order and no flaws would yield a single-valued distribution (neglecting statistically variations in bond energies) of failure strength. All practical materials have a distribution of imperfections, ranging from molecular disorders to microcracks, which cause a distribution of strengths to arise when a sample population is tested. If there is a discrete relationship between strength and flaw size, then one can postulate a distribution of flaw sizes from a distribution of strengths. The failure process is viewed as initiating when a population of flaws begins to grow in a material under cyclic loading. If the flaw growth law is known, the statistical nature of these flaw sizes dictates that a spectrum of failure lifetimes results as the population of growing flaws reaches a critical size for failure.

To describe this process mathematically, we proceed as follows. First, a sample population of specimens (~20) is tested in tension. The resulting ultimate strengths are fitted to the normal distribution function

$$f(\sigma_b) = \frac{1}{\sigma\sqrt{2\pi}} \exp\left\{-\frac{1}{2}\left[\frac{\sigma_b-\mu}{\sigma}\right]^2\right\} \tag{1}$$

where μ and σ^2 are the mean and variance of the distribution and σ_b is the breaking strength. It is a theoretical and experimentally verified fact (9) that the ultimate or breaking strength of elastomers depends on the flaw size according to

$$\sigma_b = \frac{M}{c^N} \tag{2}$$

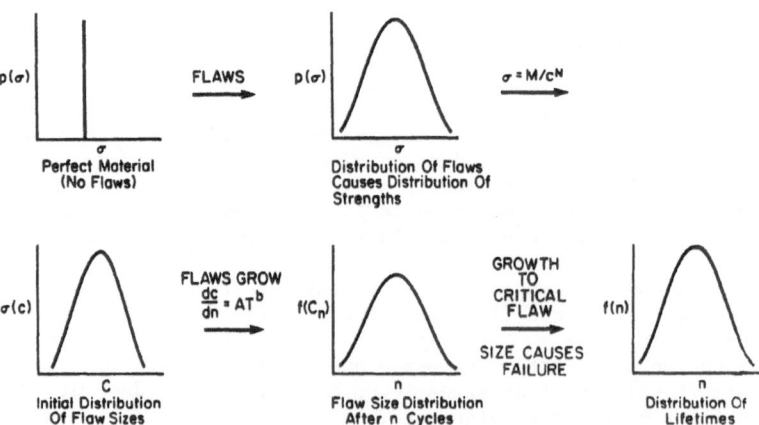

Fig. 1. Schematic of critical flaw growth approach to prediction of fatigue lifetime distribution.

188

A second population (~20) of specimens containing purposely introduced sharp flaws of known critical size is then tested in uniaxial tension and the constants M and N evaluated. Equation 2 is substituted into 1 and, after a variable transformation, we obtain

$$f(c) = \frac{MN}{\sigma\sqrt{2\pi}\ c^{1+N}} \ \exp\left\{-\frac{1}{2}\ [(\frac{M}{c^N} - \mu)/\sigma]^2\right\} \tag{3}$$

Thus, the distribution of hypothetical flaw sizes in a given specimen population is obtained. A similar procedure can be used for any other distribution function.

The growth of this flaw distribution is governed by a tearing energy concept (10)

$$\frac{dc}{dn} = AT^b \tag{4}$$

where A and b are constants depending on experimental conditions and the material. T, the tearing energy, is given for an edge crack by $T = 2KUc$ where K is a numerical constant and U is the area under the stress-strain curve. Integrating this equation between the limits c_0 to c (initial flaw size and flaw size after n cycles, respectively) and 0 to n, where n is the number of cycles, and substituting into Equation 3 yields the probability distribution of crack sizes after n cycles.

Assuming that the final flaw size is much greater than the initial flaw size, c_0, the probability distribution of fatigue lives can be directly derived (9) as

$$f(n) = \frac{MN}{\sigma\sqrt{2\pi}} (\frac{1}{b-1}) \ \frac{n^{(N+1-b)/(b-1)}}{(\frac{1}{A(b-1)(2KU)^b})^{N/(b-1)}}$$

$$\exp\left\{-\frac{1}{2}\left\{\left[\frac{M\ n^{N/(b-1)}}{(\frac{1}{A(b-1)(2KU)^b})^{N/(b-1)}} - \mu\right]/\sigma\right\}^2\right\} \tag{5}$$

Thus, Equation (5) provides the fatigue lifetime distribution functions under given experimental conditions without carrying out any fatigue tests at all, if the six constants (M, N, μ, σ, A, b) are known.

Three independent, short-term experiments provide the necessary values of the constants; namely, determination of the breaking strength in uniaxial tension with and without an introduced critical flaw, and a flaw propagation experiment in which the flaw growth is measured as a function of a small number of cycles.

Biaxial Deformation

The same format can be used to predict biaxial fatigue lifetime distributions. Determining the six constants under biaxial stress becomes experimentally much more difficult because the stress is usually hydraulically applied. Nonetheless, this technique has been developed

by Bhate and Kardos (11) using a Swanson tuned fluid oscillator system
(12). The tuned fluid oscillator may also be used in a static mode to
determine the biaxial strength distribution and the effect of a flaw on
that distribution. The six constants determined from these three biaxial
experiments can be used with Equation 5 to predict the biaxial fatigue
lifetime distribution.

EXPERIMENTAL RESULTS AND DISCUSSION

Materials and Procedures

A detailed description of the testing procedures and materials may
be found elsewhere (9,11).

Briefly, the materials used were Biomer[a], Avcothane 51[b], and
Pellethane CPR 2363-80A[c]. Biomer is a segmented polyetherurethane
polymerized by Ethicon, Inc. and centrifugally cast from solution into
sheets of 30 mils thickness by Thoratec, Inc. Avcothane 51 is a block
copolymer containing 90% polyetherurethane and 10% poly(dimethyl siloxane),
which is dip-cast from a 11–15%w solution of a 2:1 mixture of tetrahydrofuran and dioxane. The sheets are vacuum-oven dried at 50°C.
Pellethane CPR 2363-80A is an extruded thermoplastic, polyetherurethane
elastomer.

A limited number of experiments were also carried out on Hexsyn[d], a
carbon black-filled Goodyear polyolefin rubber. This material is a
terpolymer containing 95% poly(hexene-1) and 5% of a 2:3 mixture of
4-methyl-1,4 hexadiene and 5-methyl-1,4-hexadiene. The polymer is
vulcanized using a standard sulfur-based recipe and then extracted with
a solution of acetone and toluene and then vacuum-dried. Initial strength
characterization data was also obtained for a NBS butyl rubber, a carbon
black-filled compound of isobutylene-isoprene rubber designated as
National Bureau of Standards reference material 388j. The material was
synthesized, compounded, and vulcanized according to ASTM D-1388 formula 1A,
and supplied in sheet form.

Uniaxial stress-strain data was obtained on an Instron testing
machine at a strain rate of 5 min^{-1} using sandpaper-lined grips. Precise
cuts (flaws) were placed in some of the samples by using an optical
microscope and a precision microtome. The flaw propagation and fatigue
experiments were carried out on an MTS closed-loop electrohydraulic
testing machine. A traveling microscope was used to measure the length
of the flaw at various numbers of cycles. For each of three frequencies
(1, 0.1, and 0.01 Hz) experiments were done for at least four different
maximum stress levels.

For a given stress level (i.e. given extension) the flaw propagation
or fatigue specimens clearly stretch at different rates when cycling at
different frequencies. If the true stress-true strain curves change
appreciably with extension rate, different stress-strain curves must be
used for calculating U at different frequencies.

---

[a]trademark of Ethicon, Inc., Somerville, NJ.
[b]trademark of Avco Everett Company, Everett, MA; currently marketed as
 Cardiothane 51 by Contron, Inc., Everett, MA.
[c]trademark of Upjohn Company, Torrance, CA.
[d]trademark of Goodyear Tire and Rubber Company, Akron, OH.

To study these rate effects, engineering stress-engineering strain curves were obtained on an Instron testing machine at 1, 10, and 40 inches per inch per minute.

Fatigue experiments were carried out at various frequencies and maximum engineering stresses (0 psi minimum stress) under cyclic sinusoidal conditions. Each stress level results in a different value of U and so a different distribution function of fatigue lifetimes will result at each different stress level.

Static Strength Distributions

The static uniaxial strength distributions were obtained and fitted to a normal distribution function. A chi-squared goodness of fit test showed that a normal distribution fit the data accurately at a 95% confidence level. Table 1 presents the means and standard deviations for the various material populations studied. One notes two basic classes of materials, the urethane-based polymers (Biomer, Avcothane, and Pellethane) and the rubber materials (Goodyear Polyolefin and NBS Butyl). Among the urethanes, Biomer stands out as having a relatively narrow strength distribution. This probably reflects a minimum of process-induced flaws. Goodyear Polyolefin and NBS Butyl also have reasonably narrow static strength distributions. Pellethane and Avcothane, on the other hand, are notably more flawed; some of the flaws are visible to the naked eye and consist of voids and separated phases. Thus, the static strength distribution is a very sensitive measure of the flaw distribution in the material and can be used profitably to quality-control these sheet materials.

Strength Dependence on Flaw Size

Table 2 lists values for the parameters M and N in Equation 2. These were obtained from the plots of Equation 2, typical of which are those shown in Figure 2 for Avcothane and Pellethane. High values of M along with low values of N indicate relatively good resistance to strength degradation from flaws. Biomer again looks relatively good, followed by Goodyear Polyolefin. Pellethane and Avcothane show notably poorer resistance to flaw propagation. One can only speculate on reasons for this behavior. Differences in molecular weight might be responsible, but a lack of comparative molecular weight data makes this impossible to invoke. It is more likely that crosslinks are involved. The rubbers are of course 3-dimensional networks and show good resistance. Biomer is a segmented copolymer whose hard segments act as physical crosslinks and thereby give the material rubberlike characteristics including good resistance to flaw growth. Pellethane and Avcothane, while they are both copolymers and primarily etherurethanes (Avcothane contains some siloxane blocks), do not have any chemical or hard segment crosslinks.

Flaw Propagation and the Effect of Frequency

Once the strain energy density U has been evaluated from the area under the stress-strain curve (9), the effect of tearing energy on the flaw propagation rate can be evaluated using Equation 4. While the U values for Biomer were unaffected by strain rate, there was a significant effect on the values for Avcothane and Pellethane. Consequently, in the cases of both the flaw propagation and the fatigue experiments, the average strain rate during a stress cycle was calculated from the values of the maximum strain in that cycle and the frequency. The stress-strain curve obtained at that same value of strain rate was then used to

Table 1. Parameters Obtained from the Normal Static Strength Distribution

Material	Number of Samples Tested	Mean, μ (psi)	Standard Deviation σ
Biomer	31	5874	199
Avcothane	40	5639	567
Goodyear Polyolefin	40	2208	163
Pellethane	40	6518	664
NBS Rubber	40	2846	119

Table 2. Parameters M and N for the Relation Between Static Strength and Flaw Size, $\sigma_b = M/c^N$

Material	M	N
Biomer	608.0	0.39
Avcothane	14.1	1.28
Goodyear Polyolefin	167.0	0.35
Pellethane	11.1	1.24
NBS Rubber	211.0	0.44

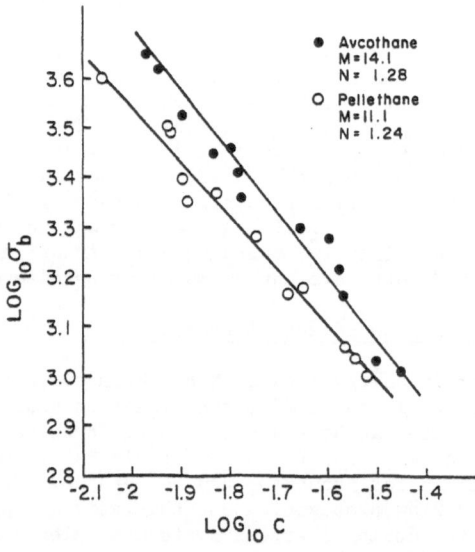

Fig. 2. Characterization of flaw growth resistance through the effect of flaw size on static strength for Avcothane and Pellethane.

calculate the strain energy density, U, for the corresponding flaw propagation or fatigue experiment. The value of K in Equation 4 reflects the stored energy dissipation near the crack and depends weakly on strain level; K does not depend on the type of material (13).

Figure 3 shows the effect of tearing energy on flaw growth for Avcothane at one Hz. From this plot the constants A and b in Equation 4 may be evaluated. Table 3 summarizes the flaw propagation constants for the materials studied over a frequency range of two orders of magnitude. Unfortunately it was not possible to accurately measure flaw growth in either the Goodyear Polyolefin or the NBS rubber materials. The crack invariably deviates by curving away from straight paths along its initial axis, making the data scatter too large for use in fatigue life prediction. This phenomenon is common in anisotropic materials or filled isotropic materials, and has been described in detail by Andrews (14).

The power law coefficient b exhibits very little change with frequency, except for the case of Biomer for which it appears to go through a minimum. For both Avcothane and Pellethane, the coefficient A decreases by a factor of about 10 for every 10-fold increase in frequency. For Biomer, A appears to go through a maximum. One can certainly conclude that the frequency behavior of Biomer is different from that for the other two urethane-based polymers. Utilizing the above fatigue prediction format for Avcothane and Pellethane, one would predict a decrease in fatigue life with an increase in frequency, an effect also noted for other polymers (15,16). However, it should be noted that the frequency range covered is much too limited and higher frequency testing in the more desirable 30 Hz range may exhibit totally different behavior.

Fatigue Lifetime Distributions - Predicted vs. Measured

At a 90% confidence level, the predicted uniaxial fatigue lifetime distribution accurately described the experimentally determined distribution for Biomer for three different maximum cyclic stress levels (9). The two distributions were compared using a Kolmogoroff statistical goodness-of-fit test (17) for small numbers of samples.

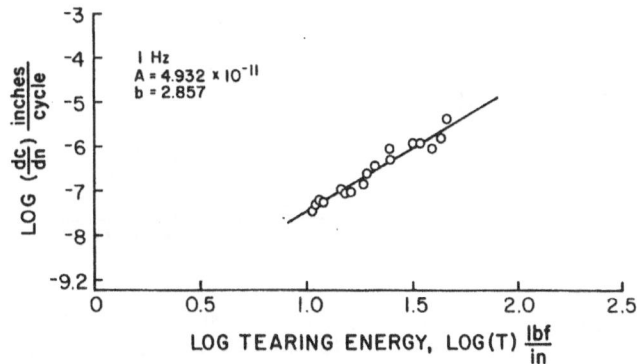

Fig. 3. Crack growth dependence on tearing energy for Avcothane at 1.0 Hz.

Table 3. Effect of Frequency on the Flaw Propagation Rate
As Correlated With the Tearing Energy

| | Material | | | | | |
| Frequency | Biomer | | Avcothane | | Pellethane | |
	A	b	A	b	A	b
0.01 Hz	1.78×10^{-8}	2.50	1.01×10^{-9}	2.27	0.33×10^{-6}	1.72
0.1 Hz	4.0×10^{-7}	1.58	2.012×10^{-10}	2.36	1.0×10^{-7}	1.66
1.0 Hz	2.51×10^{-8}	2.12	4.932×10^{-11}	2.86	5.62×10^{-8}	1.99

Figures 4-7 show the predicted cumulative lifetime distributions for
Avcothane and Pellethane at 0.01 and 0.1 Hz at a maximum stress level of
1800 psi. The solid lines indicate the predicted distributions while the
real-time data is shown as open circles. Note that the fatigue life
spectrum shifts toward higher values of n as the frequency decreases, in
agreement with the flaw propagation results. It is also obvious that
the prediction is conservative at this stress level (1800 psi is roughly
31% of the mean static strength for Avcothane and 28% for Pellethane).
With one exception, all of the data points occur at failure lifetimes
greater than predicted.

The question, of course, is whether or not the predicted curves are
correctly describing the experimental results. As in the case of Biomer
(9), Kolmogoroff statistics were used to evaluate the goodness of the fit.
Unlike the chi-square test, the Kolmogoroff test (17) can be applied to

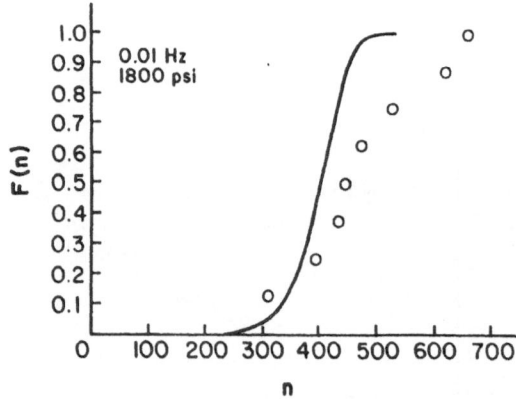

Fig. 4. Cumulative fatigue lifetime distribution for
Avcothane at 0.01 Hz and 1800 psi maximum stress.

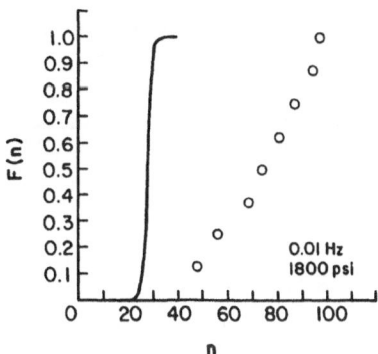

Fig. 5. Cumulative fatigue lifetime distribution for
         Avcothane at 0.1 Hz and 1800 psi maximum stress.

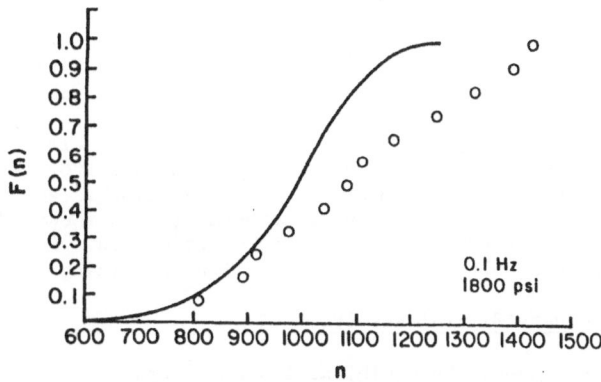

Fig. 6. Cumulative fatigue lifetime distribution for
         Pellethane at 0.01 Hz and 1800 psi maximum stress.

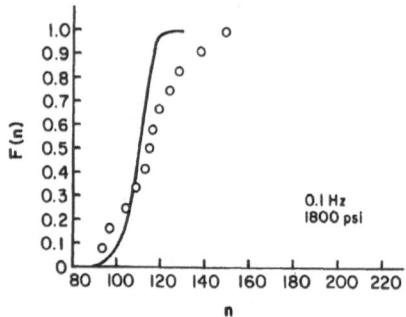

Fig. 7. Cumulative fatigue lifetime distribution for
         Pellethane at 0.1 Hz and 1800 psi maximum stress.

small numbers of samples and is computationally simpler. Briefly, the
test measures the maximum absolute deviation of the particular sample
cumulative distribution function (CDF) from the value obtained from the
postulated distribution function. Table 4 summarizes the results for
Avcothane and Pellethane at the two different frequencies. At a 90%
confidence level (α = 0.1) one calculates a critical value for the
maximum deviation between the two distributions. If all the differences
between the two distributions lie below this critical value, there is
no reason to reject the hypothesis that the predictive equation does
indeed describe the lifetime distribution. The equation provides an
adequate fit of the data at the 90% confidence level for both frequencies
of the Avcothane data and for the 0.1 Hz Pellethane data. However the
0.01 Hz Pellethane data are far beyond the predictive accuracy of the
CDF at any reasonable confidence level. It is not clear why this is the
case since the material for both frequencies came from the same sheet.
Except for Pellethane, at 0.01 Hz, the prediction is extremely good at
the low lifetime region of the distribution; in all cases the prediction
is conservative throughout the entire distribution. Both of these trends
are highly desirable in designing for life-sustaining applications.

Some Preliminary Biaxial Results

    Figure 8 shows the equibiaxial stress-strain behavior for Biomer,
which must be used to calculate the tearing energy for the flaw growth
rate experiments. The effect of tearing energy on the flaw growth rate
under equibiaxial cyclic stress is shown in Figure 9 for Avcothane at a
frequency of 26.5 Hz. The slope and intercept of this plot provide the
values for the constants b and A of Equation 4. Along with the effect
of flaw size on static biaxial strength, these data can be used to predict
the biaxial fatigue life distribution. It remains to be seen what the
effect of a performance liquid environment will be on both the uniaxial
and biaxial fatigue behavior. Clearly more high frequency data are needed
under both uniaxial and biaxial stress states before any conclusions on
accelerated testing can be drawn.

196

Table 4. Summary of Kolmogoroff Statistics Test Applied to
Cumulative Distribution Functions for Avcothane and
Pellethane

Material	Frequency	Critical Deviation	Maximum Deviation
Avcothane	0.01	0.411	0.410
Avcothane	0.1	0.338	0.302
Pellethane	0.01	---	---
Pellethane	0.1	0.310	0.302

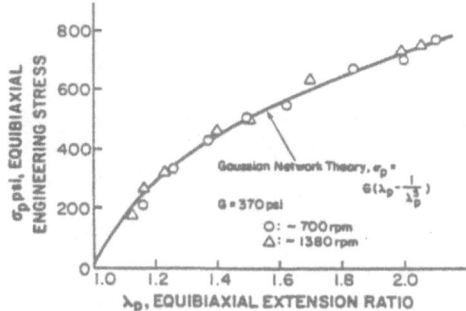

Fig. 8. High frequency equibiaxial stress-deformation
relation for Biomer.

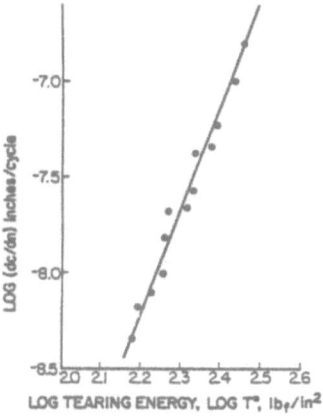

Fig. 9. Cyclic biaxial flaw growth relation for Avcothane
at 26.5 Hz.

ACKNOWLEDGMENT

    We are grateful for the support of this work by the National Heart,
Lung, and Blood Institute, Division of Heart and Vascular Diseases, Devices
and Technology Branch under Contract No. N01-HV-02910.

REFERENCES

1. J. A. Manson and R. W. Hertzberg, Crit. Rev. Macromol. Sci., 1, 433,
    (1973).
2. R. W. Hertzberg and J. A. Manson, Fatigue of Engineering Plastics,
    Academic Press, NY (1980).
3. J. M. Whitney, Fatigue of Fibrous Composite Materials, ASTM STP 723,
    American Society for Testing and Materials, Phila., 1981, p. 133.
4. A. N. Gent, P. B. Lindley and A. G. Thomas, J. Appl. Pol. Sci., 8,
    453-477 (1974).
5. G. J. Lake and P. B. Lindley, J. Appl. Pol. Sci., 8, 107-121 (1974).
6. J. P. Berry, J. Poly. Sci., 50, 107 (1961).
7. R. E. Whittacker, J. Appl. Pol. Sci., 18, 2339-2353 (1974).
8. A. T. DiBenedetto and G. Salee, Proc. 34th Antec, Soc. Plastics
    Engineers, Atlantic City, 1976, p. 103.
9. K. P. Gadkaree and J. L. Kardos, J. Appl. Pol. Sci., 29, 3041 (1984).
10. R. S. Rivlin and A. G. Thomas, J. Pol. Sci., 10, 291 (1953).
11. A. P. Bhate and J. L. Kardos, Pol. Eng. Sci., 24, 862 (1984).
12. A. P. Bhate, W. M. Swanson, and J. L. Kardos, Biomaterials (submitted).
13. M. W. Greensmith, J. Appl. Pol. Sci., 7, 993 (1963).
14. E. H. Andrews, Fracture in Polymers, Oliver and Boyd, London, 1968,
    pp. 152-153.
15. J. C. Radon, J. Macromol. Sci.-Phys., B14, 511 (1977).
16. D. C. Prevorsek and Y. D. Kwon, J. Macromol. Sci.-Phys., B12, 447
    (1976).
17. K. V. Bury, Statistical Models in Applied Science, John Wiley, 1975,
    pp. 204-206.

CHARACTERIZATION OF POLYURETHANES FOR BLOOD-CONTACTING APPLICATIONS

Stuart L. Cooper, Michael D. Lelah, and Timothy G. Grasel

Department of Chemical Engineering
University of Wisconsin
Madison, Wisconsin 53706

INTRODUCTION

Segmented polyurethanes are widely used in commercial and experimental blood-contacting applications which include vascular prostheses, blood filters, catheters, insulation for pacemaker leads, heart valves, cardiac assist devices, and chambers for artificial hearts. The use of this family of polymers for such applications is due to the physiological acceptability, relatively good blood tolerability, relative stability over extended implant periods, and excellent physical and mechanical properties that are exhibited by these materials[1].

Polyurethane block copolymers compose a large family of thermoplastic elastomers. These polymers consist of chain-extended diisocyanate "hard segments" dispersed in a macroglycol matrix of "soft segments." Due to the incompatiblity between the hard and soft chain segments, the polyurethanes exhibit microphase separation. The hard domains, which possess a relatively high glass transition temperature or melting point when compared to the soft segment material, act as multifunctional crosslinking sites and as reinforcing fillers in the soft segment matrix. The observed elastomeric behavior is a direct result of the multiphase nature of these materials, and several investigators of blood-materials interactions believe that the presence of more than one phase may contribute to the excellent blood compatibility that has been observed in these and other block copolymer systems[2,3,4]. To date, however, a successful vascular graft of less than 5 mm diameter has not been produced from a polyurethane or from any other synthetic material.

In spite of the relatively widespread use of polyurethane block copolymers in blood-contacting applications, little is known about the molecular basis for the relatively benign blood response that is observed with these materials. There is also only a limited understanding of the extent that the various components of polyurethanes affect surface-induced thrombosis.

A unique feature of the polyurethane family is the broad range of polymers that can be synthesized. Among the numerous and interrelated parameters believed to affect blood-materials interactions are hydrophilicity/hydrophobicity, surface charge, and surface chemical composition. With a given material, the choice of casting solvent and

the use of solvent extraction are thought to cause changes in the polyurethane surface chemistry. Polyurethane properties are also affected
by the type and molecular weight of the polyol and the type of
diisocyanate and chain extender used. Systematic variations of
polyurethane components and processing conditions permit studies of
factors believed to have a role in blood-material interactions.

The recent development of ion-containing polyurethanes allows for
investigations of surface charge on blood-polyurethane interactions.
When polyurethanes are synthesized with positively- or negatively-
charged groups, tensile strength and toughness generally increase,
and normally hydrophobic polymers become hydrophilic upon ionization.
Ionic moieties homogeneously distributed in the bulk polymer phase or in
ionic domains would be expected to affect both surface electrical properties and the organization and structure of water molecules at the
aqueous interface. Surface charge has been implicated as one of the
factors affecting thrombus formation on polymeric biomaterials. Many
blood components, such as red blood cells and platelets, possess
negatively-charged surfaces, and normal vascular endothelium is
negatively charged[5]. In addition, the recent development and testing of
polyelectrolyte complexes for blood contact has been based on the
hypothesis that such highly charged materials may in part simulate the
anticoagulant properties of heparin[6,7]. Thus it is of interest to
investigate the incorporation of ionic groups into a polyurethane
matrix.

The goal of this study was to carefully characterize the bulk and
surface properties of a number of commercial and specially synthesized
polyurethanes and evaluate their blood compatibility. An acute canine
ex vivo femoral A-V shunt technique developed originally by Ihlenfeld,
et al.[8] and later modified by Lelah, et al.[9] was used to study
the blood-contacting behavior of these polymers. In addition to the
effects of chemical composition, the effects of solvent casting and
extraction on the surface properties and ex vivo thrombogenicity of
several polyurethanes were examined.

EXPERIMENTAL

Polymer Materials

This study describes three series of experiments. In the first
series[10], several commercial polyurethanes in the Biomer (Ethicon Inc.,
Somerville, NJ) family were examined. Extruded grade Biomer
tubing of 0.132 in. ID was either used as received (designated
as EB) or dissolved in a two percent solution of N,N-dimethylacetamide
(DMA) and cast on a polyethylene tubing surface of the same diameter
(designated as CB). The as received 30% (in DMA) "solution grade" Biomer
(SB) was further diluted with DMA to a 2% solution for coating onto a
chromic acid-oxidized polyethylene tubing surface. The coating procedure has been described in detail previously[16]. The bulk materials
EB and SB were examined by elemental analysis, differential scanning
calorimetry, thermomechanical testing, and stress-strain testing.

In the second part of this study[5], a family of neutral and ionic
polyurethanes was studied. Details of the synthesis and physical properties of these and the other laboratory-synthesized materials have been
described [11,12,13]. Chemical structures for the repeat units of these
polymers are included in Figure 1. The soft segment in these materials
was poly(tetramethylene oxide) of molecular weight 1000. The hard
segment component was composed of 4,4'-diphenylmethanediisocyanate

Hard Segment Diisocyanate Repeat Units:

4,4'-diphenylmethanediisocyanate
(MDI or M)

4,4'-dicyclohexylmethanediisocyanate
(H_{12}MDI)

Hard Segment Chain Extender Repeat Units:

-OCH$_2$CH$_2$-N-CH$_2$CH$_2$O-
(CH$_3$)

N-methyldiethanolamine
(MDEA or MD)

-O-CH$_2$CH$_2$-CH$_2$-CH$_2$O-

1,4-butanediol
(BD)

-N-CH$_2$-CH$_2$-N-

ethylenediamine
(ED)

Soft Segment Polyols:

-O-(CH$_2$CH$_2$CH$_2$CH$_2$O)$_n$-

polytetramethylene oxide
(PTMO)

-O-(CH$_2$CH$_2$O)$_n$-

polyethylene oxide
(PEO)

-(CH$_2$-CH-O)$_n$-
(CH$_3$)

polypropylene oxide
(PPO)

Figure 1. Repeat Unit Chemical Structures for Polymers Studied

(MDI or M), chain-extended with N-methyldiethanolamine (MD), at MDI weight percentages of 21.5 and 38 percent to make materials U-21.5 and U-38, respectively. The U-38 material was reacted as previously reported[10] to form the zwitterionomer (U-38-ZW), the anionomer (U-38-AN), and the cationomer (U-38-CA). The polyurethanes were solvent-cast from DMA solutions onto the inner lumen of 3.18 mm ID polyethylene tubing. The previously described procedures for surface analysis and testing in the _ex vivo_ blood-contacting experiment were used.

The relative proportions of starting materials used in the polymers comprising Part III of this study are summarized in Table 1. Materials include a series of polymers with different soft segment type (M/MD/PTMO(2), M/MD/PPO(2), and M/MD/PEO(1.5)), a series consisting of the same material cast from different solvents (M/MD/PTMO(1)-D, M/MD/PTMO(1)-T, and M/MD/PTMO(1)-DT), a series of materials with different hard segment diisocyanates and chain extenders (M/BD/PTMO(1), H/BD/PTMO(1), and H/ED/PTMO(1)), a hard segment "analog" (M/MD), and a polyurethane zwitterionomer containing an amount of soft segment that was high relative to previously studied materials (M/MD/PTMO(1)-Z). The commercial polyurethanes studied included Extruded Grade Biomer (Ethicon Inc.) and a high soft-segment-containing version of Solution Grade Biomer (Ethicon, Inc.). Each of these two materials (EB and HSSB, respectively) was dissolved in DMA, precipitated in water, finely divided, and extracted with methanol using a Soxhlet extraction procedure for 48 hours to produce the EB-SE and HSSB-SE polymers. The polyurethanes were solvent-cast onto the inner lumen of 3.18 mm ID polyethylene tubing as previously described.

Table 1. Formulation and Notation of Laboratory-Synthesized and Commercial Polyurethanes Studied

Designation	Diiso-cyanate	Chain Extender	Soft Segment Polyol-Mol. wt.	Molar Ratio Diisocyanate/ Chain Extender/ Polyol	Casting Solvent and Comments
PART I:					
EB	-	-	-	-	tubing
CB	-	-	-	-	DMA
SB	-	-	-	-	DMA
PART II:					
U-21.5	MDI	-	PTMO-1000	1/0/1	DMA
U-38	MDI	MDEA	PTMO-1000	3/2/1	DMA
U-38-ZW	MDI	MDEA	PTMO-1000	3/2/1	DMA; zw'ionomer
U-38-AN	MDI	MDEA	PTMO-1000	3/2/1	DMA; anionomer
U-38-CA	MDI	MDEA	PTMO-1000	3/2/1	DMA; cationomer
PART III:					
M/MD/PTMO(2)	MDI	MDEA	PTMO-2000	6/5/1	DMA
M/MD/PPO(2)	MDI	MDEA	PPO-2000	6/5/1	DMA
M/MD/PEO(1.5)	MDI	MDEA	PEO-1500	4.5/3.5/1	DMA
M/MD/PTMO(1)-D	MDI	MDEA	PTMO-1000	1.3/0.3/1	DMA
M/MD/PTMO(1)-DT	MDI	MDEA	PTMO-1000	1.3/0.3/1	50% DMA, 50% THF
M/MD/PTMO(1)-T	MDI	MDEA	PTMO-1000	1.3/0.3/1	THF
M/MD/PTMO(1)-Z	MDI	MDEA	PTMO-1000	1.3/0.3/1	DMA; zw'ionomer
M/BD/PTMO(1)	MDI	BD	PTMO-1000	4/3/1	DMA
H/BD/PTMO(1)	H_{12}MDI	BD	PTMO-1000	4/3/1	DMA
H/ED/PTMO(1)	H_{12}MDI	ED	PTMO-1000	4/3/1	DMA
M/MD	MDI	MDEA	-	1/1/0	DMA
EB	-	-	-	-	DMA
EB-SE	-	-	-	-	DMA; MeOH ext'd
HSSB	-	-	-	-	DMA
HSSB-SE	-	-	-	-	DMA; MeOH ext'd

Surface Characterization

Sections of the same tubing used in the blood contact experiments were used for surface characterization. Contact angle measurements were made using underwater captive bubble techniques[14] with air and octane as the probe fluids. Calculation of surface energy parameters was based on the harmonic mean equation as described by Andrade, et al.[15]. Electron Spectroscopy for Chemical Analysis (ESCA) was performed, and the data were analyzed as previously described[5]. Attenuated total reflection infrared spectroscopy (ATR-IR) spectra were obtained on a Nicolet 7199 FTIR with a variable angle Barnes 300 ATR accessory at a resolution of 2 cm^{-1}.

Ex Vivo Single and Series Shunt Evaluation of Blood Compatibility

In the initial study of cast and extruded Biomer[10], a single shunt technique was used as described by Cooper and coworkers[8]. The schematic diagram for the series shunt animal experiment[9] used in the second and third portions of the study is illustrated in Figure 2, while the cannulation site and shunt details for polyethylene sections connected in series are shown in Figure 3. New sets of identical test

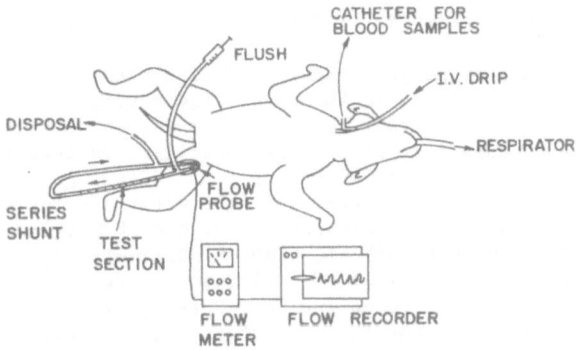

Figure 2.   Diagram of animal experiment showing shunt, catheter, and flow measuring and recording instrumentation[9].

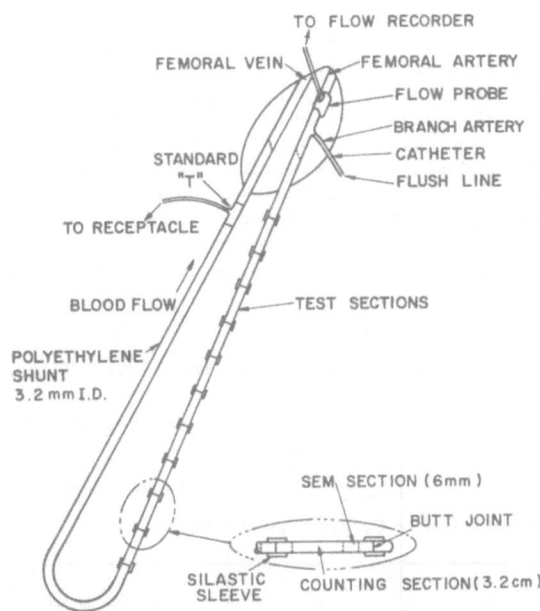

Figure 3.   Cannulation site and shunt arrangement showing details of test sections connected in series[9].

surfaces joined as a shunt section were inserted for each time period. Each material was tested in at least three different animals.

In the study of polyurethanes and polyurethane ionomers (Part II), all materials were tested in the same surgery set. In the study of polyurethanes with different hard and soft segment components (Part III), it was necessary to divide the investigation into two separate sets of animal experiments, with each set consisting of three animal surgeries. In the first surgery set, the materials tested were M/MD/PTMO(2), M/MD/PPO(2), M/MD/PEO(1.5), M/MD/PTMO(1)-D, M/MD/PTMO(1)-DT, M/MD/PTMO(1)-T, M/MD, M/MD/PTMO(1)-Z, and a polyethylene control material PE-1 (Intramedic PE-350, Clay Adams, a division of Becton, Dickinson and Co., Parsippany, NJ). The remaining polyurethanes and the same polyethylene control material (PE-2) were tested in the second surgery set. It will be later observed that the platelet deposition profiles were nearly identical for PE-1 and PE-2, and thus direct comparisons could be made between materials in different surgery sets for the Part III study.

RESULTS AND DISCUSSION

Part I: Studies of Cast and Extruded Biomer

As determined by chemical analysis and differential scanning calorimetry and mechanical testing[10], the extruded grade Biomer had a lower soft segment molecular weight than the solution grade material (650 for EB versus 2000 for SB). As a result of this higher molecular weight, the soft segment phase of SB was semicrystalline at temperatures below 10°C. Because water is used at least in part as a chain extender to form the EB polymer, a lower urea concentration in the hard segment is obtained, resulting in relatively good melt processability. Chemical structures for the two materials consistent with the testing performed were proposed by Lelah, et al.[10].

Table 2 summarizes the ESCA elemental analysis of the surface and the contact angle/surface energetics results. Both the ESCA and the contact angle data indicated that the surface concentration of polyether soft segment, which was relatively hydrophobic in comparison with the hard segment, was highest on the EB polymer and lowest on the SB polymer. Soft segment concentrations on the EB surface were higher than on the CB surface, indicating that the method of fabrication affected the composition of the surface layer.

Figures 4 and 5 show the transient platelet and fibrinogen deposi-

Table 2. Surface Properties for Biomer Polyurethanes

Designation	Peak Platelet Deposition, ($\#/1000\ \mu m^2$)	Surface Energy: γ_{sw} (dyn/cm)	$\dfrac{\gamma_s^p}{\gamma_s^d}$	ESCA $\dfrac{N}{C-O-C}$
EB	42	21	0.46	0.082
CB	85	21	0.45	0.138
SB	197	12	0.72	0.233

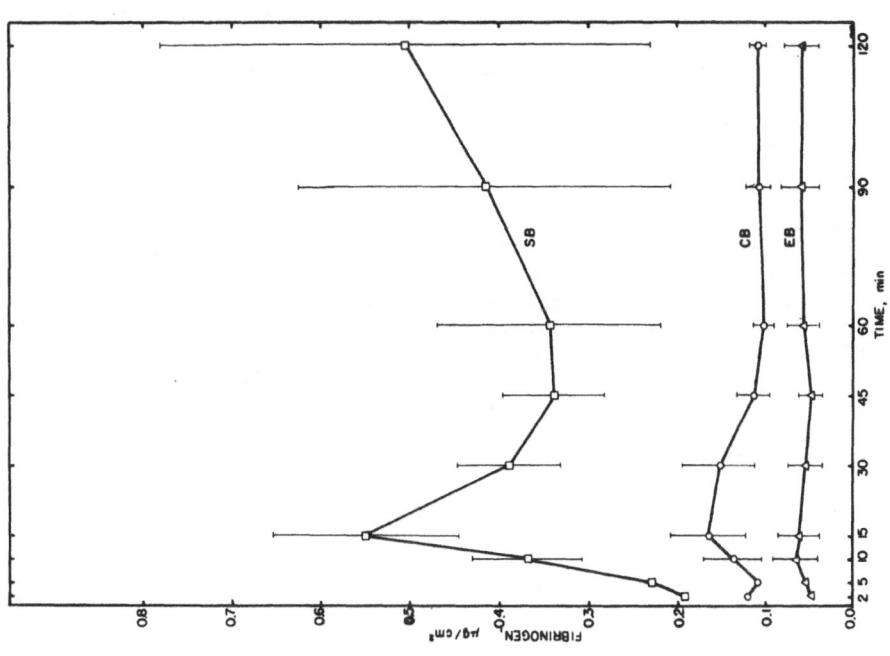

Figure 5. Fibrinogen deposition on Biomer materials[10].

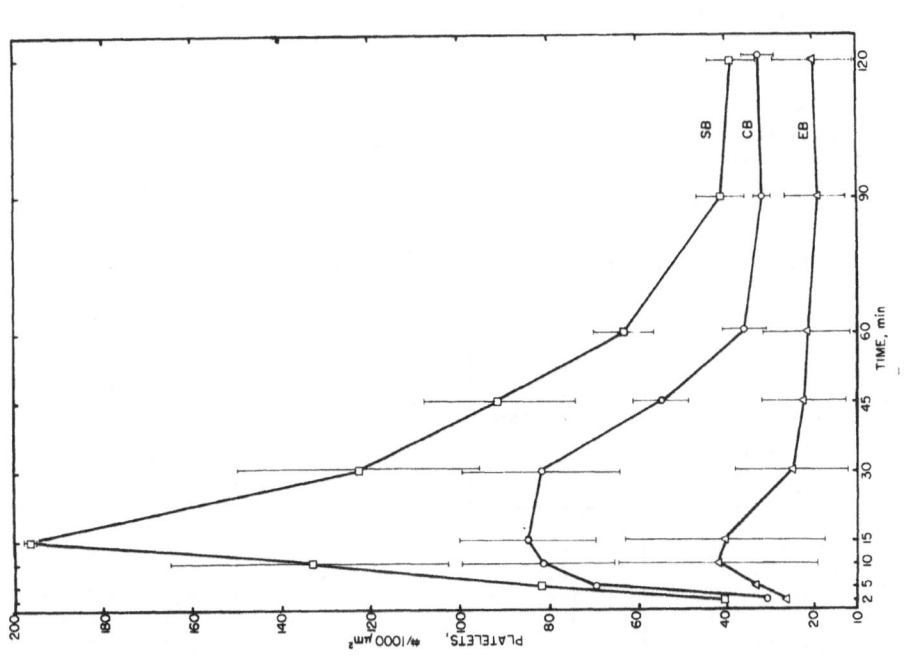

Figure 4. Platelet deposition on Biomer materials[10].

tion profiles, respectively, on the three Biomer surfaces. The platelet and fibrinogen deposition data from the ex vivo model, in combination with electron microscopy of adhered platelets and thrombi, show that thrombus deposition on SB is greater than that on either form of extruded Biomer (CB or EB). The platelet and fibrinogen peak heights provide a comparison of the thromboembolytic potential of the various surfaces. This comparison is important, since a greater peak height represents a greater degree of thrombus deposition on a surface, which can potentially occlude the artificial vessel. The peak in platelet and fibrinogen deposition is due to the competition between thrombus formation and embolization. In studies of other materials in the same single shunt model[10], polyethylene and plasticized polyvinylchloride were shown to be relatively thrombogenic, as indicated by a relatively high peak platelet and fibrinogen deposition level. These ex vivo results showed that Biomer polyurethane block copolymers may be desirable for blood-contacting applications. Also, polyurethane thrombogenicity was shown to be influenced by small changes in sample composition and method of fabrication. For this family of materials, blood compatibility was correlated with increasing concentration of polyether soft segments on the surface.

Part II: Studies of Polyurethanes and Polyurethane Ionomers

Figures 6 and 7 show the temporal sequence of platelet and fibrinogen deposition per unit area over the first sixty minutes of blood contact. Error bars represent standard errors of the mean for the three experiments. By using the peak platelet and fibrinogen deposition levels as a means of determining thrombogenicity in this ex vivo experiment, it was possible to determine that the positively charged cationomer was less blood compatible than the neutral urethanes or other ionomers. The anionomer and zwitterionomer exhibited a lower amount of platelet and fibrinogen deposition than the uncharged polyurethane upon which these ionomers were based. The U-21.5 material, which contained a relatively large concentration of bulk soft segment material, also appeared to be less thrombogenic than the U-38 polymer in this test.

Table 3 summarizes the surface analysis results and can be used to relate the surface properties to the observed blood-contact results. The ESCA N/C-O-C elemental ratio (nitrogen to oxygen-bonded-carbon) indicated that the zwitterionomer showed an enhanced surface hard

Table 3. Surface Properties for Polyurethanes and Polyurethane Ionomers

Designation Name	Peak Platelet Deposition, ($\#/1000\ \mu m^2$)	Surface Energy: γ_{sw} (dyn/cm)	$\dfrac{\gamma_s^p}{\gamma_s^d}$	ATR-IR HBI[*]	ESCA $\dfrac{N}{C-O-C}$
U-21.5	43	7.2	1.47	0.53	0.034
U-38-ZW	62	7.3	1.11	0.36	0.134
U-38-AN	124	7.0	1.01	0.77	0.068
U-38	211	13.9	0.80	1.10	0.066
U-38-CA	1078	11.7	0.77	0.83	0.064

[*] Absorbance ratios of 1710 cm^{-1} to 1730 cm^{-1} bands

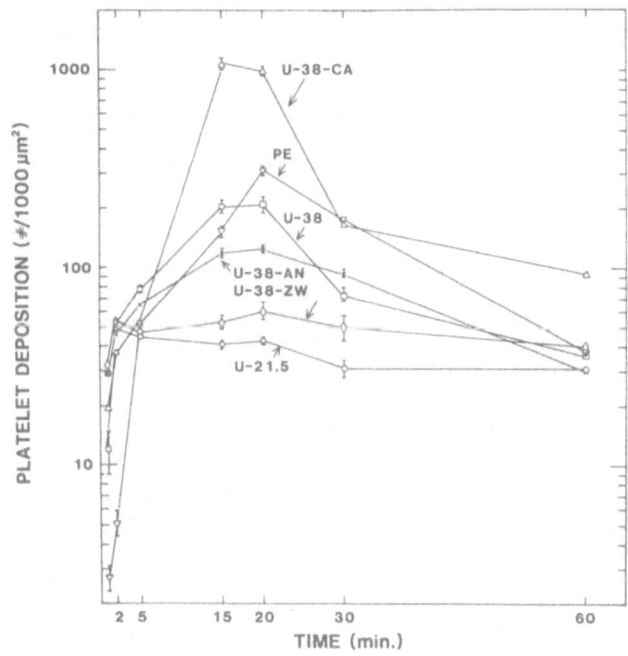

Figure 6. Platelet deposition on polyurethanes and polyurethane ionomers[5].

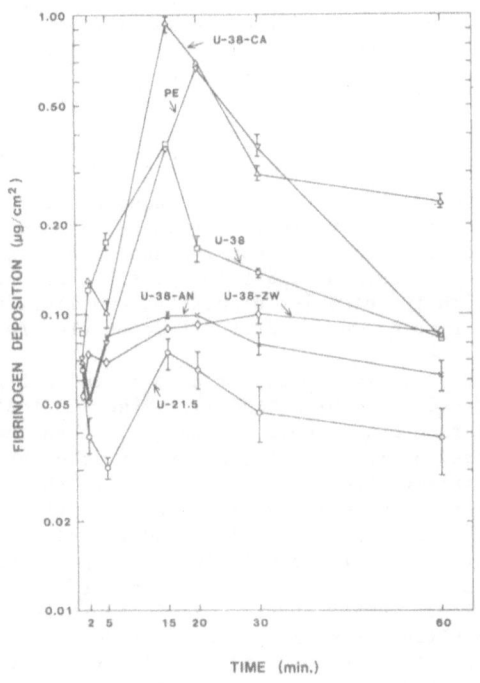

Figure 7. Fibrinogen deposition on neutral and ionic polyurethanes[5].

segment concentration when compared to the uncharged polymer. However, comparisons of the other three U-38-based materials do not show significant changes in surface hard-to-soft segment ratios.

The contact angle and surface energetics data show a relationship between solid-water interfacial tension and peak platelet deposition. The relatively hydrophilic zwitterionomer and anionomer possess an enhancement of the relatively hydrophilic and highly charged hard segment component at the surface when compared to the uncharged U-38 polyurethane. Interestingly, the increase in surface hydrophilicity is not nearly as pronounced with the cationic urethane. Surface energetics indicate that the cationomer surface was only slightly more hydrophilic than the surface of the uncharged polymer. A possible explanation is that the ionic group on the cationomer is actually on the backbone of the polymer chain and is considerably less mobile than the ionic groups of the other ionomers. The effect of the positively-charged groups is apparent, however, in the transient platelet and fibrinogen deposition profiles.

The U-21.5 material contains a relatively large amount of bulk soft segment, and the high-vacuum ESCA and air-equilibrated ATR-IR data indicate that the surface soft segment level was considerably higher than for the U-38 polymer. The pure hard segment material is considerably more hydrophilic than the pure poly(tetramethylene oxide) soft segment. The surface energetics data, however, indicate that the relatively hydrophilic U-21.5 contains a higher concentration of MDI units and urethane groups at the surface. This observation contradicts the ESCA and ATR-IR data, which were not obtained in aqueous environments. The U-38 material can be more hydrophobic than U-21.5 only if it possesses an enriched surface soft segment concentration, which could be one result of improved phase separation. Thus, phase separation may play a role in polyurethane blood compatibility. Another implication of these observations, and one which will be discussed in subsequent sections, is that a multiprobe surface characterization approach is necessary to properly characterize these polyurethane block copolymers.

Part III: Polyurethanes with Different Hard and Soft Segment Types[16,17]

Figures 8-12 show the temporal sequence of platelet deposition per unit area over the sixty minutes of blood contact on the surfaces studied. The error bars represent standard errors of the mean for three experiments. Fibrinogen deposition profiles are not shown but correlate quite well with the platelet deposition profiles, with nearly simultaneous peaks in fibrinogen and platelet deposition for most surfaces.

Figure 8 shows that platelet deposition is higher on M/MD/PEO(1.5) than on M/MD/PPO(2) or M/MD/PTMO(2). Figure 9 indicates that M/MD/PTMO(1) cast from DMA is more thrombogenic than the same material cast from THF, based on the relative number of adherent platelets at the peak platelet deposition time for each material. The platelet deposition profile of Figure 10 indicates that M/MD/PTMO(1) is relatively thrombogenic when compared to M/BD/PTMO(1) and H/BD/PTMO(1). These latter polyurethanes are relatively thrombogenic when compared to the polyurethaneurea H/ED/PTMO(1). The high soft segment-containing solution grade Biomer (HSSB) surface was found to be more resistant to platelet deposition than the extruded grade Biomer (EB) surface, as shown in Figure 11. In both cases, solvent extraction dramatically improved the blood compatibility of these materials. The hard segment analog material M/MD was the most thrombogenic material examined, as seen in Figure 12. Finally, consistent with the results in Part II of

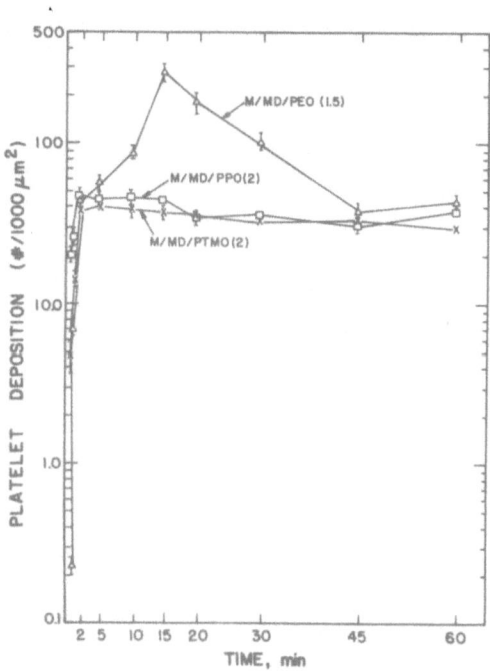

Figure 8. Platelet deposition profiles for a series of polyurethanes with different soft segment types[16].

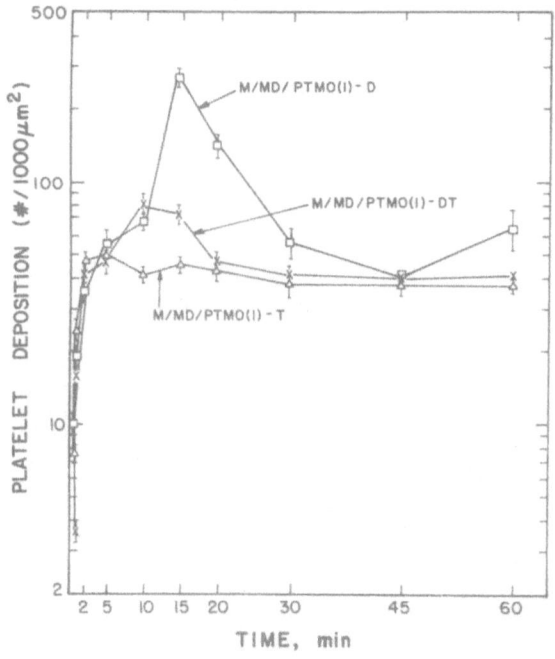

Figure 9. Platelet deposition profiles for M/MD/PTMO(1) cast from DMA, DMA/THF, and THF[16].

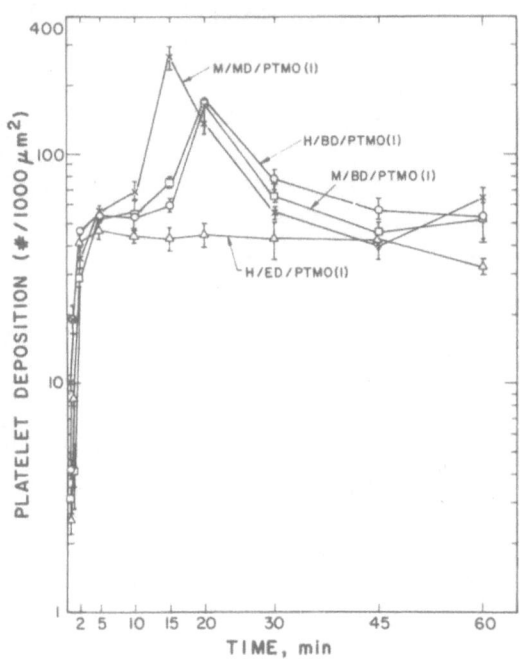

Figure 10. Platelet deposition profiles for polyurethanes with different hard segment diisocyanates and chain extenders[16].

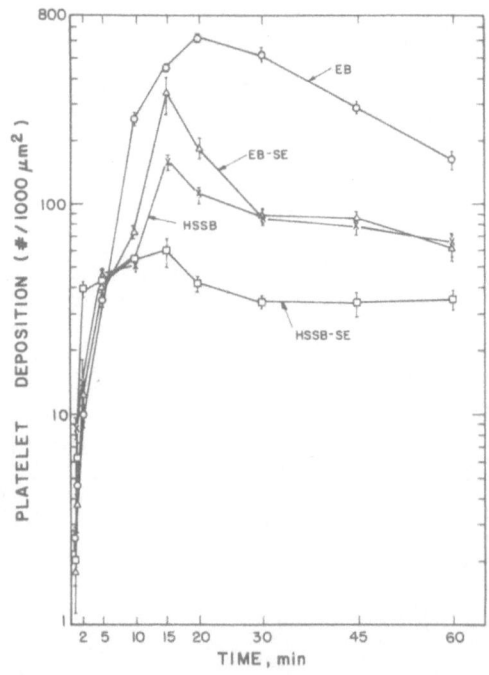

Figure 11. Platelet deposition profiles for commercial polyurethanes with and without solvent extraction[16].

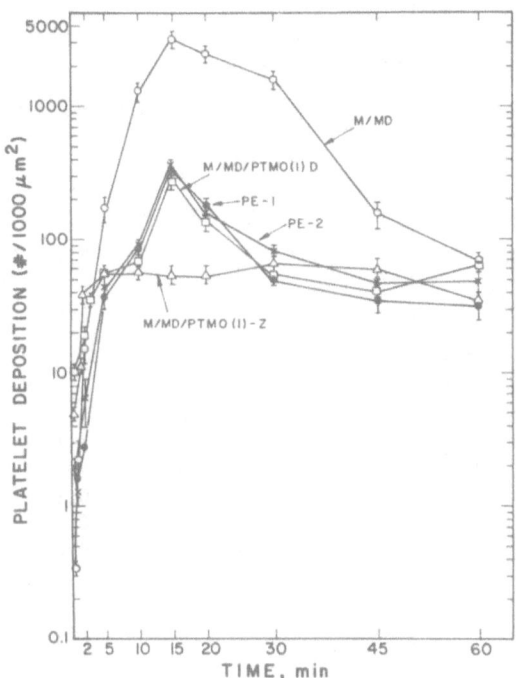

Figure 12. Platelet deposition profiles for hard segment "analog" M/MD, nonionic and zwitterionic polyurethanes M/MD/PTMO(1)-D and M/MD/PTMO(1)Z, and polyethylene control materials PE-1 and PE-2[16].

this study, the M/MD/PTMO(1)-Z zwitterionomer was found to be more thromboresistant than its nonionic counterpart.

Blood compatibility information and surface characterization data for the polyurethanes studied appear in Table 4. A more complete tabulation of the surface analysis results of this study is provided by Lelah, et al.[16,17].

At this point, it is important to point out that polymer surface characterization is difficult, because no single technique can provide the necessary physical and chemical information at the appropriate sampling depths. As an example, ESCA provides valuable surface chemical information, but it is obtained at extremely high vacuum levels. On the other hand, contact angle measurements, which in these studies were carried out underwater, allow a closer approximation to the aqueous blood environment, but the averaged surface energy data do not lead directly to chemical structure information. In many cases, it is seen that the surface energetics are particularly revealing in correlating the observed trends. The air-equilibrated (ATR-IR) and high-vacuum (ESCA) data in some cases do not lead to direct correlations with the blood contact data. These observations are not unexpected and indicate why multiple methods of surface analysis are necessary in biomaterials evaluation.

Table 4. Surface Properties for Polyurethanes with Various Hard and Soft Segment Components

Designation	Peak Platelet Deposition ($\#/1000\ \mu m^2$)	Surface Energy: γ_{sw} (dyn/cm)	$\dfrac{\gamma_s^p}{\gamma_s^d}$	ATR-IR HBI[*]	ESCA $\dfrac{N}{C-O-C}$
M/MD/PTMO(2)	43	21.7	0.53	0.50	0.18
M/MD/PPO(2)	48	8.4	0.88	0.72	0.12
M/MD/PEO(1.5)	281	1.6	1.5	0.95	0.12
M/MD/PTMO(1)-D	268	17.8	0.56	0.53	0.084
M/MD/PTMO(1)-DT	80	8.3	0.87	0.58	0.073
M/MD/PTMO(1)-T	51	0.9	1.9	0.52	0.065
M/MD/PTMO(1)-Z	65	6.3	1.1	0.20	0.226
M/MD	3229	3.4	2.1	1.10	0.051[+]
M/BD/PTMO(1)	166	54.4	0.18	1.04	0.120
H/BD/PTMO(1)	170	57.1	0.22	0.54	0.288
H/ED/PTMO(1)	47	8.2	0.89	0.90	0.467
EB	622	24.5	0.44	1.15	0.109
EB-SE	336	8.8	0.84	1.04	0.096
HSSB	160	10.6	0.79	0.58	0.067
HSSB-SE	60	7.0	1.14	0.67	0.023

[*] Absorbance ratio of 1710 cm^{-1} to 1730 cm^{-1} bands

[+] N/C ratio

The relative thrombogenicity of the relatively hydrophilic PEO-based polyurethane (Figure 8), is notable, because Merrill, et al.[18,19] noted that PEO-based polyurethanes retain relatively few platelets in their in vitro platelet retention test. However, such in vitro tests typically do not correlate with measures of ex vivo blood compatibility[20].

The effect of fabrication processes on surface properties and blood-contacting behavior has been demonstrated[10], and the results of Figure 9 show that solvent coating conditions are important in the determination of surface properties and blood compatibility. The material cast from THF was more hydrophilic and thus probably contained a larger surface concentration of atoms associated with the relatively hydrophilic hard segment than the polymer cast from DMA. The ESCA data, however, do not indicate that the surface of the THF-cast polymer is enriched in hard segment when compared to the DMA-cast material. This case also demonstrates the necessity for multiple surface characterization techniques, because it is apparent that there are different gradients in relative hard and soft segment concentrations or different molecular orientations for these materials in vacuo and underwater.

In a limited study of the effect of hard segment diisocyanate, little difference is noted between the platelet deposition profiles of M/BD/PTMO(1) and H/BD/PTMO(1) in Figure 10. This observation is in agreement with the nearly identical surface energy values determined by the contact angle studies. A comparison between the ED- and BD-chain-extended polyurethanes (Figure 10) indicates that the ethylene diamine-extended material was more thromboresistant and more hydrophilic than the butanediol-based material. The more thromboresistant polymer (H/ED/PTMO(1)) was found to have a higher surface hard to soft segment ratio, as determined by both contact angle studies and ESCA. Thus, hard segment components are also seen to be important in determining the blood response of polyurethane block copolymers.

In the present study, the hard segment analog M/MD was found to be very polar (high $\gamma_s P/\gamma_s d$) and also rather hydrophilic (low γ_{sw}). As shown in Figure 12, this material was also found to be the most thrombogenic polymer studied. It is interesting to compare this result with the other observations in this study which show that increased hard segment levels on the surface often result in decreased thrombogenicity. It is apparent that the microphase separation in these block copolymers is important in determining the extent of the observed blood response.

The transient platelet deposition profiles for the commercial polyurethanes before and after extraction are shown in Figure 11. The surface characterization data (Table 4) show an increase in surface hydrophilicity upon extraction, and thus a surface enrichment of hard segment is indicated for the extracted polymers. In this study, the polymers were re-cast after extraction, and the bare surfaces were shown to be smooth by scanning electron microscopy. Thus, the changes in blood contact behavior were due to modifications in surface chemistry and not in surface topography or roughness.

CONCLUSIONS

The blood-contact and surface characterization results have indicated that the blood compatibility of polyurethane block copolymers is affected by the nature of both the hard and soft segment components. The extent and direction of the observed variances in blood compatibility, however, were not found to follow simple rules. In the original study of cast and extruded Biomer[10], for instance blood compatibility improved with increasing concentration of polyether soft segments at the surface. The later studies, however, indicated that in many instances, blood compatibility improved as the surface-water interfacial tension decreased, indicating that the presence of hard segment on the surface is of some importance in determining blood response.

Incorporation of ionic groups into the bulk polyurethane results in a hard domain structure that may be significantly different than in a neutral polyurethane. Ionization of polyurethanes is found to be a useful technique to study the role of surface chemistry in artificial surface-induced thrombosis. The thromboresistance of the zwitterionomers and the anionomer tested is related to a high concentration of the mobile side chain ionic sulfonate groups at the surface. Ion incorporation appears to strongly influence the blood response of these materials by making them more hydrophilic and enhancing surface hard segment levels. Effects of ionic mobility, charge type, and surface ionic concentration are topics for further studies. The presence of a positively charged quaternary amine group causes a polyurethane to become considerably more thrombogenic. This effect may be due to the nature of the positive charge or because of its low mobility caused by its attachment to the polymer backbone. The mechanisms causing these effects are not yet known and also merit further investigation.

Fabrication conditions were found to affect both polymer surface properties and blood response. These results have important implications in blood-contacting applications. Polyurethanes fabricated under

one set of conditions for blood compatibility testing may not perform
similarly when fabricated using a different technique for a specific
device or implant application. Extraction of polyurethanes was also
shown to markedly affect the blood response.

As a family of polymers, the polyurethanes were found to be
relatively thromboresistant in the acute ex vivo test used in these
studies. The results support continued investigation of polyurethanes
for blood-contacting applications.

REFERENCES

1. M. D. Lelah and S. L. Cooper, "Polyurethanes in Medicine," CRC
   Press, Boca Raton, FL (1985).
2. G. J. Picha, D. F. Gibbons, and R. A. Auerbach, Effect of Polyurethane Morphology on Blood Coagulation, J. Bioeng., 2:301 (1978).
3. T. Okano, S. Nishiyama, I. Shinohara, T. Akaike, Y. Sakurai,
   K. Kataoka, and T. Tsuruta, Effect of hydrophilic and hydrophobic
   microdomains on mode of interaction between block polymer and blood
   platelets, J. Biomed. Mater. Res., 15:393 (1981).
4. A. Takahara, J. Tashita, T. Kajiyama, and M. Takayanagi, Blood
   compatibility and microphase-separated structure of segmented
   poly(urethaneureas) with various soft segment components, Rep.
   Prog. Polym. Phys. Jpn., 25:841 (1982).
5. M. D. Lelah, J. A. Pierce, L. K. Lambrecht, and S. L. Cooper,
   Polyether-urethane ionomers: surface property/ex vivo blood
   compatibility relationships, J. Colloid Interface Sci., 104:422
   (1985).
6. M. Jozefowicz and J. Jozefonvicz, Antithrombogenic polymers, Pure
   Appl. Chem., 56:1335 (1984).
7. L. C. Sederel, L. van der Does, J. F. van Duijl, T. Beugeling, and
   A. Bantjes, Anticoagulant activity of a synthetic heparinoid in
   relation to molecular weight and N-sulfate content, J. Biomed.
   Mater. Res., 15:819 (1981).
8. J. V. Ihlenfeld, T. R. Mathis, L. M. Riddle, and S. L. Cooper,
   Measurement of transient thrombus deposition on polymeric
   materials, Thromb. Res., 14:953 (1979).
9. M. D. Lelah, L. K. Lambrecht, and S. L. Cooper, A canine ex vivo
   series shunt for evaluating thrombus deposition on polymer
   surfaces, J. Biomed. Mater. Res., 18:475 (1984).
10. M. D. Lelah, L. K. Lambrecht, B. R. Young, and S. L. Cooper,
    Physicochemical characterization and in vivo blood tolerability of
    cast and extruded Biomer, J. Biomed. Mater. Res., 17:1 (1983).
11. K. K. S. Hwang, C. Z. Yang, and S. L. Cooper, Properties of poly-
    ether-polyurethane zwitterionomers, Polym. Eng. Sci., 21:1027
    (1981).
12. C. Z. Yang, K. K. S. Hwang, and S. L. Cooper, Morphology and
    properties of polybutadiene- and polyether-polyurethane
    zwitterionomers, Makromol. Chem., 184:651 (1983).
13. J. A. Miller, K. K. S. Hwang, and S. L. Cooper, Properties of poly-
    ether-polyurethane anionomers, J. Macromol. Sci. Phys., B22:321
    (1983).
14. W. C. Hamilton, A technique for the characterization of hydrophilic
    solid surfaces, J. Colloid Interface Sci., 40:219 (1972).
15. J. D. Andrade, S. M. Ma, R. N. King, and D. E. Gregonis, Contact
    angles at the solid-water interface, J. Colloid Interface Sci.,
    72:488 (1979).
16. M. D. Lelah, T. G. Grasel, J. A. Pierce, and S. L. Cooper, Ex vivo
    interactions and surface property relationships of polyether-
    urethanes, J. Biomed. Mater. Res., submitted for publication (1985).

17. M. D. Lelah, Ph.D. Dissertation, University of Wisconsin-Madison (1984).
18. V. sa da Costa, D. Brier-Russell, E. W. Salzman, and E. W. Merrill, ESCA studies of polyurethanes: blood platelet activation in relation to surface composition, J. Colloid Interface Sci., 80:445 (1981).
19. E. W. Merrill, V. sa da Costa, E. W. Salzman, D. Brier-Russell, L. Kuchner, D. F. Waugh, G. Trudel, S. Stopper, and V. Vitale, A critical study of segmented polyurethanes, in: "Biomaterials: Interfacial Phenomena and Applications," S. L. Cooper and N. A. Peppas, eds., Adv. Chem. Ser. 199:95 (1982).
20. R. S. Wilson, M. D. Lelah, and S. L. Cooper, Blood-material interactions: Assessment of in vitro and in vivo test methods, in: "Techniques in Biocompatibility Testing," D. F. Williams, ed., CRC Press, Boca Raton, FL (1985).

THE EFFECT OF VARYING COMPLIANCE UPON THE LONG-TERM PATENCY OF A

POLYURETHANE ARTIFICIAL ARTERY

D. Annis, A.C. Fisher, T.V. How and L. de Cossart

The Institute of Medical and Dental Bioengineering
University of Liverpool, P.O.Box 147, Liverpool L69 3BX, UK

At the Ist International Conference on "Polymers in Medicine" we
described a compliant microfibrous polyetherurethane arterial graft of
small luminal cross-sectional area, the wall of which was composed of fibres
between 1.0 m and 2.0μm in diameter. At that time we reported our results
of the implantation into animals of artery grafts of this construction
having luminal diameters of 6mm and larger. We also reported the commence-
ment of a study in which we implanted grafts of similar construction having
an internal diameter of 3.7mm. In this present publication we follow the
performance of these grafts during the last two years.

Amongst the many novel features of our graft is the measure of its
compliance. By manipulation of the electrostatic spinning process we are
able to manufacture grafts having a range of compliance which lies within
the rather wide range of compliance recorded in the literature for the
canine carotid artery. When it became apparent that a sufficient number
of these implanted grafts were remaining patent for periods in excess of
340 days it became possible for us to compare the long-term patency of
grafts of different compliance. The compliance was changed without
changing the gross physical characteristics of the grafts, their length,
wall thickness and luminal diameter remaining the same. The fibrous
nature of their surfaces was also unchanged. A reduction in compliance
was achieved in manufacture by increasing the compaction of the fibres
within the thickness of the wall. Thus we were able to compare the long-
term patency of grafts that differed in no way other than in the measure
of their compliance.

Although there is a widely held belief that a small diameter vascular
prosthesis should have compliance that approaches that of the natural
vessel to which it is attached, it must be stressed that this view is, as
yet, unproven. Much of the available evidence upon which it is based is
unsatisfactory, offering comparisons of patency rates that follow the
implantation of a wide range of grafts having widely differing compliance,
but also having widely differing structure. The observation that an almost
rigid woven Dacron grafts performs less well than a more compliant
autologous vein graft may indicate no more than that the vein graft is
superior, not because of its compliance but because of its being composed
of living tissue. A major difficulty that has been encountered in the
study of the effect of compliance upon patency of small bore grafts has
been the generally poor performance of all synthetic arteries smaller than

6mm internal diameter and the extremely poor results in both man and in the larger mammals where synthetic vessels of 4mm internal diameter have been implanted.

The structure and method of manufacture of our graft have been fully described elsewhere (Annis et al. 1978). Briefly, the graft is spun from a solution of Biomer polyetherurethane in dimethyl acetamide. Fibres are drawn electrostatically towards a rotating steel mandrel of appropriate size. The build up of these fibres on the mandrel produces a tube having a microfibrous porous wall of controlled anisotropy. The fibres are between 1.0 and 2.0μm in diameter. The wall of the graft, though micro-porous, is grossly smooth to the naked eye. The graft has good surgical handling characteristics, cutting without fraying and holding fine stitches to within half a millimetre of the cut end of the wall of the graft.

THE DESIGN OF THE EXPERIMENT

In a series of 18 consecutive implantations, 4cm lengths of the dog carotid artery were excised and replaced by 3.8cm lengths of grafts of our manufacture, having a luminal diameter of 3.7mm. Twelve of the grafts had circumferential compliance of 2.4%, the remaining six having a compliance of 0.7%.

The definition of compliance as we use it and the method of its measurement have been fully described by us elsewhere (How et al. 1984). Briefly, the graft is slowly pressurized at a constant rate from 0-200mmHg, such that the resultant circumferential strain rate is not greater than 0.03min.$^{-1}$ Pressure diameter and longitudinal force are monitored continuously and compliance is expressed as the percentage change of external diameter of the graft occurring between 80 and 120mm Hg pressure with respect to the diameter at 100mm Hg. The graft is extended in length by 6% during measurement of changing diameter.

All grafts were implanted by the same surgeon using a precisely controlled technique. For three days before surgery the dogs were given 75mg of aspirin (ASA) and 200mg of dipyridamole. During the time whilst the carotid artery was clamped the dogs received heparin. ASA and dipyridamole therapy was continued throughout the post-operative period and in all animals therapy has been continued to the present time.

THE METHOD OF ASSESSMENT OF PATENCY

Two weeks after surgery the grafts were scanned using a Toshiba duplex ultrasound system with a real-time B mode scanner (SAL/50A) and a 2.4MHz pulsed wave Doppler (DLS/01A). Using this system the graft was first imaged using the B mode scanner and, having been precisely located, the presence or absence of blood flowing within was detected by targetting the pulsed Doppler sampling volume onto the image of the lumen of the graft. Flow in the graft was confirmed by the appearance of Doppler shift frequency waveforms arising from the movement of red blood cells within the graft. In addition to confirming the presence of flow, the instrument determined the direction of flow, its magnitude and the degree of turbulence within the graft. During our early experience of the use of the instrument it was necessary to resolve doubtful cases by surgical exploration and the direct observation of the graft using a hand-held continuous-wave Doppler probe in contact with the outer wall of the graft. With increasing experience of the use of the instrument, exploration of the grafts is no longer necessary.

RESULTS

<u>Subset 1</u> :   Grafts of 2.4% compliance: ASA and Dipyridamole daily since the week before implantation.

Of the 12 grafts in Subset 1, (Figure 1) 7 remain patent at times ranging between 602 days and 468 days since implantation (an average patency of 540 days). Five occluded; however, of these, 3 occluded within the first few days following surgery, no flow being detected by ultrasound scanning at any time. We describe these as 'acute failures' and ascribe their failure to small but important technical errors of suturing in the making of the end-to-end anastomosis. In all three the site of occlusion was at one or other of the anastomoses where a thrombus formed attached to the line of anastomosis. With accretion of more thrombus it grew until it was large enough to occlude the lumen of the graft. In none was thrombus attached elsewhere to the inner surface of the graft.

There were 2 late failures. One occurred on the 98th day after assessing patency by direct exposure of the graft. This graft was found to be patent, but exploration was followed by the development of a large haematoma in the exploration wound which discharged and became infected. The second late failure occurred 301 days after implantation. On opening the explanted graft, fresh thrombus lay within a thin, intact tubular membrane, attached at each anastomosis but not attached to the inner surface of the graft along its length.

<u>Subset 2</u> : Stiffer grafts (compliance 0.7%) : ASA and Dipyridamole given daily since the week before implantation.

None of the 6 grafts in Subset 2 failed acutely (Figure 2). Five of the grafts remain patent at times ranging from 344 - 342 days (an average of 343 days) after surgery. One graft failed on the 59th post-operative day. It lay within a cavity containing old, altered blood - the result of a partially absorbed wound haematoma. Within the graft was a fresh thrombus, not itself attached to the wall of the graft but lying within a tubular membrane which was attached to the graft near each anastomosis but was not attached along the length of the graft.

SUBSET 1   Compliance 2.4%

Aspirin and dipyridamole given throughout the experiment

12 GRAFTS WERE IMPLANTED

^{3}/12 occluded within the first 2 weeks after surgery
                  (before the first ultrasound scan)

^{2}/12 occluded later   -   on the 98th and 301st day
                         after surgery

^{7}/12 ARE PATENT NOW   -   602 DAYS
                         554 "
                         551 "
                         551 "
                         536 "
                         515 "
                         468 "   -   AFTER IMPLANTATION

Fig. 1. The Patency of 3.7mm I.D. Carotid Replacement Grafts

SUBSET 2   Compliance 0.7% (stiffer grafts)

Aspirin and dipyridamole given throughout the experiment

6 GRAFTS WERE IMPLANTED

^{0}/6 occluded within the first 2 weeks after surgery
              (before the first ultrasound scan)

^{1}/6 occluded later - on the 59th day after surgery

^{5}/6 ARE PATENT NOW - 344 DAYS
                       343 "
                       343 "
                       343 "
                       342 " - AFTER IMPLANTATION

Fig. 2. The Patency of 3.7mm I.D. Carotid Replacement Grafts

The Gross and Microscopical Features of Longer-term Patent Grafts

Some grafts from an earlier trial have been removed whilst still
patent, but none, of course, from this ongoing trial. They show both
anastomoses healed and covered on the luminal surface only by an almost
transparent extension of the natural intima which, having crossed the line
of lumen, extends for about one quarter of the length of the graft from
each end. Histological examination shows a thin intima securely attached
to the wall of the graft. The intervening inner surface, though smooth
and glistening, is not covered by new intima. There is no evidence of
intimal thickening either on the inner surface of the graft or on the
adjacent natural artery.

DISCUSSION

In the assessment of the effect on the patency of our graft of a
changed wall stiffness, the first and overriding need was for a rigidly
controlled, precisely reproducible surgical technique and a surgeon with
the skill to perform it. The same surgeon performed all the operations.

Scrutiny of our results has shown that our technique has been
rewarded with good and long-maintained patency of the majority of the
grafts implanted. It also has shown that, almost without exception,
failure can be traced back to some small error of surgical technique.

Of 18 grafts in the combined series of Subsets 1 and 2, three failed
acutely. The mode of failure in all was thrombus formation on the suture
line - often limited in its attachment to the site of a single suture.
Three of the 18 grafts failed later (i.e. after the second post-operative
week). Both were related to the accumulation of blood around the graft
which became encysted, leaving a collection of old blood engulfing the
graft. This was due to our failure to leave a dry wound at the end of
the surgical operation. This experience leaves us in no doubt that good
surgery is a prerequisite of any study of graft patency.

With a sufficient number of grafts still patent in excess of 340 days
after implantation we are able to compare a set of grafts of compliance
2.4% with a set of stiffer grafts of similar construction having compliance
of 0.7%. Of the 6 grafts of 0.7% compliance implanted, none has failed

acutely. Five of the six are patent now, 344 days after implantation. One graft failed 59 days after surgery. A direct comparison can be made between this subset and those having compliance of 2.4% where 7/9 are still patent now, 602-468 days after surgery. It is apparent from these figures that, at least for 344 days after implantation, an increase in compliance from 0.7% to 2.4% did little or nothing to affect the long-term patency of our graft.

Despite the difference in compliance of our two sets of grafts, it is not possible to relate either of them to the compliance of the natural carotid artery in the dog because of the uncertainty surrounding the values for the natural artery. Values of compliance of the dog carotid artery reported in the literature, when converted to our expression of compliance, vary from 1.74% to 7.37%. These figures were obtained from the elastance ($\Delta PD/\Delta D$) values determined by Peterson et al. (1960) and from the pressure-diameter curve of a carotid artery subjected to a continuous increase in internal pressure (Cox, 1975). In both studies the vessel diameter was measured by means of a gauge which requires mechanical contact with the vessel wall. Hutchison (1974) used a non-contacting displacement transducer to measure diameter and quoted a mean elastance value at 100mm Hg pressure of 1.125×10^6 dynes/cm^2. This corresponds to a compliance according to our definition of 4.7%.

Peterson et al. carried out their measurement in vivo while Cox and Hutchison both used excised vessels. The wide range in compliance is most probably due to the effect of the smooth muscle tone on the mechanical properties and the unknown changes in tone caused by handling the vessel during its exposure.

There is no yardstick which we can use to match the compliance of our graft to that of the natural artery. Nevertheless, we have demonstrated that there is a range of compliances within which we can expect good longer-term patency. To know the full span of the range it will be necessary to produce and implant a set of grafts of very much reduced compliance - similar in compliance to that of woven Dacron grafts or grafts of extended PTFE which, in a biological context, are rigid. Only then shall we know how wide a range of compliance can be tolerated by our graft without jeopardizing its longer-term patency.

REFERENCES

Annis, D., Bornat, A., Edwards, R.O., Higham, A., Loveday, B. and Wilson, J., 1978, An elastomeric vascular prothesis, Trans.Am.Soc. Artif. Intern. Organs, Vol. XXIV, 209-213.
Cox, R.H., 1975, Anisotropic properties of the canine carotid artery in vitro, J. Biomechanics, 8 : 293-300.
How, T.V., Bhuvaneshwar, G.S. and Annis, D., 1984, Infrared diameter gauge for in vitro mechanical testing of vascular grafts, J.Biomed.Eng. 6 : 195-199.
Hutchison, K.J., 1974, Effect of variation of transmural pressure on the frequency response of isolated segments of canine carotid arteries, Cir. Res.35 : 742-751.
Peterson, L.H., Jenson, R.E. and Parnell, J., 1960, Mechanical properties of arteries in vivo. Cir.Res. 8 : 622-639.

SMALL DIAMETER BLOOD VESSEL PROSTHESES FROM POLYETHERS

J.G.F. Bots, L. van der Does and A. Bantjes

Dept. of Chemical Technology, Biomaterials Section
Twente University of Technology
P.O. Box 217
7500 AE Enschede, The Netherlands

ABSTRACT

Studies on the relation between blood platelet adhesion and type
and amount of polyether segments in copolyether-urethanes report
a reduced platelet adhesion with increasing polyether content.
We therefore assumed that combinations of polyethylene oxide
(PEO) and polypropylene oxide (PPOX) might give materials with a
good blood compatibility. Water-soluble PEO was attached to PPOX
by UV initiated crosslinking. Films were tested on hydrophilicity, mechanical properties, protein adsorption and blood compatibility. The hydrophilicity was determined by swelling experiments. A compromise between hydrophilicity (PEO) and mechanical
strength (PPOX) was met at a swelling of 0.5 (PPOX/PEO ratio :
90/10). In protein adsorption studies only small amounts of adsorbed proteins were found. Three blood material interaction in
vitro tests, - kallikrein generation (factor XII activation),
activated partial thromboplastin time and blood platelet adhesion
-, gave good results: a low platelet adhesion and kallikrein
generation and a high APTT value. Porous tubings (I.D. 1.3 mm)
were fabricated by spinning from solution and implanted in the
abdominal aorta of rats. Stress-strain diagrams were comparable
to those reported for natural blood vessels.

INTRODUCTION

A special type of block copolymers, consisting of alternating hard
and soft segments, is widely known and two representatives, -Biomer[R,1] and
Avcothane[R,2] -, are used as biomaterials in several blood contacting devices such as catheters, artificial blood vessels and heart valves. Kim et
al.[3] and Lyman et al.[4] have reported that a segmented copolyether-ure-
thane-urea (PEUU) compared to silicone rubber (SR) and a fluorinated ethy-
lene-propylene copolymer (FEP) shows an increased albumin adsorption together with a low gammaglobulin and fibrinogen adsorption and a low platelet adhesion. This behaviour is generally accepted as being indicative for
a blood material interaction without severe reactions like thrombus formation.

Indications were obtained by Lyman that protein is more strongly adsorbed onto the hard polyurethane-urea segments than on the soft polypropylene oxide blocks. He concluded from ESCA measurements that the air side of a cast film was richer in polyether than the glass mold side of the film.

Interesting observations have also been reported by Lelah et al.[5], who found that extruded Biomer[R] had a higher content of soft polytetramethylene oxide segments on the surface than cast Biomer[R], resulting in a lower fibrinogen adsorption and platelet adhesion for the former. Results from these and other studies[6-7] indicate that the polyether segments in copolyurethanes are probably responsible for the preferred adsorption of albumin and the reduced platelet adhesion.

Apart from this the ratio of hydrophilic and hydrophobic areas on the polymer surface is known to have a marked influence on the blood compatibility. For instance, a different behaviour in protein adsorption[8] and platelet adhesion [9-10] was found in contact with hydrophilic or hydrophobic surfaces.

The importance of the morphology and texture of the blood contacting polymer surface has also been reported by Yui et al.[11] and Hess et al.[12]. Yui synthesized polypropylene oxide segmented polyamides with various polyamide segment length and found the platelet adhesion to be strongly dependent on the size of the crystalline and amorphous domains. The platelet adhesion was significantly minimized on a copolymer, in which the average diameter of the crystalline and amorphous domains were 6.42 nm and 5.18 nm respectively. Hess concluded from in vivo experiments that neointima formation in small diameter (1.5 mm) prosthetic blood vessels is strongly stimulated by a porous fibrillar inner structure.

Sa da Costa[13] extensively studied the relationship between platelet adhesion and type and amount of polyether segments in copolyether-urethanes and she found a reduced platelet adhesion with increasing polyether content. Polyethylene oxide (PEO) was reported to be most active in suppressing platelet adhesion compared to polypropylene oxide (PPOX) and polytetramethylene oxide (PTMO).

From these results we derived the hypothesis that combining PEO with PPOX or PTMO might lead to materials with a good blood compatibility: PEO as the blood compatible, hydrophilic component and PPOX or PTMO as the more hydrophobic and mechanically stronger one.

PEO was attached to high molecular weight atactic PPOX by UV initiated crosslinking in the presence of dicumylperoxide (DCP). Swelling experiments were performed with the resulting films in order to determine the equilibrium water content. Blood material interactions were studied with three tests: kallikrein generation (factor XII activation), activated partial thromboplastin time (APTT) and blood platelet adhesion. Adsorption of albumin, fibronectin, fibrinogen and gammaglobulin was investigated using an ELISA test[8]. For in vivo experiments blood vessel prostheses were spun from solution and implanted in the abdominal aorta of rats. The mechanical properties were compared to those of natural blood vessels.

EXPERIMENTAL

Materials

High molecular weight PEO, $\bar{M}_w$ = 1, 3 and 6 x 10^5 -, was purchased from Polysciences Inc. (Warrington, USA). The crystalline and water-soluble polymers showed a broad molecular weight distribution: $\bar{M}_w/\bar{M}_n$ = 6.1 (GPC-LALLS).

Atactic PPOX (DSC: 95% atactic) was synthesized[14] from propylene oxide (Merck) with a catalyst prepared from diethyl zinc (Schering AG, Berlin, Germany, 20% (w/w) in toluene) and triphenyltinhydroxide (M&T, Vlissingen, The Netherlands). $\bar{M}_w$ varied between 0.8 and 2.5 x 10^5, depending on the monomer/catalyst ratio and for all the PPOX batches $\bar{M}_w/\bar{M}_n$ was < 1.3. Semi-crystalline PPOX was prepared[15] with $FeCl_3$ as a catalyst and the crystalline fraction (DSC: 95% crystallinity) was separated from the amorphous one by recrystallisation (twice) from methanol, which dissolves the atactic component.

Semi-crystalline PTMO, - DSC: 70% crystalline -, was synthesized from tetrahydrofuran (Merck) with $SbCl_5$ as a catalyst[16]. Molecular weights of 2 to 4 x 10^4 ($\bar{M}_w$) were obtained: $\bar{M}_w/\bar{M}_n$ = 1.2 (GPC-LALLS). For crosslinking experiments dicumylperoxide (DCP) (Schuchardt, Germany), recrystallised twice from methanol, was used. The solvents, used for the catalyst preparation for the polymerisation of propylene oxide (xylene, toluene and hexane (Merck)) were distilled from sodium. Tetrahydrofuran and propylene oxide were purified by distillation from lithium aluminium hydride (Merck) and calciumchloride (Merck), respectively. All purifications were performed in a nitrogen atmosphere.

Methods

Film preparation and swelling experiments

Mixtures of PEO, PPOX and DCP were dissolved in dichloromethane or chloroform and films were cast from viscous 5 - 10% (w/w) solutions. After slow evaporation of the solvent the films were dried in vacuum (0.2 - 0.5 mmHg) at room temperature and subsequently UV irradiation (Hanau lamp TQ 81, wavelengths 254 and 366 nm) proceeded in a nitrogen atmosphere (the temperature during irradiation was kept constant). The irradiated films were washed with water for 24 hrs. to remove non-crosslinked PEO and with acetone (24 hrs.) for removal of non-crosslinked PPOX as well as DCP and decomposition products like acetophenone and 2-phenylisopropanol. Weight losses were determined and the equilibrium water content after 24 hrs. of swelling was measured. Swelling was calculated using the relationship S = $(W_{eq}/W_{dry} - 1)$.

Films of PEO, PPOX and mixtures containing 4.8% DCP (w/w) for measurements of mechanical properties and protein adsorption were cast from CH_2Cl_2, irradiated for 1 hr. at 50°C and water washed (no acetone washing).

Samples for thermal analysis (DSC) were cast from CH_2Cl_2 without adding DCP. The coatings of PPOX/PEO, PPOX, PTMO and Biomer[R] (Ethicon Inc., Sommerville, N.J., USA) on the inside of glass tubes for the kallikrein generation and APTT test and on glass slides for the blood platelet adhesion test were obtained by covering the glass surfaces with a thin layer of a slightly viscous solution of Biomer[R] (DMF), PTMO (THF), or PPOX, PPOX/PEO (CH_2Cl_2) and allowing the solvent to evaporate. The PPOX containing coatings (with an initial DCP content of 4.8% (w/w)) were irradiated for 1 hr. at 50°C and subsequently washed with water (no acetone washing since this loosens the coating).

Blood vessel prostheses

From a 4% (w/w) solution of PPOX/PEO/DCP : 85.8%/9.5%/4.7% (w/w) in dichloromethane/dichloroethane 70/30 (v/v) fibres with a diameter ranging from 5 to 50 μm were spun and by winding on a rotating axis porous "woven-like" tubings were fabricated (and subsequently UV irradiated during 60 min.). Inner and outer diameter could be varied by changing the axis diameter (0.6 - 2.0 mm) and the spinning time (30 min. - 2 hrs.). The pore size (5 - 200 μm) as well as the fibre diameter could be varied depending on the flow rate of the polymer solution.

Mechanical properties

Stress-strain measurements were made with porous tubings, - length 40 mm, I.D. 1.3 mm, O.D. 1.7 mm -, and dumb-bell shaped test strips, - 6 x 40 mm -, using a tensile testing device (Instron floor model TT-CM, High Wycombe, UK) with drawing rates of 5 mm/min.

Differential scanning calorimetry (DSC)

In the temperature range from 20°C to 100°C thermograms were recorded using a Du Pont DSC 990 at a heating rate of 10°C/min. Indium was used to calibrate the temperature scale and to determine the heat of fusion of PEO, atactic and crystalline PPOX and PTMO. Transition temperatures were determined of PEO, PPOX and PTMO with samples weighing 10 ± 2 mg.

Protein adsorption

Crosslinked films of PPOX/PEO 90/10 were washed with PBS buffer and screened on adsorption of albumin (BSA), fibrinogen, fibronectin and gammaglobulin with an ELISA test[8]. Equilibrium protein adsorption was measured after 100 min. human plasma incubation. Tissue culture polystyrene (Costar, Cambridge, Mass., USA) was taken as a reference material and Biomer[R], cast from DMF (10% (w/w), air side) was used for comparison.

Kallikrein generation[17]

Kallikrein generation could be detected using a chromogenic substrate (AB KABI Diagnostic, type S2302). The acetic acid catalysed release of p-nitroanilin was measured by spectrophotometric analysis at 405 nm. Aliquots of 1 ml 10% citrated human plasma (deluted to 10 ml with Tris buffer, Merck) were incubated at 37°C in polymer coated glass tubes (see film preparation). Within a period of 15 min., – at t = 1,2, 3.5, 5, 7.5, 10, 12.5 and 15 min. –, samples of 100 μl were taken and added to the chromogenic substrate (300 μl Tris buffer, 100 μl S2302 stock solution, 37°C). After addition of 200 μl 50% (v/v) acetic acid solution the coupling reaction of the chromogenic substrate was stopped and the released p-nitroanilin was measured. The reference materials were silica borate glass and SR[19] (RTV Adhesive sealant, General Electric, Bergen op Zoom, The Netherlands), the latter as a coating on glass tubes (from a 5% (w/w) solution in THF, see film preparation).

APTT test[18]

0.5 ml citrated human plasma was incubated at 37°C in polymer coated glass tubes. After 1 min. 50 μl cefalin solution (Sigma RBC, 1/3 stock suspension; 2/3 0.85% (w/w) NaCl solution) was added, followed after 1 min. by 50 μl 0.2 M $CaCl_2$ solution (Merck). The interval between $CaCl_2$ addition and the beginning of coagulation was determined (references: silica borate glass and SR).

Blood platelet adhesion[20]

A cell, consisting of two glass slides, separated by a PVC ring (I.D. 10 mm) was filled without air bubbles with citrated human plasma containing ± 5 x 10^4 platelets/μl. One of the glass slides was coated with the polymer to be tested (see film preparation), the other with SR as a reference. The cell was centrifugated twice, the first time in a centrifugal force towards the polymer coated test surface and afterwards to the reference by turning the cell 180°. The adhered platelets on the polymer surfaces were counted using a microscope (magn. 400x).

RESULTS AND DISCUSSION

In order to study the crosslinking process, films were prepared from PEO ($\bar{M}_w$ = 6 x 10^5) differing in initial DCP content and UV radiation time. Fig. 1 shows a decrease of S (swelling) with increasing DCP concentrations and longer irradiation time. The increasing crosslink density lowers S and the amount of non-crosslinked PEO, which could be determined by washing

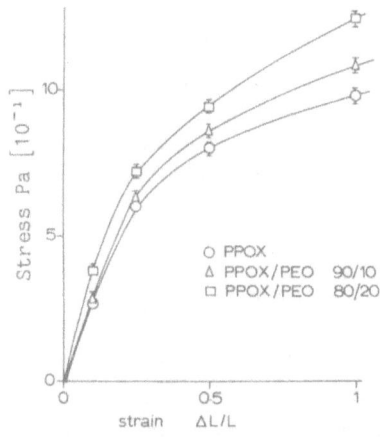
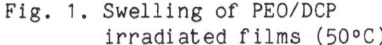
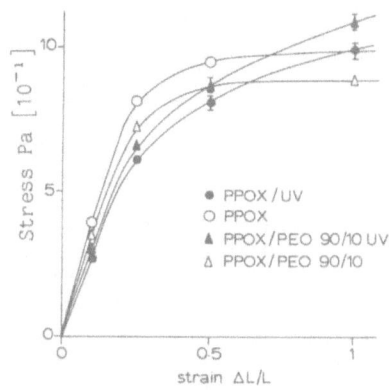

Fig. 1. Swelling of PEO/DCP
irradiated films (50°C)

Fig. 2. Swelling of PPOX/PEO `
irradiated films (50°C)
DCP: 9% (w/w)

with water after irradiation (Table 1). After washing with acetone in order to remove DCP and decomposition products it could be concluded from NMR measurements of the acetone fraction that not all the DCP had been used, - even after 150 min. irradiation -, and that a certain amount of peroxy radicals was probably incorporated in the polymer network.

The temperature dependence of the curing process was investigated by irradiating 10% DCP (w/w) containing films at 20°C, 35°C, 50°C, 75°C and 90°C for 90 min. with UV. At 75°C and 90°C the polymer films deformed and melted (DSC: T_m = 66°C). The 50°C sample gave the highest crosslink density: S = 6.8, whereas at 35°C and 20°C S values of 7.8 and 10 were found respectively, illustrating the influence of the temperature on the crosslinking process. The resulting films had no mechanical strength due to the high water content.

Table 1. Weight losses of PEO/DCP films
in water and acetone

UV Irr. (min)	DCP (%)	Weight losses (%)	
		water	acetone
30	4.8	55	4
90	4.8	30	3
150	4.8	20	2
30	16.7	23	14
90	16.7	13	9
150	16.7	1	6

Attempts were made to reduce the swelling and to improve the mechanical properties by preparing films of PEO and PPOX in the presence of DCP. A linear relationship between the PEO content and S was found for irradiation times of 30, 90 and 150 min. at 50°C and it is obvious from the data in Fig. 2 that the presence of PPOX had resulted in a substantial decrease of the swelling. Weight losses of 15 - 25% after washing with acetone, which dissolves PPOX, DCP and decomposition products were observed but no relationship between weight loss and irradiation time could be determined. Extraction of the films with water to remove non-crosslinked PEO gave 0 - 5% decrease in weight, indicating that nearly all of the PEO has been crosslinked.

Unlike the PEO/DCP samples the PPOX containing films showed elastomeric properties. However, for the preparation of these films a rather high DCP content and UV radiation times of 30 - 150 min. were used. For possible application as biomaterials the concentration of the toxic DCP should be as low as possible and long UV radiation times should be avoided to prevent polymer degradation and incorporation of peroxy radicals in the polymer network. By lowering the DCP content the same S values can be obtained by longer UV irradiation. Films were made from PPOX/PEO with a low DCP content (4.8% (w/w)). Although a short UV radiation (30 min.) was applied, films were obtained with not too high S values.

Stress-strain diagrams (Fig. 3) of these films having a PPOX/PEO ratio of 100/0, 90/10 and 80/20 (irradiated at 50°C) show that the stress for a particular strain increases with an increase in PEO content. Fig. 4 shows the stress-strain curves of PPOX and PPOX/PEO 90/10 films before (open symbols) and after (closed symbols) UV irradiation. The behaviour of the irradiated films differs from the classical theory, which predicts an increase in stress at any particular strain due to crosslinking.

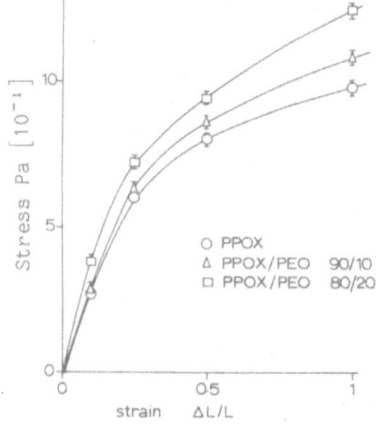

Fig. 3. Stress-strain diagrams of irradiated and water washed PPOX/PEO films

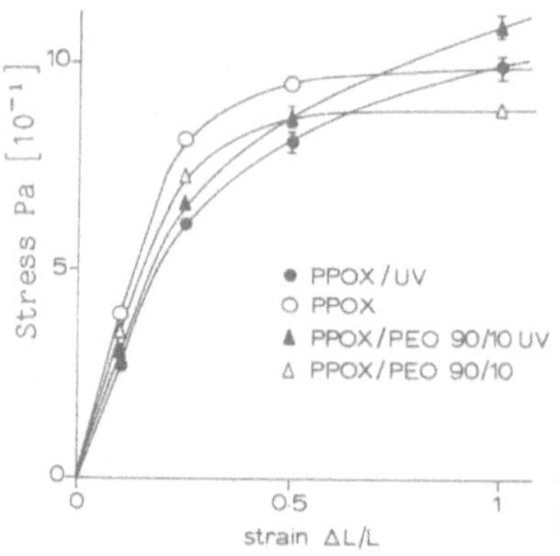

Fig. 4. Stress-strain diagrams of PPOX/PEO films. Influence of the irradiation and water-washing

For the interpretation of the results of the blood material interaction tests information about the morphology of the polymeric films is necessary. It has been reported[21] that the miscibility of PPOX and PEO is restrained to very low molecular weight polymers: PPOX: $\bar{M}_w$ = 425 and PEO: $\bar{M}_w$ = 300. Mixtures of higher molecular weight are incompatible and phase separation occurs. In the compatible mixtures Rastogi et al.[22-23] observed surface activity, which resulted in a migration towards the surface of the component having the lowest surface tension and for binary blends of PPOX and PEO this resulted in an increased surface concentration of PPOX.

In order to get more information about the morphology of our high molecular PPOX/PEO blends, differential scanning calorimetry and ESCA studies are in progress.

Blood material interactions of the polymeric films as coatings on glass were studied with three in vitro tests. The materials were cross-linked films of PPOX and PPOX/PEO 90/10, containing 4.8% DCP (w/w), irradiated with UV for 1 hr at 50°C, and PTMO and Biomer[R] for comparison. The PPOX/PEO 80/20 film (S=0.9) could not properly be attached to glass and

has therefore not been tested. The in vitro tests were performed with
coatings still containing DCP and/or decomposition products, because
washing with acetone could not be applied owing to loosening of the
coatings.

Fig. 5 shows the results of the kallikrein generation test, normalis-
ed to silica borate glass as a strongly activating surface and compared to
silicone rubber, which is non-activating. Kallikrein generation reflects
the activation of factor XII, which generally has been accepted to occur
at negatively charged surfaces. Although no net negative charge is present
on the polyether surfaces, a generation of kallikrein was observed for
these polymeric coatings, possibly resulting from a certain degree of
polarity in the carbon-oxygen bonds. In discussing kallikrein generation
by the polyethers it should also be mentioned that PPOX coatings with and
without DCP gave about the same results, indicating only a small effect of
remaining DCP and/or decomposition products.

The polyethers studied differ in hydrophilicity and crystallinity,
which might result in different factor XII activation. The polymers PPOX
and PTMO are both hydrophobic, - S values of 0.07 and 0.04, respectively
(Sa da Costa found for PTMO: 0.05). From DSC measurements it was concluded
that PPOX was highly atactic, whereas PTMO had a crystallinity of about
70%. Because of the comparable, slightly hydrophilic character of PTMO and
PPOX, the difference in factor XII activation may be attributed to the
cyrstallinity of PTMO.

If crystallinity is assumed to stimulate factor XII activation, the
presence of crystalline PEO in PPOX/PEO 90/10 would have been expected to
result in an increase of kallikrein generation. Merrill et al.[24] have
reported, however, that equilibration of a copolyether-urethane with water
at 25°C depressed the crystal-amorphous transition of the soft PEO segments to temperatures below 0°C, thus creating an "amorphous" hydrophilic
PEO phase.

The presence of PEO in PPOX/PEO 90/10, probably as an amorphous
phase, had a marked effect on the hydrophilicity, with S values from 0.3 -
0.5, depending on the irradiation time. The reduction of kallikrein generation, compared to PPOX and PTMO, seems therefore to indicate a relationship between hydrophilicity and factor XII activation.

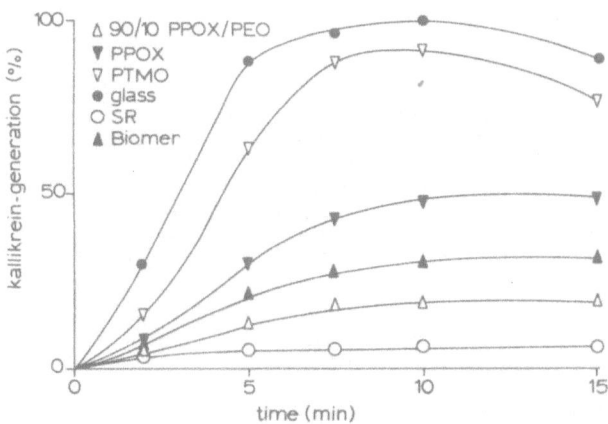

Fig. 5. Kallikrein generation of PPOX/PEO 90/10, PPOX, PTMO, Biomer[R],SR
   and glass.

The results of the APTT test (Table 2), displaying the activation of both the intrinsic and extrinsic coagulation system, show the same trend as the kallikrein generation, giving the highest coagulation time for PPOX/PEO 90/10. Blood platelet adhesion tests (Table 3) confirmed the findings of Lyman[4] and Sa da Costa[6-7], giving the lowest platelet adhesion for PPOX/PEO 90/10.

From the results of the kallikrein generation test, the APTT determination and the platelet adhesion experiments it can be concluded that the blood compatibility of the polyethers, in particular the PPOX/PEO 90/10, is as good as or even better than Biomer[R].

Table 2. APTT values of polyethers, Biomer[R], SR and glass

Polymer	APTT (sec)
Glass	150 ± 20
PTMO	245 ± 40
Biomer[R]	425 ± 20
PPOX	490 ± 25
PPOX/PEO 90/10	580 ± 30
SR	650 ± 30

Table 3. Blood platelet adhesion on polyether, Biomer[R] and SR surfaces

Polymer	platelets/4 x 10^4 μm^2
PTMO	90 ± 20
Biomer[R]	120 ± 30
PPOX	12 ± 4
PPOX/PEO 90/10	5 ± 2
SR	600 ± 70

In order to elucidate the difference in behaviour between the polyethers and Biomer[R], in contact with blood or plasma, protein adsorption was studied using an ELISA test[8]. Fibrinogen deposition (Table 4) was not significant on the polyether surfaces (PPOX/PEO 100/0, 90/10 and 80/20) in contrast to the adsorption on Biomer[R], cast from DMF. Lelah et al.[5] found considerably less fibrinogen adsorption on the surface of extruded Biomer[R] than on Biomer[R], cast from DMA. Fibrinogen adsorption on the polyether surfaces is lower than on extruded Biomer[R], which has been reported by Lelah to be the most thromboresistant.

Table 4. Protein adsorption[a] (A_{450}) on polyether, Biomer[R] and tissue culture polystyrene surfaces

Polymer	Albumin	Fibrinogen	Fibronectin	Gammaglobulin
PPOX	12 ± 10	15 ± 12	10 ± 2	19 ± 15
PPOX/PEO 90/10	8 ± 6	6 ± 1	13 ± 1	9 ± 2
PPOX/PEO 80/20	9 ± 7	6 ± 1	14 ± 10	9 ± 7
Biomer[R]	450 ± 10	480 ± 50	10 ± 10	11 ± 10
TCPS	1900 ± 400			

[a] A_{450} x 10^{-3}

The very low protein adsorption on the polyethers suggests a certain inertness towards protein induced reactions like fibrin and thrombus formation, however, more detailed information has to be derived from in vivo experiments.

For this reason tubings, having a certain porosity, were spun from solution (PPOX/PEO 90/10, 1 hr. UV irradiation at 50°C, DCP: 4.8% (w/w)). Histiopathologic observations of White et al.[25] on porous elastomeric polymer samples in dogs illustrate the influence of porosity on tissue ingrowth: small pores (18 - 45 μm) allow ingrowth of histiocytic tissue whereas larger pores (60 - 120 μm) are ingrown with an organised, fibrous tissue, which provides more structural support to the implant than the former. Effects of the lack of pores in vascular grafts are reported by Lyman et al.[28], who observed fibrotic reactions in solid wall Biomer[R] grafts, also induced by the mismatch in mechanical properties between the prosthesis and the natural vessel.

Hess et al.[12] found a continuous and permanent neo-intima on highly fibrillar surfaces with pore sizes of 20 - 50 μm and claimed that this structure provides the essential anchoring facilities to the cells invading the prosthesis from the anastomotic areas.

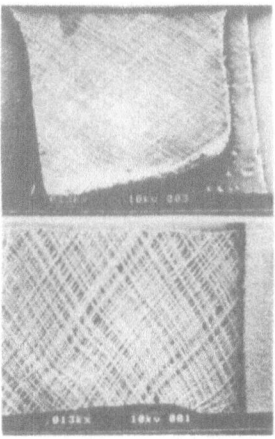

Fig. 6. Lumen (upper) and outside (lower) of a pros-
thesis. (magn. 8x)

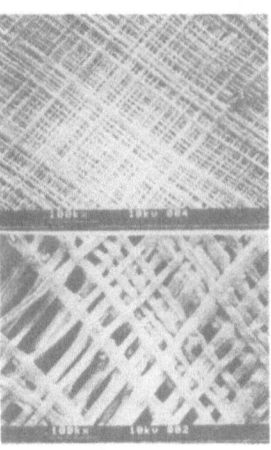

Fig. 7. Inner (upper) and outer
(lower) fibre morpholo-
gy. (magn. 40x)

From these experiments we designed a small diameter blood vessel prosthesis, - inner diameter: 1.3 mm -, having pores of 20 - 40 μm and fibres with diameters of 5 - 20 μm on the lumen side, which both gradually increase to pores of 100 - 200 μm and fibre diameters of 35 - 50 μm at the outside of the prosthesis. This concept allows formation of neo-intima beginning at the anastomoses using the anchoring positions of the fibrillar structure as well as ingrowth of tissue through the prosthesis wall to further support the endothelial layer.

Fig. 6 shows the lumen and outside of the prosthesis at the same magnification, indicating the difference in pore diameter. A closer look (Fig. 7) also illustrates the different fibre morphology. The large fibres are flattened after the spinning due to the presence of high boiling solvent (dichloroethane), which allows some deformation to take place but this is not observed for·smaller fibres, from which the solvent evaporates more easily.

The presence of a high boiling solvent in the solvent mixture is necessary to obtain optimal confluency at the fibre crossings (Fig. 8), which is essential for good mechanical properties like tear strength.

Fig. 8. Confluency at fibre
        crossings (magn. 300x)

Fig. 9. Winding angles of 45°
        (upper) and 60° (lower)

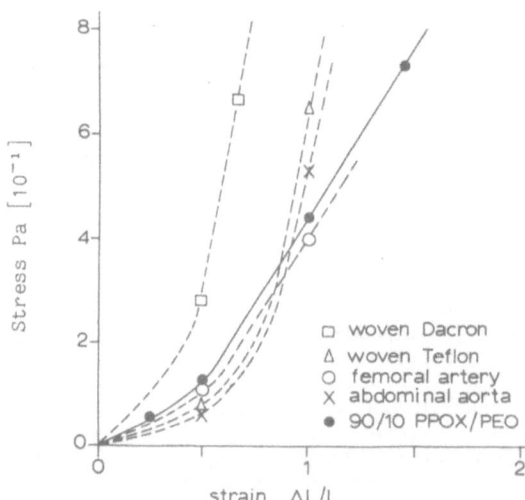

Fig. 10. Longitudinal stress-strain diagram of PPOX/PEO. Winding angle 45°

By changing the winding angle (Fig. 9) the mechanical properties can be influenced. Stress-strain diagrams (Fig. 10) of PPOX/PEO 90/10 (winding angle 45°) indicate the resemblance between the prosthesis and natural vessels[26] like the abdominal aorta and femoral artery: the same low stress for small elongations, when predominantly changes in fibre orientation takes place, is observed. Since for natural blood vessels the longitudinal distensibility tends to be less than the circumferential one, a winding angle of 35 - 40° will result in optimal compliance in both directions, which according to Leidner et al.[27] is advantageous since lower winding angles show less local fibre deformation caused by stress of sutures.

This study shows that by a simple procedure elastomeric products with a reproducible hydrophilicity can be made and in combination with the spinning from solution method, which gives any opportunity to vary the pore size and fibre thickness of the prosthesis, this allows us to fabricate "tailor-made" small diameter blood vessel prostheses.

The encouraging results from the in vitro tests and the good mechanical properties justify more research on the in vivo behaviour, which is performed at the moment with rats (abdominal aorta). The results of this in vivo study and the characterisation of the blood contacting polyether surfaces with ESCA and DSC will be published.

ACKNOWLEDGEMENTS

The authors are indebted to Dr. Olijslager of KRI-TNO, Delft for the blood platelet adhesion tests and to Drs. Buys of TNO Utrecht for assisting in the PPOX synthesis.

This study was subsidized by the Netherlands Organization for the Advancement of Pure Research (ZWO-SON).

REFERENCES

1. J. V. Ihlenfeld, T. R. Mathis, L. M. Riddle and S. L. Cooper, Measurements of transient thrombus formation on polymeric materials, Thromb. Res. 14:953 (1979).
2. S. W. Graham and D. M. Hercules, Surface spectroscopic studies of Avcothane, J. Biomed. Mat. Res. 15:349 (1981).
3. S. W. Kim, R. G. Lee, H. Oster, D. Coleman, J. D. Andrade, D. J. Lentz and D. Olsen, Platelet adhesion to polymer surfaces, Trans. Am. Soc. Artif. Int. Org., vol. XX:449 (1974).
4. D. J. Lyman, K. Knutson, B. McNeill and K. Shibatani, The effect of chem. struct. and surf. prop. of synthetic polymers on the coagulation of blood, Trans. Am. Soc. Artif. Int. Org., Vol. XXI:49 (1975)
5. M. D. Lelah, L. K. Lambrecht, B. R. Young and S. L. Cooper, Physico-chemical characterisation and in vivo blood tolerability of cast and extruded Biomer, J. Biomed. Mat. Res. 17:1 (1983).
6. V. Sa da Costa, D. Brier-Russel, E. W. Salzman and E. W. Merrill, ESCA studies of polyurethanes, J. Coll. Interf. Sci. 80:445 (1981).
7. V. Sa da Costa, D. Brier-Russel, G. Trudel III, D. F. Waugh, E. W. Salzman and E. W. Merrill, J. Coll. Interf. Sci. 76:596 (1980).
8. W. Breemhaar, E. Brinkman, D. J. Ellens, T. Beugeling and A. Bantjes, Preferential adsorption of high density lipoprotein from blood plasma onto biomaterial surfaces, Biomat. 5:269 (1984).
9. Y. Ykada, H. Iwata, F. Horii, T. Matsunaga, M. Taniguchi, M. Suzuki, W. Taki, S. Yamagata, Y. Yonekawa and H. Handa, Blood compatibility of hydrophilic polymers, J. Biomed. Mat. Res. 15:697 (1981).
10. D. L. Coleman, D. E. Gregonis and J. D. Andrade, Blood-materials interactions, J. Biomed. Mat. Res. 16:381 (1982).
11. N. Yui, J. Tanaka, K. Sanui, N. Ogata, T. Okano and Y. Sakurai, Characterisation of the microstructure of PPOX-segmented polyamide and its suppression of platelet adhesion, Polym. J. 16:119 (1984).
12. F. Hess, C. Jerusalem, P. Grande and B. Braun, Significance of the inner surface structure of small caliber prosthetic blood vessels in relation to the development, presence and fate of a neo-intima, A morphological evaluation, J. Biomed. Mat. Res. 18:745 (1984).
13. V. Sa da Costa, Thesis, 1979, Mass. Inst. Techn., USA.
14. H. C. W. M. Buijs, US. pat. 3.798.249.
15. W. R. Sorenson and T. W. Campbell, Preparative methods of polymer chemistry, Interscience, New York, 1968, USA.
16. E. L. Muetterties, US. pat. 2.748.145, 2.856.370.

17. W. E. Hennink, Thesis, 1985, Techn. Univ. Twente, The Netherlands.
18. L. Sederel, Thesis, 1982, Techn. Univ. Twente, The Netherlands.
19. M. J. Gallimore and P. Friberger, Simple chromogenic peptide substrate assays for determining prekallikrein, kallikrein inhibition and kallikrein "like" activity in human plasma, Thromb. Res. 25:293 (1982).
20. J. M. George, Direct assessment of platelet adhesion to glass, Blood 40:862 (1972).
21. S. Krause, Polymer-polymer compatibility, in: "Polymer Blends", D. R. Paul and S. Newman, Academic Press, New York (1978).
22. K. Rastogi and L. E. St. Pierre, Interfacial phenomena in macromolecular systems V, J. Coll. Interf. Sci. 35:16 (1971).
23. K. Rastogi and L. E. St. Pierre, Interfacial phenomena in macromolecular systems III, J. Coll. Interf. Sci. 31:168 (1969).
24. E. W. Merrill, E. W. Salzman, S. Wan, N. Nahmud, L. Kushner, J. N. Lindon and J. Curme, Platelet-compatible hydrophilic segmented polyurethanes from polyethylene glycols and cyclohexane diisocyanate, Trans. Am. Soc. Artif. Int. Org. Vol XXVIII: 482 (1982).
25. R. A. White, F. M. Hirose, R. W. Sproat, R. S. Lawrence and R. J. Nelson, Histiopathologic observations after short-term implantations of two porous elastomers in dogs, Biomat. 2:171 (1981).
26. M. Hasegawa and T. Azuma, Mechanical properties of synthetic arterial grafts, J. Biomechn. 12:509 (1978).
27. J. Leidner, E. W. C. Wong, D. C. MacGregor and G. J. Wilson, J. Biomed. Mat. Res. 17:229 (1983).
28. D. J. Lyman, F. J. Fazzio, H. Voorhees, G. Robinson and D. Albo, Compliance as a factor effecting the patency of a copolyurethane vascular graft, J. Biomed. Mat. Res., 12:337 (1978).

IMMOBILIZATION OF LIVING CELLS IN BIOCOMPATIBLE SEMIPERMEABLE MICROCAPSULES:

BIOMEDICAL AND POTENTIAL BIOCHEMICAL ENGINEERING APPLICATIONS

Mattheus F.A. Goosen*, Geraldine M. O'Shea,
Hrire M. Gharapetian and Anthony M. Sun

Connaught Research Institute
1755 Steeles Avenue West
Willowdale Ontario, Canada
M2R 3T4

*Department of Chemical Engineering
Queen's University
Kingston, Ontario, Canada
K7L 3N6

INTRODUCTION

At the present time, the clinical and industrial development of immobilized cell technology is constrained not by a lack of opportunity but rather by a lack of suitable techniques for immobilizing viable plant, animal and microbial cells. In the biomedical engineering area, for example, several methods have been developed for microencapsulating biologically active molecules, tissues and cells so that they remain viable and in a protected state within a semipermeable membrane which permits passage of low molecular weight substances, such as nutrients and oxygen, but not of cells and high molecular weight proteins (1-8). However, there have been no successful attempts to develop semipermeable microcapsules which have long-term biocompatibility with the body tissues and which are impermeable to the components of the immune system. As a result, survival times of transplanted microencapsulated tissue or cells in-vivo have consistently been less than three weeks,(7, 9-12) severely limiting the usefullness of the encapsulation procedure in the treatment of diseases such as diabetes.

The industrial development of immobilized cell technology, on the other hand, is constrained by a lack of suitable bioreactors for handling bacterial, plant and animal cells and by a lack of suitable bioseparation procedures for handling the highly complex (and high value) products such as proteins and alkaloids (13, 14). In addition to the technology limitation there is the concern in Canada, the U.S. and Europe that the number of appropriately-trained biochemical engineers (in fermentation, cell culture and bioseparation) will be inadequate to meet the needs of emerging Biotechnology industries.

Recent studies have indicated that a novel alginate-poly-1-lysine (PLL) microencapsulation procedure, has great clinical potential in the protection of transplanted living cells and tissue (9-12,15-17). In experimental animals, for example, a single intraperitoneal transplant of encapsulated islets reversed the diabetic state for more than one year. In contrast, a single injection of unencapsulated islets was effective for less than two weeks. In addition, the encapsulation technique also has great industrial potential in the development of new membrane-bioreactors for the production of high-value biologicals and pharmaceuticals from immobilized living cells (15).

MICROENCAPSULATION OF LIVING CELLS

The microencapsulation of viable cells such as islets of Langerhans within an alginate-PLL-alginate membrane has been described in previous studies (9,12,15). Briefly, islets were suspended in 2 mL of 1.5 % (w/v) sodium alginate (Kelco-Gel LV, Kelco Specialty, Colloids Ltd., Toronto, Canada) in 0.85 % NaCl at a concentration of 1000 islets/mL. Spherical droplets of this suspension were formed by an air jet-syringe pump droplet generator. With this apparatus, the cell-sodium-alginate suspension is extruded through a needle located inside a sheathed tube through which air flows at a controlled rate. As liquid droplets are forced out of the end of the needle by the syringe pump, the droplets are pulled off by the shear forces set up by the rapidly flowing air stream.

In some experiments a specially designed electrostatic droplet generator was used in place of the air-jet droplet generator. With this device, the sodium alginate liquid, with or without cells, is place in a syringe to which is attached a stainless steel needle, bevelled on all sides. The syringe is attached to a syringe pump and a high voltage wire is attached to the needle. A second wire of opposite polarity is attached to the collecting vessle containing calcium chloride. The needle is located at a specified distance, usually 10 mm, from the top of the liquid in the collecting vessel. The electrostatic generator itself consists of a power supply, logic circuitry and a console panel for controlling pulse voltage, pulse frequency and pulse length. The pulse voltage, for example, determines the strength of the force pulling the droplets from the end of the needle. The pulse frequency determined how many pulses can be applied to the droplet per unit time, and the pulse length determines the length of time for which the droplet-forming force can be applied. Unless otherwise indicated, the air-jet droplet generator was used for the formation of microcapsules.

Spherical droplets were collected in 1.5 % $CaCl_2$ where they gelled. Following washing steps with 30 mL volumes of 0.1 % CHES [2 (N-cyclohexyl-amino) ethanesulfonic acid] and 1.1 % $CaCl_2$, the calcium alginate droplets were reacted with PLL, a positively charged poly-electrolyte, by suspension in 30 mL PLL solution for six minutes [0.05 % (w/v) Mv = 1.7 x 10^4; Sigma, St. Louis, MO)]. The resulting capsules were washed with 30 mL volumes of 0.1 % CHES, 1.1 % $CaCl_2$, and 0.85 % NaCl and suspended for four minutes in 0.03 % sodium alginate which formed the outer layer of membrane. Further washing with 0.85 % NaCl was preceded by the treatment of the capsules with 0.05 M sodium citrate, pH 7.4, for six minutes which liquified the gel inside the capsules. The excess citrate was removed by two NaCl washes and the encapsulated islets cultured at 37 °C. About 50 % of the capsules (900 µm diameter) contained islets (1 to 2 islets/capsule). Compared to an earlier microencapsulation procedure employing an alginate-PLL-polyethyleneimine (PEI) capsule membrane (7,9) several modifications were made; the initial PLL/alginate reaction time was increased from three to six minutes, the PLL concentration was increased from 0.03 to 0.05 % (w/v), and the PEI was replaced by alginate. In addition, specially purified sodium alginate (Kelco-Gel LV, Kelco Specialty Colloids Ltd., Toronto, Ont. Canada) was used for all microcapsule preparations.

PHYSICO-CHEMICAL PROPERTIES OF MICROCAPSULES

The shape of the calcium alginate droplets and hence the final microcapsules was assessed by extruding sodium alginate solutions of varying concentrations 0.3 to 1.5 % into a 1.1 % w/v calcium chloride hardening solution. The gel droplets which formed were examined microscopically for the presence of surface imperfections such as tails and striations. The concentration of the alginate solution was related to the viscosity using a viscosity/concentration curve (Kelco Algin Book, Second Edition).

The shape of the calcium alginate gel droplets was found to be critically dependent on the initial sodium alginate concentration (or viscosity) used in the microencapsulation procedure. At sodium alginate concentrations of 1.2 % (w/v) or greater, the gel droplets were perfectly spherical showing no surface roughness or irregularities. However, at a concentration of 1.1 % about 75 % of the capsules had very small tails. At lower concentrations, all of the capsules had tails and the size of these tails along with surface irregularities and striations progressively increased with a decrease in alginate concentration. While in all instances, the droplets could be broadly described as spheroidal, it was only at concentrations of sodium alginate solution of 1.2 % and above, i.e. viscosities of 30 cps and above, that perfect spheres were formed (Table 1). In one experiment, the purified alginate was replaced by the cruder alginate solutions 0.6 - 0.8 % w/v (Sigma) used in previous studies (7-9). The resultant gel droplets not only showed tails and striations, but also had a granular texture with greater surface irregularities.

In order to assess the effect of the viscosity average molecular weight ($\bar{M}v$) of the PLL on the permeability of the capsule membrane, seven types of microcapsules (with or without cells) were prepared using seven different PLL preparations ($\bar{M}v$ of 4,000; 17,000; 25,000; 40,000; 90,000; 190,000 and 400,000). Bovine serum albumin ($\bar{M}n$ 67,000) was added to a saline solution containing 5 mL of empty microcapsules to give a total volume of 7 mL. The change in absorbance of the solution at 280 nm was measured at room temperature as a function of time. In another experiment, 0.05 M sodium citrate buffer was added to the microcapsules containing red blood cells causing the red blood cells to rupture thus releasing haemoglobin ($\bar{M}n$ 66,000) into the interior of the microcapsules. The diffusion of haemoglobin out of the microcapsules was assessed by measuring the change in absorbance (540 nm) of the supernatant.

Bovine serum albumin was observed to diffuse readily into microcapsules prepared with high molecular weight PLL as demonstrated by the rapid decrease in absorbance of the albumin solutions containing the empty microcapsules (Fig. 1). The higher the PLL molecular weight used in the microcapsule preparation, the more rapid the absorbance decrease. Similarly, the microcapsules prepared with high molecular weight PLL were found to be very permeable to haemoglobin as demonstrated by the rapid increase in absorbance of the solution containing the microencapsulated ruptured red blood cells. The higher the PLL molecular weight used in the microcapsule preparation, the more rapidly the absorbance of the solution increased.

This study, therefore, showed that the permeability of the microcapsule could be controlled by varying the molecular weight of the PLL. The permeability of the microcapsules, as measured with bovine serum albumin

Table 1. Dependence of microcapsule morphology on viscosity of
sodium alginate solution

Sodium alginate		Fraction of microcapsules which are perfectly spherical and smooth (%)
Concentration [% (w/v)]	Viscosity (cps)	
1.5	51	100
1.4	43	100
1.3	36	100
1.2	30	100
1.1	25	<25
1.0	20	0
0.9	16	0
0.7	11	0
0.3	4	0

and haemoglobin, decreased with a decrease in the PLL used in the
microcapsule preparation. Only the microcapsules prepared with PLL of
Mv 17,000 were impermeable to albumin, haemoglobin and immunoglobulins
indicating a molecular weight cut-off below 67,000. Since no
microcapsules could be formed with PLL of 4,000 and since the
microcapsules prepared with PLL of 40,000 were permeable to both albumin
and haemoglobin, the ideal PLL molecular weight for transplantation
purposes appears to be close to 17,000.

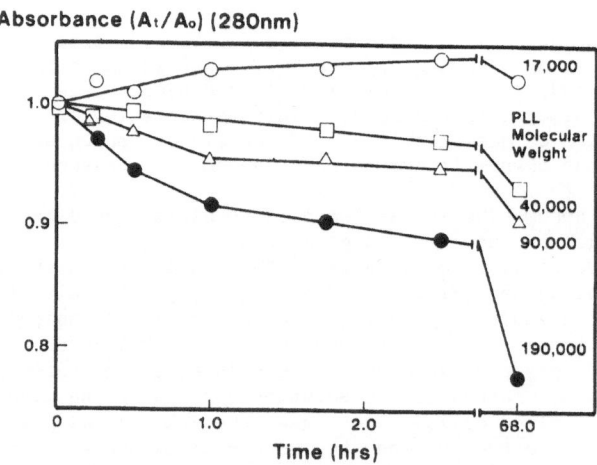

Figure 1. Diffusion of bovine serum albumin (molecular weight of
6.7×10^4 into microcapsules (without cells) prepared with different
molecular weights (Mv) of PLL; (At/Ao) absorbance of solution at any
time/initial absorbance.

When the microcapsules were placed into the sodium citrate buffer in order to liquify the calcium alginate gel core, they expanded; the increase in microcapsule diameter being directly proportional to the PLL molecular weight used in the microcapsule preparation; the higher the PLL molecular weight the greater the increase in microcapsule diameter (Fig. 2). This effect was observed for both empty microcapsules and microcapsules containing cells. The extent of expansion of the capsule membrane, as a function of PLL molecular weight, can be explained as follows: the longer the PLL chains, the more difficult it is for them to penetrate the alginate gel matrix. This results in fewer PLL/alginate ionic interactions or cross-links for the higher molecular weight PLL chains. Therefore, when the unbound alginate gel inside the capsule is liquified, there would be a larger increase in capsule diameter for the microcapsules prepared with higher molecular weight PLL.

In order to demonstrate that the capsule wall was impermeable to the components of the immune system, normal serum immunoglobulin (Connaught Labs. Ltd., Toronto) was added to a solution containing 5 mL empty microcapsules prepared with PLL of 17,000 or 25,000 to give a total volume of 7 mL. The change in absorbance of the solution at 280 nm was measured as a function of time. The microcapsules were found to be impermeable to immunoglobulins since no change was observed, over a 24 hour period, in the absorbance of the immunoglobulin solution containing the microcapsules.

The evaluation of the surface finish of the microcapsules by scanning electron microscopy (SEM) revealed essentially smooth interior and exterior capsule surfaces. Dry capsule wall thickness ranging from 0.2 to 0.4 μm were observed. Using interferometry, for the microcapsules prepared with PLL of Mv of 25,000, the two-wall thickness of individual dried microcapsules was determined to be 1.72 ± 0.48 microns. Using an image shearing method, the wet wall thickness was found to be 5.23 ± 0.77 microns (Table 2).

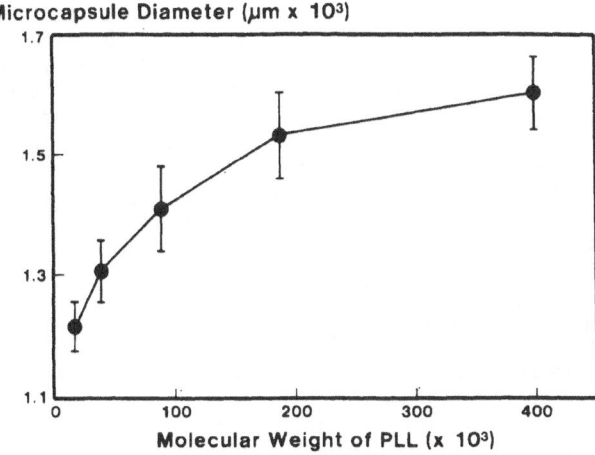

Figure 2.   Effect of molecular weight of PLL on the final size of the micro-capsules.   The microcapsules expanded in sodium citrate;   this expansion was directly in proportion to the PLL molecular weight used in the micro-encapsulation preparation.

Table 2. Dependence of Microcapsule Wall Thickness and
Water Content on PLL Molecular Weight

PLL		CAPSULE MEMBRANE		
$\overline{M}v$ (x10³)	CONC. (%)(w/v)	WALL THICKNESS μm		WATER CONTENT (%)
		Dry*	Wet	
25	0.05	1.72±0.48	5.23±0.77	83.5
90	0.05	0.51±0.03	6.24±0.27	91.3
25 4 }	0.10	1.24±0.31	5.11±0.52	87.9

*two wall thickness

Using the electrostatic droplet generator the size of the microcapsules could be reduced from 900 microns to less than 500 microns with the same gauge needle (Table 3).

Based on these results the wet wall thickness of the microcapsules was found to be 14 times greater than the dry wall thickness indicating that the capsule wall, 5.0 μm thick, is a hydrogel containing approximately 85 % w/w water. Many hydrogels have gained general acceptance as being biocompatible materials since, as implants, they are tolerated by the body fluids (18-22). It appears that the low interfacial tension between the swollen gel surface and the aqueous biological environment minimizes protein interaction. A strong protein/polymer interaction may serve as a trigger mechanism for the rejection of implanted polymer by the body (20). The soft rubbery consistency (high compliance) of most hydrogels may also contribute to their biocompatibility by reducing frictional irritation to surrounding tissues (21).

The SEM studies showed that the alginate-PLL-alginate microcapsules have a relatively smooth surface. This smoothness could be due to the outer alginate membrane and/or to the expansion (stretching) of the microcapsules in the citrate incubation step. The smoothness of the outer surface would inhibit surficial cell attachment, enabling the microcapsules to remain semipermeable and effective inside the body for extended periods of time. The high water content and negative charge of the outer alginate membrane would also inhibit surficial cell growth. Based on animal experiments, very little cell growth has been observed on the surface of alginate-PLL-alginate microcapsules recovered from diabetic animals up to 10 months post-implantation (12).

240

BIOMEDICAL APPLICATIONS: TRANSPLANTATION OF MICROENCAPSULATED ISLETS IN DIABETIC ANIMALS

Compared to alginate-PLL-PEI microcapsules, transplantation of islets of Langerhans encapsulated in alginate-PLL-alginate membranes produced surprisingly different results (Fig. 3). A single transplant of 4.5×10^3 encapsulated islets restored normoglycemia in all recipients within two days. In a different experiment (12), of the five animals which received transplants, three remained normoglycemic for more than 100 days. One of these, three animals still remained normoglycemic 240 days post-transplantation and one had not returned to the diabetic state when sacrificed 365 days post-transplantation. Free capsules were recovered from the abdominal cavity of individual animals at 90, 156 and 365 days post-transplantation. The capsules were intact and were shown by histological studies to contain viable islet cells. Furthermore, these recovered islets secreted insulin in culture in response to a glucose challenge. In only one case, did the transplant fail within one month; on autopsy, it was found that the implanted capsules were enclosed in fibrous tissue at the implantation site. This may have been caused by the introduction of some contaminant with the implant. Transplant recipients gained weight steadily during their normoglycemic period, had normal urine volumes and their eyes showed no evidence of cataract development, indicating good control of the diabetic state. In contrast, untreated diabetic control rats developed cataracts within a few months. Diabetic control rats receiving single injections of unencapsulated islets were normoglycemic for less than 10 days.

These studies demonstrated for the first time that islet allografts could survive in vivo for up to 12 months when encapsulated in a protective, durable and biocompatible membrane. Using earlier encapsulation procedures, allografts survived for no longer than three weeks (7) and no intact capsules or viable islets were recovered from the recipients. Microencapsulation of the islets obviates the need for any immunosuppressive therapy for the recipients or the pre-treatment or culturing of the islets prior to transplantation. Furthermore, if diabetes has an auto-immune etiology, the capsule membrane would protect the islets from destruction in the event of a recurrence of the initial disease. Experiments with higher animals and xenografts are currently underway.

POTENTIAL BIOCHEMICAL ENGINEERING APPLICATIONS: IMMOBILIZED CELL BIOREACTORS AND NOVEL BIOSEPARATION PROCEDURES

In biochemical engineering one of the major research efforts over the the past two decades has been the development of immobilized enzymes for catalyzing specific reactions (23). Free enzymes were not ideal as industrial catalysts for biotransformation reactions due to their instability in, for example, organic solvents or at high temperatures. However, for the immobilization of enzymes, it is first necessary to extract the enzymes from the cells. In order to avoid this procedure the technique for the direct immobilization of whole microbial cells was developed. B. Subtilis cells entrapped in polyacrylamide or K-carrageenan, for example, were used for the increased production of the enzyme α-amylase (24,25). Z. mobilis cells immobilized on an ion exchange resin were used to give higher ethanol productivities (26). However, there are several drawbacks to the currently-used whole-cell immobilization techniques such as loss of cells due to leakage in the carrier-binding method, the increase in mass transfer resistance in the gel entrappment method and the unsuitability of the chemical cross-linking method for the immobilization of living cells (23). On the

Table 3. Effect of Needle Size and Distance from Needle Tip to Calcium Chloride Solution the Diameter of Alginate Gel Droplets Produced by the Electrostatic Droplet Generator

NEEDLE GAUGE	DISTANCE FROM NEEDLE TIP TO $CaCl_2$ SOLUTION (mm)	CALCIUM ALGINATE GEL DROPLETS	
		Fraction Spherical (%)	Diameter (μm)(± 5.0)
22	30	100	1275±79
22	15	100	579±72
22	7.5	100	495±73
22b	15	100	420±42
26	15	100	313±45
26b	15	100	173±18

b = bevelled on all sides

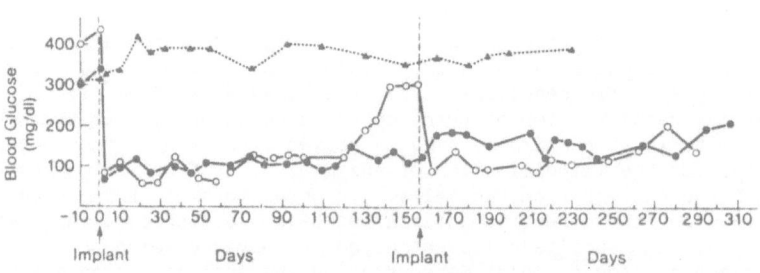

Figure 3. Fasting plasma glucose concentrations of diabetic rats after intraperitoneal transplantation of microencapsulated islets. Three Weistar Rats were made diabetic by the injection of streptozotocin (65 mg/kg). Two weeks after the development of diabetes, two of them received a single intraperitoneal injection of 4.5×10^3 encapsulated islets. One of the rats received a second transplant two weeks after regressing to the diabetic state.

other hand, the alginate-PLL microencapsulation technique which involves very mild reaction conditions and results in the formation of a hydrogel capsule membrane, overcomes many of the drawbacks of previous cell immobilization procedures (15). In addition, due to the semi-permeable nature of the capsule wall, the technique has the potential of providing a new and improved procedure for the separation and recovery of the final product. For example, it should be possible to develop new bioseparation procedures by exploiting the semipermeability of the capsule membrane, to selectively entrap and concentrate high molecular weight products. The microcapsules could then be separated from the medium and ruptured, thus releasing the concentrated product.

Plant cells in free suspension or in immobilized form also show great promise for the production of a variety of biochemicals such as pharmaceuticals, food colours and fragrances (27, 28). Aside from providing better control over product supply and quality, immobilized plant cells have shown increased product yield and a prolonged production phase compared to free cells in suspension (29). On the other hand, in other studies involving the hydroxylating capacity of alginate gel immobilized plant cells (32), the hydroxylation reaction had to be stopped after 60 days due to the disintegration of the gel. In addition, due probably to mass transfer limitations, the hydroxylating capacity of the immobilized cells was found to be only one-half that of free cells in suspension. By encapsulating the cells it may be possible to obtain increased product yield, longer reaction times and greater reaction capacities as a result of the greater stability and lower mass transfer resistance of the microcapsule compared to the alginate gel matrix.

While small quantities of immobilized cells may be readily grown in shakeflasks, larger quantities require a fermentor. The scale-up of bioreactors, however, usually results in additional problems, such as, for example, increased pressure drop due to cell over growth in packed-bed reactors (30, 31). Compared to microbial cells, plant cells are relatively sensitive to shear forces, due to their large size and fine cell wall, and therefore, aside from air-lift fermentors, most conventional fermentors cannot be used (29). Through immobilization, plant cells, may be protected from shear forces by the polymeric matrix (32-34). The major disadvantage of immobilized plant cells, however, is that most plant products are stored intracellularly and efficient product release is therefore essential to the performance of an immobilized-cell reactor. Product release maybe obtained, for instance, through variation in pH and the use of permeabilizing agents such as dimethyl sulfoxide (DMSO), (29, 34). Brodelius in a recent article has indicated that alginate-entrapped C. roseus cells may be suitable for intermittant release of ajamlicine by permeabilization with 5 % DMSO (29). Intermittant product release and bioconversion using microencapsulated plant cells have not, as yet, been studied.

The development of microencapsulated living cell bioreactors for the production of high value pharmaceuticals, biologicals and biochemicals would help to demonstrate the advantages of microencapsulation over standard gel immobilizaton. It should be possible, for example, to demonstrate that the semipermeable capsule membrane can not only protect shear-sensitive plant cells against high shear forces caused by rapidly flowing medium in a bioreactor but also against bacterial contamination. By controlling the permeability of the capsule membrane, mass transfer resistances would be minimized. Encapsulated cells could be used for biotransformation reactions of high value drugs such as β-methyl-digitoxin (a cardiac drug) and for the semi-continuous production of high-value biochemicals in moving-bed column bioreactors, through intermittant release of intracellular product by permeabilization. In the latter case, microcapsules and medium would be added to the bottom of the column bioreactor, the microcapsules would move slowly up the column. Medium would be continuously recirculated throught the slower moving microcapsule bed (to reduce mass transfer limitations). At the top of the reactor the capsules would be removed and transferred to 5 % DMSO to release the product. The encapsulated cells would then be removed from the DMSO/product solution and returned to the bottom of the reactor.

In summary, it has been shown that by optimizing the alginate PLL microencapsulation parameters, durable, semipermeable capsules may be produced which are perfectly smooth and spherical and which are effective in treating diabetic animals . Experiments are presently underway with labelled and unlabelled proteins to detemine the exact molecular weight cut-off of the different PLL capsule membranes. The microencapsulation of living cells in protective, semipermeable membranes has great clinical potential as a new form of treatment for diseases such as diabetes. Furthermore, the procedure also has great industrial potential in the development of new types of bioreactors for the production of high-value pharmaceuticals and biochemicals from immobilized living cells.

Acknowledgements

We wish to thank Ms. H. van Rooy and Mrs. A. Wood for their technical assistance. Dr. S. Chou has been helpful throughout the studies. Dr. P. Anderson of KMS Fusion Inc. must be acknowledged for his help. Ms. S. Peters must be thanked for typing the manuscript. This work was supported in part by the Medical Research Council of Canada, the Canadian Diabetes Association, the Juvenile Diabetes Foundation International and the Natural Sciences and Engineering Research Council of Canada.

References

1. T.M.S. Chang, Semipermeable microcapsules, Science 146, 545, (1964).

2. K. Mosbach and R. Mosbach, Entrapment of enzymes and microogranisms in synthetic cross-linked polymers and their application in column techniques, Acta Chem. Scand. 20: 2807, (1966).

3. T.M.S. Chang, F.C. MacIntosh, and S.G. Mason, Semi-permeable aqueous microcapsules, Can. J. Physiol. and Pharm., 44: 115, (1966).

4. V. Hackel, J. Klein, R. Megret and F. Wagner, Immobilization of microbial cells in polymeric matrices, Europ. J. Appl. Microbiol., 1: 291, (1975).

5. M. Kierstan and C. Bucke, The immobilization of microbial cells, subcellular organelles, and enzymes in calcium alginate gels, Biotech and Bioeng., 19: 387, (1977).

6. M.F. Sefton and R.L. Broughton, Microencapsulation of erythrocytes, BBA, 717, (1982).

7. F. Lim and A.M. Sun, Microencapsulated islets as bioartificial pancreas, Science, 210: 908, (1980).

8. F. Lim and R.D. Moss, Microencapsulation of Living Cells and Tissues J. Pharmacol. Sci., 70: 351, (1981).

9. A.M. Sun, G.M. O'Shea and M.F.A. Goosen, "Development and in-vivo testing of an artificial endocrine pancreas", in: Biocompatible Polymers, Metals and Composites. Szycher M. (Ed.) Technomic Publ.; Chapter 40: 929, (1983).

10. A.M. Sun, G. O'Shea, H. van Rooy and M. Goosen, Microencapsulation d'ilots de langerhans et pancreas artificial, J. Ann. Diabetologie (Hotel-Dieu) 20: 161, (1982).

11. Y.F. Leung, G.M. O'Shea, M.F.A. Goosen and A.M. Sun, Microencapsulation of crystalline insulin or islets of Langerhans: an insulin diffusion study, Artificial Organs, 7: 208, (1983).

12. G.M. O'Shea, M.F.A. Goosen and A.M. Sun, Prolonged survival of transplanted islets of Langerhans encapsulated in a biocompatible membrane, BBA, 804: 113-118, (1984).

13. A.E. Humphrey, Comercializing Biotechnology: Challenge to the Chemical Engineer, Chemical Engineering Progress, December 7-12, 1984.

14. J.F. Kennedy, A Future for Immobilized Cell Technology, Nature, 299: 777-778 (1982).

15. M.F.A. Goosen, G.M. O'Shea, H.M. Gharapetian, S. Chou and A.M. Sun, Optimization of microencapsulation parameters: semi-permeable microcapsules as a bioartificial pancreas, Biotechnology and Bioengineering, 27: 145-150 (1985).

16. A.M. Sun, G.M. O'Shea and M.F.A. Goosen, Injectable, biocompatible islet microcapsules as a bioartificial pancreas, Proc. 4th Congress I.S.A.I.O., Kyoto, Japan, (1983).

17. M.F.A. Goosen, G.M. O'Shea, S. Chou, A.M. Sun, Physicochemical properties of injectable microcapsules as bioartificial organs, Proc. 33rd Can. Chem. Eng. Conf., Toronto, Canada, 2: 441-446, (1983).

18. A.S. Hoffman, S.R. Hanson, L.A. Harker, B.D. Ratner, T.A. Horbett and L. Reynolds, In-vivo evaluation of artificial surfaces using a baboon model of arterial thrombosis, A.I.Ch.E. Annual Meeting, Chicago, Nov. 16-20, (1980).

19. D.E. Gregonis, G.A. Russell, J.D. Andrade and A.C. de Visser, Preparation and properties of stereoregular poly (hydroxyethyl methacrylate) polymers and hydrogels, Polymer, 19: 1279: 1978.

20. A.S. Hoffman, G. Schmer, G. Harris and W.G. Kraft, Covalent binding of biomolecules to radiation-grafted hydrogels on inert polymer surfaces, Trans. Amer. Soc. Artif. Intern. Organs, 18:10, (1972).

21. H. Scott, P.L. Kronick and E.E. Hillman, "Active-vapour grafting of hydrogels in medical prosthesis", Annual Report, PB206 499, National Technical Information Service, Springfield, Va., August (1971).

22. M.F.A. Goosen and M.V. Sefton, Properties of a heparin-poly (vinyl alcohol) hydrogel coating, J. Biomed. Mater. Res., 17: 359, (1983).

23. I. Chibata, T. Tosa and M. Fujimura,"Immobilized Living Microbial Cells" in Ann. Report on Ferm. Proc., 6: 1-22 (1983).

24. P.S.J. Cheetham, C.E. Imber and J. Isherwood, The Formation of Isomaltulose by Immobolized Erwinia Rhapontici Nature, 299: 628-631, (1982).

25. A. Shiumyo, H. Kimura and H. Okada, ---, Eur. J. Appl. Microb. Biotech. 14: 7, (1982).

26. T.A. Krug and A.J. Daugulis, Ethanol Production Using Zymononas Mobilis Immobilized on an Ion Exchange Resin, Biotechnol. Letters, 5: 159-164, (1983).

27. O. Shahai and M. Kmith, Commercializing Plant Tissue Culture Processes: Economics, Problems and Prospects. Biotechnol. Progress, 1: 1-9, (1985).

28. W.G.W. Kurz and F. Coustabel, ---, Adv. in Applied Microbiology, Perlman, (ed) A.P., New York, 25: 209-240 (1979).

29. P. Brodelius, Production of biochemical with immobilized plant cells: possibilities and problems, Annals New York Acad. Sciences, 413: 383-393, (1983).

30. A.J. Daugulis, T.A. Krug and C.E.T. Choma. Filament Formation and Ethanol Production by Zymomonas mobilis in Adsorbed Cell Bioreactors, Biotech. and Bioeng. XXVII, (1985).

31. A.J. Daugulis, T.A. Krug and C.E.T. Choma, "Scale Up of an Adsorbed Cell Bioreactor for Ethanol Production" in Fifth Can. Bioenergy R and D Seminar, March 26-28, Ottawa, S. Hasnain (ed), Elsevier Sci. Publ., 499-502, (1984).

32. J.E. Prenosil and H. Pedersen, Immobilized Plant Cell Reactors, Enzyme Microb. Technol., 5: 323-331, (1983).

33. I.A. Veliky and A. Jones, Bioconversion of gitoxigenin by immobilized plant cells in a column bioreactor, Biotechnology Lett., 3: 551-554, (1981).

34. M.L. Shuler, O.P. Sahai and G.A. Hallsby, Entrapped Plant Cell Tissue Cultures, Ann. N.Y. Acad. Sci., 413: 373, (1983).

CHARACTERIZATION OF MEMBRANES FOR ARTIFICIAL ORGANS

Erhan Pişkin

Chemical Engineering Department
Hacettepe University
Ankara, Turkey

INTRODUCTION

Exceptional advancement in polymer science and membrane technology during the past decade has made possible an enormous variety of biological applications for membrane-moderated devices. Conventional membrane separation techniques, i.e., microfiltration, ultrafiltration, reverse osmosis, dialysis, electrodialysis and gas permeation, have already captured the attention of several industries (e.g., chemical, pharmaceutical, medical, biological and food industries, biotechnology, etc.). Many novel applications for today's membranes are opening new horizons in the research, clinical and processing fields.

Today, a suitable membrane separation device is available for almost all kinds of separation problems. The young membrane seperation processes are nicely employed in concentration, purification, separation, fractionation of effluents which contains valuable and especially heat sensitive (e.g. biological species) substances. The literature bearing upon the membrane separation processes is enormous and widely spread over different fields. There are excellent reviews and books dealing with these processes. This paper, does not attempt to give any further details of the conventional membrane processes and neither their applications. Instead, the emphasis of this section is on the description of another novel and very exciting applications of membrane moderated systems, i.e. in artificial organs technology,which have already emerged in clinical medicine. Selected applications of polymeric membranes in the field of artificial organs including membrane oxygenators, hemodialysis, hemofiltration, plasmapheresis, hemoperfusion, microcapsules (or so-called artificial cells) are considered by giving some insight into the subject.

MEMBRANES

Over the years, polymeric films have been employed as barriers to the free transmission of the migrating species. Polymers in film form in which not only over-all impermeability (or permeability) is important, but also their ability to discriminate is employed in the transmission of species on the basis of several factors, e.g. size, shape, chemistry,charge, etc. Membranes have been considered as specific polymeric films, or more likely as imperfect barriers between two environments which impede the permeation

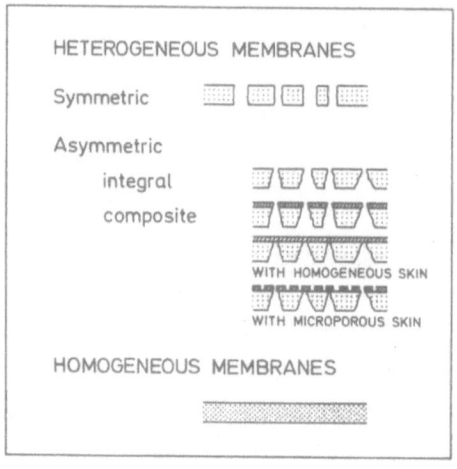

Fig. 1. Membrane structures.

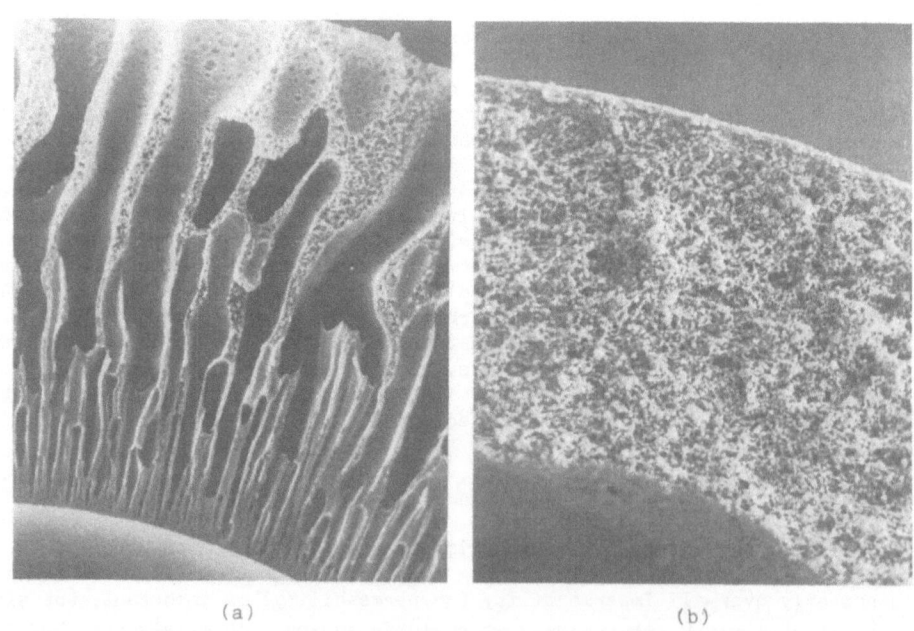

Fig. 2. Two limiting types of membrane structures:(a) Finger type;
(b) sponge type. (Gambro Dialysatoren KG, with permission).

of some species while permits the passage of others.

Synthetic membranes may consist of a large variety of materials including macromolecules (high polymer) with natural or synthetic origin, metals, ceramics, glass, carbons, etc. High polymers are used widely in the construction of membranes. Because, they offer a variety of properties for specific applications, including readily available and low in cost, easy of fabrication, favorable mechanical properties, relatively inert and chemically stable, acceptable stability and compatibility within biological environment and many others. Polymers which enjoy the greatest use in the biological applications today are: Cellulose and its derivatives; silicone rubber; polytetrafluoroethylene (Teflon); polyamids (Nylons); polyvinyl chloride; polymethylmethacrylate; acrylic copolymers; polysulfones; polycarbonate; polyelectrolytes; and others.

Synthetic polymeric membranes may be characterized both by structure and by function. The former describes what they are and the later how they perform. Structural features may be illustrated in various ways such as chemical composition, molecular configurations and conformations, segmental and molecular mobility (or stiffness), crystallinity and microcrystalline morphology, orientation, crosslinking and other intermolecular attractions, macroheterogenity due to pore structure or additives (e.g. filters, reinforcement materials, plasticizers, antioxidants, stabilizers, etc.). The most important properties of membranes are, of course, their functional properties, i.e. permeability and permselectivity, which are closely related to their basic structural properties. By emphasizing the structural features, membranes can be tailor-made to suit a particular requirement. When a membrane is considered for an artificial organ application, not only the primary properties (i.e. permeability and permselectivity) must be concerned, but also its stability and compatibility in vivo conditions must be clearly outlined. First of all, the membrane should provide adequate mass transfer rates with acceptable selectivity, and should likely maintain its performance in vivo over a desired period. It should also be nontoxic, nonpyrogenic, and blood or tissue compatible, and thus not inducing any undesirable host reactions such as blood clotting, tissue necrosis, carcinogenesis and allergic responses. Additionally, it should be purified, fabricated and sterilized easily, and have adequate mechanical and thermal stabilities without resulting any physical integrity failure during storage and usage.

Synthetic membranes may be classified according to their structure, as heterogeneous and homogeneous. As it is examplified in Figure 1, heterogeneous membranes exhibit a readily porous structure, and can be symmetric or asymmetric; integral or composite; with or without a skin layer which is a dense polymer film and induces high permselectivity. In contrast, homogeneous membranes are likely to be considered as a continuous medium without any pores in it.

Heterogeneous Membranes

Heterogeneous membranes, so-called microporous membranes can be made from a wide variety of polymers by following various specific ways and means. One of the most common procedure is solvent-casting (phase inversion) in which a polymer solution is cast into a film on a proper surface the polymer is precipitated in a non-solvent while the solvent is evaporated [1]. Symmetric or asymmetric membranes can be prepared by this technique. Two limiting types of structures are achieved, namely finger type which gives high mass transfer rates, but have low permeability and low mechanical stability, and sponge (or foam) type which induces high permselectivity and high mechanical integrity, but low permeation rates (Fig.2). Cellulose and its esters, PVC, PC, Nylons, PVF, polyacrylates, polyacetals, and many other polymers are used to prepare heterogeneous membranes by this technique. Symmetric microporous membranes are also made by sintering of finely devided polymers (e.g. PE, PP or PTFE powders or fibers) [2]. Stretched membranes which are also symmetric microporous membranes and exhibit good chemical, thermal and mechanical stabilities, are prepared by extrusion from polymer particles, such as PE, PP or PTFE [3]. The homogeneous films obtained by extrusion is then stretched perpendicular to the direction of extrusion, thus highly porous structure is achieved. Many other interesting approaches have been also proposed to prepare, both integral and composite, heterogeneous membranes, including track-etching [4], interfacial polymerization [5], ionotropic gel formation [6], plasma deposition [7].

Transpor Properties. As mentioned above a heterogeneous membrane has interconnected voidages or pores in which species can pass through. Species that are larger than even the largest pores of the pore size distribution of a microporous membrane are retained upon its surface (Figure 3). This mode is likely characterized as sieve-type retention. The particle size, shape, charge and maybe chemistry of the permeate molecules, and also pore size and structure and pore size distribution determine the selective permeability within heterogeneous membranes [8].

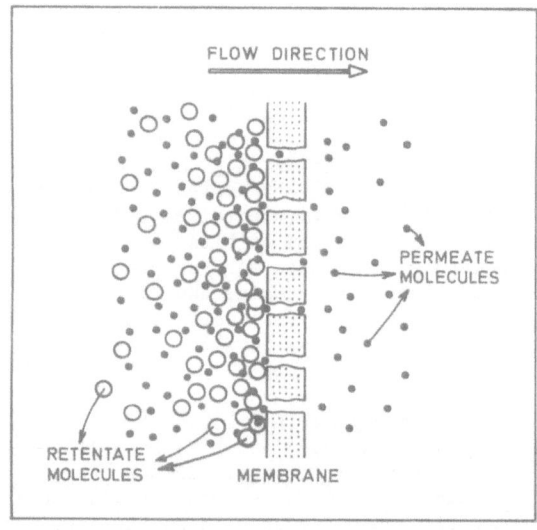

Fig. 3. Transport through heterogeneous membranes.

Homogeneous Membranes

Homogeneous membranes, i.e. non-porous membranes, as mentioned above, are so called because they have no identifiable voidages at colloidal level and species pass through them as if by diffusion through a solid phase (Figure 4).

Homogeneous membranes can be made from mostly polymers such as silicone rubber, PVC, PAN, PS, PVA, PE, PP, Nylons, etc. These membranes are generally prepared by casting from a solution or from a polymer melted by extrusion, blow and press molding. Different procedures may be applied depending on the membrane configuration desired (e.g. films, tubes, hollow fibers) [10]. In addition to fabrication procedure, chemical nature of the polymer might be considered in order to obtain different microcrystalline structures.

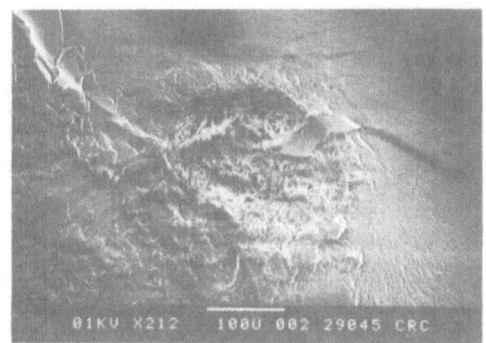

Fig. 4. Cross section of a reservoir
type polymeric drug carrier
system which consist of PHEMA,
PEG and Mitomycin-C 9 .
(a scanning electron micrograph)

By changing the polymer type (chemistry of backbone and side chains, substituents, etc.), polymer molecular weight and its distribution, the type
of cross-linking and its density; by using mixtures of hydrophobic and
hydrophilic polymers; and by including additives, membranes with different
microstructures, thus leading a wide variety of properties, may be obtained
for specific applications.

Transport Properties A homogeneous membrane consists of a dense polymer
film where there are no pores in the structure. Therefore, the mass transfer
rates are rather low in these membranes, but, they exhibit high selectivity
(which means here solubility). Thus, two species with different solubilities
but identical diffusivities can be separated by utilizing these type of
membranes.

It should be pointed out that these membranes must have also some kind
of openings which permit the transfer of the permeating molecules. This
mode of processes is highly complicated and not understood thoroughly. In
the free volume theory of diffusion through homogeneous polymeric structures
this mode is examined in terms of the probability of finding enough local
volume, while the activation energy approach deals with the creation of
the free spaces [11].

Transmission of species through homogeneous membranes depends on their
solubility and diffusivity. The nature of the permeate molecules (i.e. size.
shape, charge, chemistry), the nature of the polymeric membrane (i.e. chemical
nature, chain flexibility, cohesive energy, cross-linking and its density,
degree of crystallinity and its type and distribution, compounding materials,
and their shape, size, distribution, concentration, concentration distribution, orientation, topology, etc), and temperature are markedly effective
on both solubility and diffusivity, thus on transport rates through homogeneous membranes [12].

APPLICATIONS

Artificial Kidney Systems
The kidneys are the chief excretory organs of the body with the

respiratory system and the skin. The kidneys excrete waste products of metabolism, e.g. urea, uric acid, creatinine and others, and adjust the loss of water and electrolytes from the body in order to keep body fluids relatively constant in amount and composition. The kidneys are responsible for maintaining the pH of the blood and, thus, the body fluids at a relatively constant value. In renal failure, as kidney function decreases, various substances, which are normally excreted by the healthy kidneys, will accumulate, and thus leading morbidity and disability. At this point, the loss in excretory capacity has to be matched by exogeneous means, i.e. artificial kidney systems, in order to maintain the overall mass balance in the body.

Today, artificial kidney systems are not yet able to compensate for normal renal functions. Instead, they are applied to the patients with renal insufficiency in order to maintain the concentrations of potentially harmful endogeneous or exogeneous substances below certain levels to avoid organ toxicity.

There are three artificial kidney systems that are in use today, namely peritoneal dialysis, hemodialysis and hemofiltration. Since this paper emphasises on synthetic membranes, we would not deal with attitudes of peritoneal dialysis, in which the semipermeable dialysis membrane is the peritoneum which is the membrane lining the wall of the abdomen.

<u>Hemodialysis</u> This is an extracorporeal system where arterial blood from patient is pumped through a dialysis unit consist of a semipermeable synthetic dialysis membrane (Figure 5). The blood is passed over one side of the dialysis membrane while a dialysate solution containing a suitable solution of ions (Na^+, Ca^{++}, Mg^{++}, acetate or bicarbonate, arbitrarily K^+, and dextrose) is circulated on the other side. Hence, waste or toxic products are removed from blood by dialysis. The dialysate composition is adjusted to maintain the correct electrolyte concentration in blood. Excess water can be ultrafiltered by applying a hydrostatic pressure difference between blood and dialysate side.

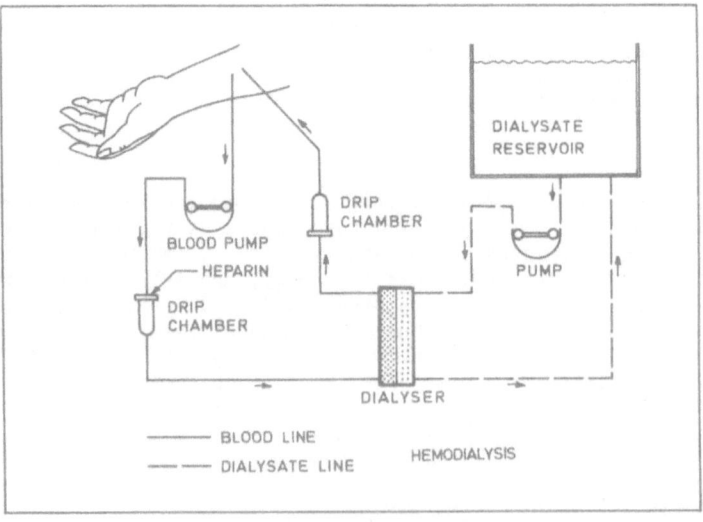

Fig. 5. Hemodialysis.

Cellulose nitrate (collodion) is the first membrane used in hemodialysis
[13]. Cellophane which is a type of regenerated cellulose. was then utilized
in early devices [14]. These membranes were then totally replaced by another
type of regenerated cellulose, so-called Cuprophan which is made of cotton
linters by Cuoxam process [15].Different kinds of Cuprophan membranes,
i.e.. in the form of sheets and tubes in earlier hemodialyzers and as hollow
fibers in recent devices, have been used to a large extent. With a variety
of pore structure, cellulose acetate membranes based on the phase inversion
technique are also considered for hemodialysis, and marketed in the form
of mostly hollow fiber units [16].

After the "middle-molecules" hypothesis proposed by Babb and Scribner
[17]. investigations on production of hemodialysis membranes have focused
on improving the membrane permeabilities for middle molecules, and also
ultrafiltration rates. Several such membranes made of polysulfone, PC, PAN
and PMMA have already been commercialized [18-21]. A variety of block copolymer
membranes (e.g. PEO-PET. PEO-PC, polyether-urethane) have been investigated
for possible hemodialysis applications [22-24]. Other novel membranes such
as cross-linked PVAL, PEG or PHEMA [25], polyelectrolytes [26], polyamino-
acid [27], collagen [28], membranes with high blood compatibilities, good
mechanical properties, and improved dialysis rates would of course trigger
the future of hemodialysis. However, hollow fiber Cuprophan units seem to
be the material of choice still today.

Hemofiltration This procedure can be considered as an alternative to
hemodialysis for the treatment of patients suffering from renal diseases.
Unlike hemodialysis which relays upon the diffusive transport of metabo-
lites and electrolytes, hemofiltration remove solutes by predominantly a
convective flow under a transmembrane pressure gradient (Figure 6). With
a filtration rate of up to 40 % of the blood flow, blood is filtered through
the membrane. The ultrafiltrate formed contains microsolutes, but no proteins
and cellular elements of the blood, and is discarded. In order to maintain
patient fluid balance a corresponding volume of sterile physiological solution
is reinfused.

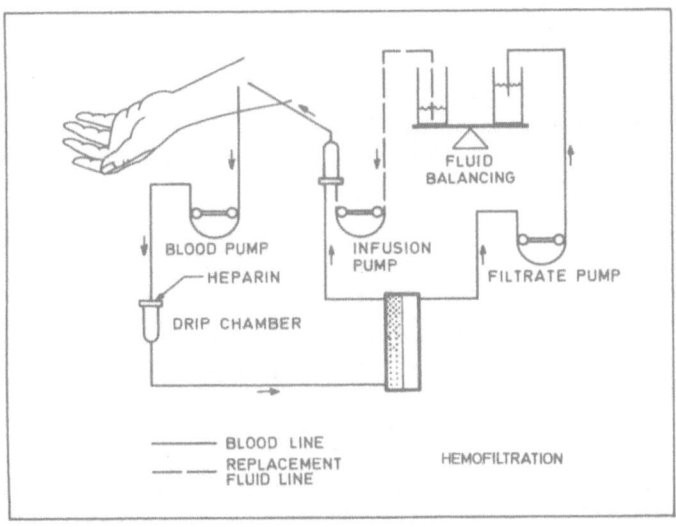

Fig. 6. Hemofiltration.

The characteristics of hemofiltration membranes is quite different from that of those used for conventional hemodialysis membranes, with their high hydraulic permeability and their retentivity. To date, regenerated cellulose, CA, PC, PAN, PMMA, polyamid, polysulfone membranes with different structure and thickness, and thus with different hydraulic and diffusive permeabilities are available commercially, and many others are in varying stages of development or undergone clinical trials [29-31].

Despite the advantages over hemodialysis in respect to elimination of intermediate or middle molecular weight substances with higher rates as a result of convective transport, hemofiltration did not spread widely owing to the high cost of commercial sterile replacement solution, together with the risk of its contamination during storage.

Plasmapheresis

Membrane plasmapheresis is a young extracorporeal treatment where blood from the patient is circulated through a membrane filter, and roughly 30 % of the whole blood is filtered (Figure 7). The plasma separated from blood contains macromolecules and micromolecules (or so-called crystalloids). In the simple plasmaseparation the fluid phases is discarded, and fresh frozen plasma containing valuable plasma proteins (e.g. albumin) is rein-fused, which is one of the major drawbacks of therapeutic plasmapheresis as practiced today. Recent developments, such as cryofiltration, double filtration plasmapheresis, and plasmapheresis combined with specific ad-sorption of the pathological plasma components has triggered this new thera-peutic technique [32-35]. These developments allow further improvements in respect to reduction of the pathological plasma constituents, but saving the whole plasma with albumin and other functional proteins, and thus sti-mulating the widespread applications of plasmapheresis.

To date a number of plasma separators utilizing membranes with improved properties, such as low or no rejection of high molecular weight proteins, without loss of filtration efficiency, have been released by several com-panies. Membranes made of PP, CA, PMMA, PE, PVC, PVAL, PC- cellulose-di-acetate, and copolymer of ethylene and vinylalcohol are employed in commercially available hollow fiber or flat-plate modules [36,37].

Eventhough very high cost of the procedure, plasmapheresis is accept-ed as an useful tool today in the therapy of many multisystem diseases. The pathogenic role of the pathological plasma components, e.g. antibodies, immunocomplexes, paraproteins, and toxic substance, has not been well defined in most cases, plasmapheresis is employed with the hope of that, the remo-val of these molecules may be of therapeutic benefit. Better understanding of mechanisms, reduction of the cost of the system, and improvements in existing techniques will, of course, make the plasmapheresis more accept-able in the future.

Hemoperfusion

Hemoperfusion is another, relatively new extracorporeal treatment method in which the blood from the patient is circulated through adsorption columns in order to remove exogenous and endogenous toxins (Figure 8). Today, it is well-established procedure to treat patients who are suffering drug intoxication. Others areas of application including chronic renal failure, hepatic failure, and many other diseases of unknown origin, e.g. schizophrenia, psoriasis and diseases related to immunological disorders are still contraversial but promising.

Activated carbons and resins are the most widely used sorbents in commercially available hemoperfusion cartridges. Other sorbents, e.g. poly-aldehydes, silica gel, active alumina and magnesia, aluminium gels and various immunosorbents, or more complex sorbent systems containing these sorbents incorporated with biofunctional agents (e.g. antigens, antibodies,

254

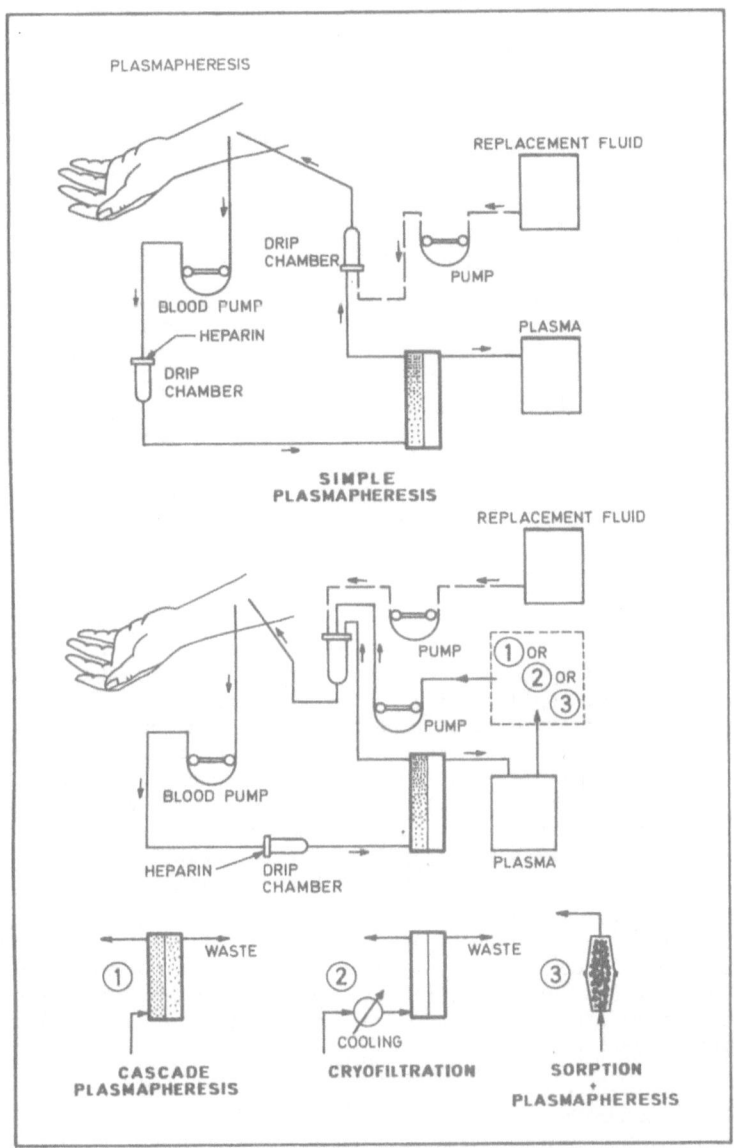

Fig. 7. Plasmapheresis

enzymes, etc.) are also utilized for clinical medicine [38-40].

In earlier hemoperfusion applications, sorbents as granular forms, i.e. mostly un-coated charcoals have been employed [41] . However, the occurrence of problems associated with these early devices have limited their usage. The fine particle generation resulting in embolism to various organs, and the severe depletion of the blood cells, i.e. platelets and leucocytes, have been reported. In order to eliminate these side effects, various approaches have been proposed including novel configuration in which

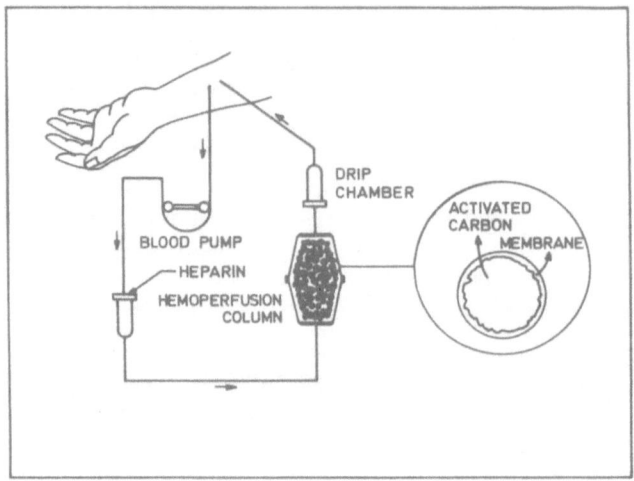

Fig. 8. Hemoperfusion.

activated carbon particles are entrapped within carriers. Fine charcoal powders are fixed by being sprayed onto tapes with an adhesive [42]. Carbon particles are entrapped within, mostly, Cuprophan hollow fibers [43]. Powdered charcoal is immobilized within polymeric gel carriers [44]. Most popular approach amongst, of course is coating, or so-called microencapsulation of sorbent granules with semipermeable membranes [45]. In this later case, i.e. a membrane moderate device, the selection of the coating material and method is the most important subject in obtaining suitable hemoperfusion systems which have high adsorptive capacity, adsorption rate and blood-compatibility and are free from fine particle generation.

Various types of polymer coatings and different encapsulation procedures have been proposed and evaluated at different stages of application [46,47]. It seems that in commercially available hemoperfusion cartridges, cellulose, cellulose acetate, cellulose nitrate and polyHEMA are the membrane material of chose. Many other polymers with promising preliminary results are still underinvestigation.

Membrane Oxygenator

All living cells require to get oxygen and to get rid of carbon dioxide. In man, the lungs are chiefly responsible for the natural process of gas exchange, or so-called blood oxygenation. The lungs can oxygenate up to 30 liters of venous blood per minute. Up to 2400 ml (STP) of oxygen can be taken up. Basal conditions for a resting healthy adult man at 37 ^{o}C are:

Blood flow rate = 5 liters/min
O_2 required = 250 ml (STP)/min
CO_2 produced = 200 ml (STP)/min

These are also nominal design requirements for total respiratory support with an artificial oxygenation.

The concept of artificial oxygenation originated as early as 1868, but, it took more than a century to develop an artificial gas exchange device at commercial level for oxygenation of blood in an extracorporeal circuit. Even today, blood oxygenators are not considered as prolonged rep-

lacements for failing lungs, instead, they are employed in open-heart surgery for a short period.

An artificial cardiorespiratory device (Figure 9), i.e. the heart-lung
machine, which consist of; a mechanical pump for extracorporeal circulation; a blood oxygenator for gas exchange; and a heat exchanger for maintaining the blood temperature at a constant value, is used during open-
heart surgery. The most important functional component of this extracorporeal circuit is the oxygenator. A succesful oxygenator must be able to
oxygenate up to 5 liters of blood per minute to nearly 100 % saturation.
Simultaneously, it must remove a proper amount of CO_2 to maintain pCO_2
levels within physiological range without causing respiratory acidosis or
alkalosis. An ideal oxygenator should have also low priming volume; should
not create harmful effects such as blood trauma, denaturation of plasma
proteins and micro-embolisation; and should be of simple design, easy handling and reliable sterilization.

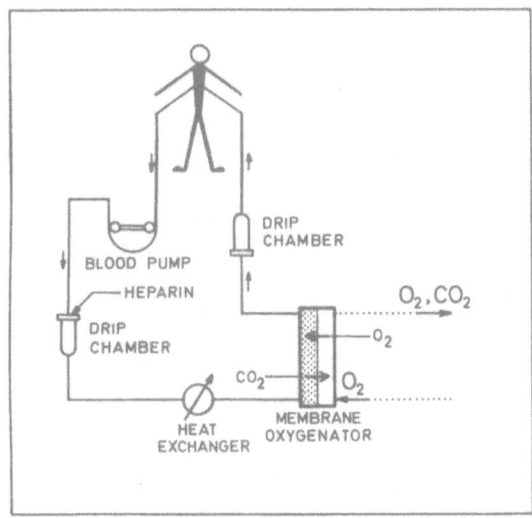

Fig. 9. The heart-lung machine.

A large variety of devices have been developed for blood oxygenation.
Amongst them, the membrane oxygenators have been given an increasing credit. In these devices, a permselective membrane is interposed between the
blood and gas phases, and thus reducing blood trauma and presumably denaturation of blood proteins found in direct blood-gas contact devices.
Membrane oxygenator design decreases the risk of microemboli formation.
But, the membrane, itself also drastically reduces the gas transfer efficiency both by adding an additional resistance in series and by causing a
relatively stagnant blood region near the wall. Thus the majority of efforts
for developing a better membrane oxygenator have been directed toward
fashioning more efficient devices that would utilize membranes with high
permeation rates for O_2 and CO_2, and would provide a thin layer of blood
over the membrane by breaking up the boundary layer.

Various membranes have been evaluated for membrane oxygenators. Cellophane. Polyethylene, ethyl cellulose, and polytetrafluoraethylene (Teflon) membranes were the earliest ones [49-51]. Silicone rubber with far favourable gas permeabilities was then the material of choose. Most silicone polymers have little strength, thus they have been incorporated with fillers [52], or have been reinforced with textiles [53], or have been made as copolymers [54]. It should be noted that even these membranes were inefficient in terms of CO_2 elimination. Heterogeneous hydrophobic membranes (Figure 10) made of Teflon and polypropylene have been, then released [55, 56]. Only these few porous membranes have so far yielded high enough gas transfer rates both for O_2 and CO_2. However, some disadvantages of these novel membranes have been also dictated such as far greater blood trauma, higher risk for micro bubbles emboli, possible ultrafiltration, etc. [57].

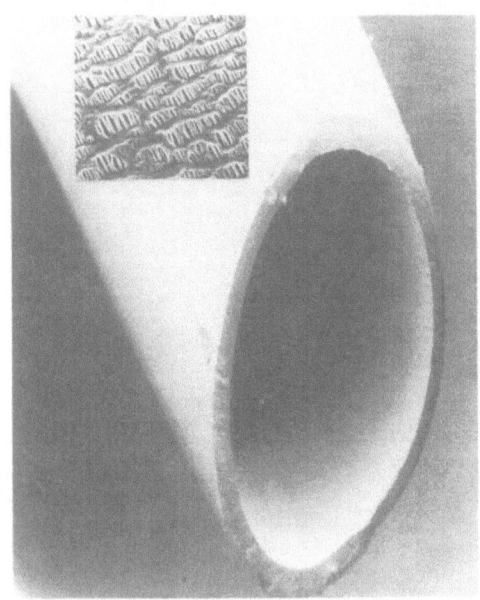

Fig. 10. Polypropylene hollow
          fiber oxygenator
          membrane (Celanese
          Corp., with permission)

As conclusion, it can be said that today there are nicely designed disposible membrane oxygenators modules in the form of spiral-wound, plate-and-frame, and hollow fibers, which fulfill many of the basic requirements for blood oxygenation in short term applications, such as open-heart surgery. But, it seems that much more efforts must be done to investigate the concept of long-term respiratory assistance in man.

Microcapsules

Biologically active substances, such as enzymes, whole cells, cell extracts, antigens, antibodies, contraceptives, and many drugs are widely used in clinical medicine for diagnostic and therapeutic applications. There are some important drawback when these species are employed in free forms. They are not sufficiently stable under operation conditions, especially in both extremes of temperature and pH. They usually result immunological reactions, and rapid removal and inactivation.

Immobilization is a breakthrough to eliminate the problems associated

with the use of bioagents in free forms. During the last decade many novel
approaches for immobilization of biologically active entities have been
released. They may be, covalently bound to a carrier; or adsorbed ionically
or physically to various kinds of support materials; or covalently crosslinked to each other; or entrapped inside the polymeric matrices, i.e. mostly
in the form of membranes: or encapsulated into the lumen of microcapsules,
nanocapsules or hollow fibers (Figure 11) [58-60].

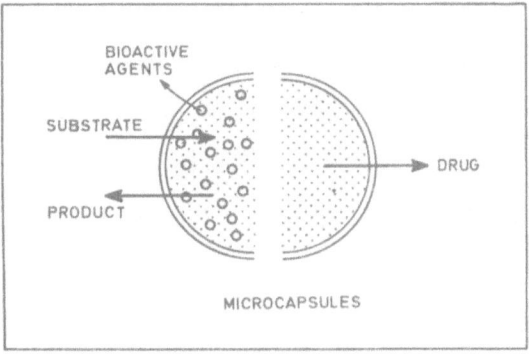

Fig. 11. Microcapsules.

The process of microencapsulation have been described several years
ago for the production of carbonless copy papers. Since then numerous applications in agriculture, pharmacy, medicine and other related fields have
been elaborated. Several techniques including phase separation methods
(e.g. aqueous or organic phase separation, meltable dispersion, spray-
drying, pan coating, etc.) and interfacial condensation or addition polymerization have been evaluated in order to prepare microcapsules [61-63].

The list of polymers that have been tried out to obtain microcapsules.
Among these, cellulose base polymers (i.e. cellulose nitrate, cellulose
acetate, ethylcellulose, cellulose acetate phthalate, carboxymethylcellulose, methyl cellulose, and others), polydimethylsiloxane and other silicones, polyurethanes, polyamids, polyHEMA, and other acrylates, lipids,
collagen, polypeptites, etc. have given most credit and have already emerged
in commercially available systems.

Biofunctional substances like sorbent system, enzymes, cells, organelles, antigens and antibodies immobilized by microencapsulation have
been used in medicine; for the treatment of patients suffering with renal
diseases, drug overdose, liver failure and other metabolic diseases [64-
67]; in enzyme replacement therapy [68, 69]; removing of immunogenic substances via specific sorbent-microcapsule systems, as an artificial pancreas
systems in order to maintain the glucose levels in diabetics [70]; and as
a reservoir type drug delivery device in order to provide controlled release or to target drugs by specific means [71-74], and in many other novel
applications.

REFERENCES

1. S.Loeb and S.Sourirajan, Adv. Chem. Ser., 38:117 (1962).
2. M.E.Nordberg, I.Amer Chem. Soc., 27:299 (1944).
3. H.S.Bierenbaum, R.B.Isaacson, M.L.Druin and S.G.Plovan, Ind. Eng. Chem. Prod. Res. Develop., 13:2 (1974).
4. R.L.Fleischer, P.B.Brice and R.M.Walker, Science, 149:383 (1965).
5. J.E. Cadotte, R.S.King, R.J.Majerle and R.J.Peterson, J.Macromol. Sci. Chem., A15:727 (1981).
6. H.Thiele and K.Hallich, Kolloid Z., 163:115 (1959).
7. H.Yasuda and C.E. Lamaze, J.Appl. Polymer Sci., 17:201 (1973).
8. N.Lakshminarayanaiah, "Transport Phenomena in Membranes", Academic Press, London (1969).
9. E.Pişkin, M.Kiremitçi and A.Denizli, in: "Proc. 12th. Int. Sym. Controlled Release of Biactive Materials". (in press).
10. R.E. Kesting, "Synthetic Polymeric Membranes", McGraw-Hill. New York (1971).
11. J.Crank and G.S.Park, "Diffusion in Polymers", Academic Press, London (1968).
12. R.McGregor, "Diffusion and Sorption in Fibers and Films". Academic Press, London (1974).
13. J.J.Abel, L.G.Rountree and B.B.Turner, J.Pharm. Exp.Ther.. 5:275 (1914).
14. W.J.Kolff and H.T.J. Berk, Geneesk Gids, 21:409 (1943).
15. W.Bandel, Chem. Z. Jahrgang, 65 (1964).
16. B.J.Lipps, R.D.Stewart, H.A.Perkins, G.W.Holmes, E.A.McLain, M.R.Rolfs and P.D.Oja, Trans. Amer. Soc. Artif. Intern. Organs, 13:200 (1967).
17. A.L.Babb, P.P.Popovitch. T.G.Christofer and B.H.Scribner, Trans. Amer. Soc. Artif.Intern. Organs, 17:81 (1971).
18. M.Luttinger, C.W.Cooper, R.I., Trans. Amer. Soc. Artif. Intern. Organs, 14:5 (1968).
19. B.S.Fisher, W.S.Higley, P.A. Cantor and W.J.R. Stone, Trans. Amer. Soc. Artif. Intern. Organs, 19:429 (1973).
20. I.O.Salyer, G.L.Ball and G.L.Beemsterboer, in: "Membrane Processes in Industry and Biomedicine", M.Bier, ed., Plenum Press, New York (1971).
21. K.Ota, J.Okazawa, E.Kumagaya, J.Agishi, N.Sugino. N.Mitani, Y.Fugii, M.Kimura, Y.Nagao, H.Tsukamoto, H.Tanzawa, Y.Sakaj, Próc.Eur. Dial. Transplant. Assoc., 12:559 (1975).
22. D.J.Lyman, B.H.Loo and R.W.Crawford, Biochemistry, 3:985 (1964).
23. P.Konstantin, H.Goehl and C.Gullberg, Artif. Organs (Suppl), 5:691 (1981).
24. D.J.Lyman and B.H.Loo, J,Biomed. Mater. Res., 1:17 (1967).
25. D.J.Lyman, in "Replacement of Renal Function by Dialysis", W.Drukker, F.M.Parsons, J.F.Maher, eds.. Martinus Nijhoff Publ., The Hague (1979).
26. A.Micheals, Ind. Eng. Chem., 57(10):32 (1965).
27. C.H.Bamford. A.Elliott and W.E.Hanby. "Synthetic Polypeptides", Academic Press, New York (1956).
28. T.Nishihara, A.L.Rubin and K.H.Stenzel, Trans. Amer. Soc. Artif. Intern. Organs, 14:169 (1968).
29. F.F.Holland, E.Klein, R.P.Wendt and K.Eberle, Trans. Amer. Soc. Artif. Intern. Organs, 24:662 (1978).
30. W.Henne, G.Duenweg and W.Bandel, Artif. Organs (Suppl), 3:466 (1969).
31. H.Göhl, P.Konstantin and C.A. Gullberg, Contr. Nephrol., 32:20 (1982).
32. W.Samtleben, M.Blumenstein, L.Liebl and H.J.Gurland, Trans. Amer. Soc. Artif. Intern. Organs, 26:12 (1980).
33. T.Agishi, I.Kaneko, Y.Hasua, et.al., Trans. Amer. Soc. Artif. Int. Organs, 26:406 (1980).

34. P.S.Malchesky, Y.Asanuma and M.Blumenstein, Mitt. Klin. Nephrologie, 10:44 (1981).
35. H.J.Gurland, M.J.Lysaght and W.Samptleben, in: "The Past, Present and Future of Artificial Organs", E.Pişkin and T.M.S.Chang, eds., Meteksan, Ankara (1983).
36. W.Henne, K.Gerlach, J.Tretzel and M.Pelger, "Membrane Technology for Plasmapheresis" presented at "Int. Workshop on Plasma separation and Plasma Fractionation, Rottach-Egern, March (1983).
37. Y.Nose, P.S.Matchesky and J.W.Smith, "Plasmapheresis", ISAO Press, Cleveland (1983).
38. V.Bonomini and T.M.S.Chang, "Hemoperfusion", Hemisphere, Washington (1980).
39. C.Giardano, "Sorbents and Their Clinical Applications", Academic Press, London (1980).
40. E.Pişkin, Life Support Systems, 2:47 (1984).
41. H.Yatzidis, Proc.Eur.Dial. Transplant. Assoc., 1:83 (1964).
42. J.B.Hill, F.L.Palaia and C.R.Horres, in "Arficial Organs", J.M.Courtney, J.D.S. Gaylor and T.Gilchrist, eds., The Macmillan Press, London (1977).
43. H.J.Gurland, L.A.Castro, G.Hillebrand and B.Schmidt, in: "Hemoperfusion: Kidney and Liver Support and Detoxification", S.Sideman and T.M.S. Chang, eds., Hemisphere, Washington (1980).
44. M.Kiremitçi and E.Pişkin, Life Support Systems (Suppl), 2:192 (1984).
45. T.M.S. Chang, Kidney Int., 10:218, (1976).
46. J.D. Andrade, R.Van Wagenen, C.Chen, K.Kopp and W.J.Kolff, Proc. Eur. Dial. Transplant. Assoc., 9:210 (1972).
47. E.Pişkin, in "The Past, Present and Future of Artificial Organs", E.Pişkin and T.M.S. Chang, eds., Meteksan, Ankara (1983).
48. W.J.Kolff and R.Balzer, Trans. Amer. Soc. Artif. Intern. Organs, 1:39 (1955).
49. G.H.A.Jr. Clowes, A.L. Hopkins and T.Kolobow, Trans. Amer. Soc. Artif. Intern. Organs, 1:23 (1955).
50. G.H.A.Jr.Clowes, A.L.Hopkins and W.E.Neville, J.Thorac. Surg., 32:630 (1956).
51. E.C.II.Pierce, Arch. Surg., 77:938 (1958).
52. W.G.Esmond and N.R.Dibelius, Trans. Amer. Soc. Artif. Intern. Organs, 11:325 (1965).
53. J.A. Thomas, Arch. Klin. Chir., 289:286 (1958).
54. E.C.Pierce II and N.R.Dibelius, Trans. Amer. Soc. Artif. Intern. Organs, 14:220 (1968).
55. K.Esatc and B.Eiseman, J.Thorac. Cardiovascular Surg., 69:690 (1975).
56. Product information, Celgard Microporous Film, Celanese Corp., Charlotte, North Carolina (1981).
57. P.D.Richardson and P.M.Galletti, in: "Pysiological and Clinical Aspects of Oxygenator Design", S.G.Dawids and H.C. Engell, eds., Elsevier, Luxembourg (1976).
58. D.R.Zaborsky, "Immobilized Enzymes" CRC Press, Cleveland (1973).
59. D.Thomas and J.P.Kernevez, "Analysis of Control of Immobilized Enzyme Systems", North-Holland, Amsterdam (1976).
60. B.Mattiason, "Immobilized Cells and Organelles", CRC Press, Boca Raton, Florida (1983).
61. T.M.S. Chang, "Artificial Cels", Charles C. Thomas, Springfield, Illinois (1972).
62. M.H.Gutcho, "Microcapsules and Other Capsules", Noyas Data Corp., Park Ridge, New Jersey (1979).
63. F.Lim, "Biomedical Applications of Microencapsulation" CRC Press, Boca Raton, Florida (1980).
64. T.M.S. Chang, Kidney Int., 10:218 (1976).
65. T.M.S. Chang, "Artificial Kidney, Artificial Liver and Artificial Cells", Plenum Press, New York (1978).

66. T.M.S. Chang, in: "The Past, Present and Future of Artificial Organs", E.Pişkin and T.M.S. Chang, eds., Meteksan, Ankara (1983).
67. T.M.S. Chang, Nephron, 36:161 (1984).
68. K.Mosbach, "Methods in Enzymology -Immobilized Enzymes", Academic Press, New York (1976).
69. T.M.S. Chang, "Biomedical Applications of Immobilized Enzymes and Proteins, Vol.I,II, Plenum Press, New York (1977).
70. F.Lim and A.M.Sun, Science, 210:908 (1980).
71. A.C. Tanquary and R.E.Lacey, "Controlled Release of Biologically Active Agents, Plenum Press, New York (1974).
72. A.F.Kydonieus, "Controlled Release Technologies: Methods, Theory and Application", vol.I,II., CRC Press, Boca Raton, Florida (1980).
73. R.Baker, "Controlled Release of Bioactive Materials", Academic Press, New York (1980).
74. S.D. Bruck, "Controlled Drug Delivery", vol.I,II, CRC Press, Boca Raton, Florida (1983).

TOTALLY BIORESORBABLE COMPOSITES SYSTEMS FOR INTERNAL FIXATION OF BONE FRACTURES

M. Vert[1], P. Christel[2], H. Garreau[1], M. Audion[3], M. Chanavaz[4], and F. Chabot[1]

1.UA CNRS 500 - University of Rouen, LSM, INSCIR, BP8, 76130 Mont-Saint-Aignan, France

INTRODUCTION

Metals have been used extensively for internal fixation of bone fractures. However, shortcomings of metallic plating or nailing are now well identified and the interest for internal fixation devices made of more flexible materials is growing fast. In this respect, plastic-based systems appear attractive, particularly carbon fiber reinforced composites.[1] However, the most suitable solution remains the use of bioresorbable plastic systems due to their combining various potential advantages: - better bone remodeling because of better matching of mechanical properties with respect to bone, - no need for reoperation to remove the appliance after bone-healing, and - possibility for incorporating antibiotics in view of preventing infection through sustained delivery of the entrapped drug.

In this regard, poly[α-hydroxy acids], especially polylactic acid stereocopolymers, polyglycolic acid and poly[lactic-co-glycolic] copolymers have been considered as good candidates because of their biocompatibility, their degradability with adjustable rates of degradation and their resorbability through degradation products normally involved in biological processes (lactic and glycolic acids). However, the main advantage of polyesters of the poly[α-hydroxy acids]-type is the possibility for adjusting their mechanical properties through structural changes (copolymerization or configurational structure of lactic acid chiral residues) instead of using additives that can leach out resulting in non-controlled aging and sometimes unwanted toxicity due to the released additives.[2]

In 1975, when some of us became interested in these problems, the potential of poly[α-hydroxy acids] for bone fracture fixation was known but no thorough investigations nor clinical experimentation was reported.

Our first personal contacts with home made poly[α-hydroxy acids] were somewhat disappointing. The degradation rates were close to those reported in literature and were thus much too high to allow the fitting of the time-requirement for long bone fracture healing normally evaluated to 4 months.

2. University of Paris VII, Faculté de Médecine Lariboisière St-Louis, LRO 10 ave de Verdun, 75010 Paris, France
3. Hopital Saint-Roch, 06000 Nice, France
4. Clinique Implantologique, 66 quai Cavelier de la Salle, 76000 Rouen France

Improvements at the different stages of the processing of ready-for implantation appliances in pure poly[α-hydroxy acids] led to standard procedures which allowed us to decrease very much the rate of degradation of massive devices made of the semi-crystalline members of the series of polylactic acid stereocopolymers with high content in L-units and low contents in D-units.[2] The decrease of molecular weights of semi-crystalline polylactic acid stereocopolymers in vivo, as monitored by gel permeation chromatography and histology, showed degradation rate slow enough to allow fracture healing before degradation started.[2] On the other hand, mechanical properties of polylactic acid stereocopolymers have been found slightly above those of poly[methyl methacrylate], PMMA, but with the same shortcomings : low impact strength and low deformation at break.[3] In particular, unexpected breakings were observed at the level of the holes in the case of metallic screwing of PLA X bone plates after 4 months implantation on the cortex of unbroken sheep femurs. These failures were probably due to the high flexibility of the bone with respect to the flexibility and the elongation at break of the plastic bone plate, especially when the plates were fixed with metallic screws. This was a source of worry for uses in osteosynthetic surgery.

When polymerists are faced with a problem of failure because of insufficient inherent mechanical properties of a polymer, either they switch to another polymer with better characteristics or they try to reinforce the polymer by making a composite system. The second possibility has been selected in order to preserve the benefits of poly[α-hydroxy acids]' biological properties. However, the search for bioresorbable systems precluded the use of carbon or glass fibers or any other non-degradable reinforcing material. As semi-crystalline poly[L-lactic acid] and polyglycolic acid melt at approximately 170-180 and 225-230°C respectively, it was attempted to take advantage of this 50°C difference between the melting temperatures of the two polymers to fabricate a totally bioresorbable composite material by compression moulding. The composite was made of a matrix of a semi-crystalline polylactic acid reinforced by poly[glycolic acid] fibers or multistrand threads as those used to make PGA commercial sutures.[3] Impact strength and resistance to metallic screwing were significantly improved.[4]

In spite of these stimulating findings which resulted from efforts focused on the evaluation of mechanical, physical and biological properties of poly[α-hydroxy acids], the problem of clinical osteosynthesis of bone fracture with totally bioresorbable plates, screws, nails, etc... still remained. Experimental devices were tentatively used for the fixation of bone after osteotomy in animals and, especially, in sheep as most of the characterization work was carried out with this animal. However, it was soon appreciated that the experimentation would be difficult owing to the "non-cooperative" behavior of the animals which supported poorly any post operative complications and, in particular, the presence of a plaster cast which was required to secure the fixation. Therefore, three years ago, we decided to change our strategy and to turn to clinical experimentation under the condition of limiting the trials to easily accessible bone sites and slightly loaded fractures instead of trying to fix long bone fractures in animals.

In this paper, we present complementary data concerning the fate of long-lasting poly[L-lactic acid] after four year implantation in sheep, preliminary results of efforts to characterize dynamic mechanical properties of our bioresorbable composite combination and some data on the clinical experimentation of poly[α-hydroxy acid] systems which is currently underway in the maxillo-facial sphere. For the sake of clarity, we recall first some of the main characteristics of these polyesters insofar as structural, physical and biological properties are concerned. The last part of this paper deals with recent clinical findings which show the property of fast degrading polymers being gradually replaced by young bone cells. Although it has not yet been tested for the composite material this property may appear to be of special interest for bone fixation and bone surgery.

POLY[α-HYDROXY ACIDS]

Poly[α-hydroxy acids] constitute a particular class of polyesters whose repeating units $--[O-CO-CHR]_n-$ are derived from α-hydroxy acids, HO-CHR-COOH. The various polymers of the series are presented in Table 1 together with the corresponding acronyms used in the rest of the text.

Table 1

Poly[α-hydroxy acids]

$+O-CH_2-CO+_n$ PGA

$+O-\overset{H}{\underset{CH_3}{C}}H-CO+_n$ PLA 100

$+O-\overset{H}{\underset{CH_3}{C}}H-CO-]_n+O-\overset{CH_3}{\underset{H}{C}}H-CO+_p$ PLA X (X=100n/n+p)

$+O-\overset{H}{\underset{CH_3}{C}}H-CO-]_n+O-CH_2CO+_q$ PLA (100-X) GA Y (Y=100q/n+q)

$+O-\overset{H}{\underset{CH_3}{C}}H-CO-]_n+O-\overset{CH_3}{\underset{H}{C}}H-CO-]_p+O-CH_2-CO+_q$ PLA X GA Y (X=100n/n+p+q) (Y=100q/n+p+q)

Though higher homologues are known, the more common poly[α-hydroxy acids] are macromolecular compounds derived from glycolic (R=H) and lactic (R=CH$_3$) acids, the latter being chiral, i.e. having two enantiomeric forms L and D with similar intrinsic chemical properties but opposite configurational structures. The most efficient method to produce high molecular weight poly[α-hydroxy acids] is the ring opening polymerization of cyclic diesters. The structure of the resulting polymers is under control as far as configurations of lactic acid residues are concerned. In contrast, L and D stereosequence distributions in stereocopolymers, and repeating unit distributions in copolymers, are somewhat out of control because of transesterification side-reactions occuring during the chain growing.[5] Anyhow, these polymers, copolymers and stereocopolymers can now be prepared with reproducible characteristics when proper processes are used. PLA 100 and PGA are semi-crystalline compounds with melting points in the range of 175-180° and 225-230° C respectively. It is well known that, among the series of poly[α-hydroxy acids] the crystallinity and thus the melting points decrease with structural irregularities, in particular D units or comonomer repeating units, introduced in PLA 100 or PGA polymer chains.

Table 2 shows the average mechanical characteristics of semi-crystalline PLA X stereocopolymers. These mechanical characteristics do not depend very much (20 to 30 %) in general) on the configuration and the composition of PLA X at room temperature. In contrast, physical and biological properties are very sensitive to the configuration and to the respective

265

amounts of the various repeating units present in polymer chains.[6]

Table 2

Mechanical characteristics of the semi-crystalline
PLA X stereocopolymers

Tensile modulus	4-5 GPa
Flexural modulus	5-6 GPa
Elongation at break	2-3%
Tensile strength	50-60 MPa
Impact strength (notched IZOD)	30-40 $J.m^{-1}$

The fates of poly[α-hydroxy acids] were determined after implantation in the cortex of the tibiaes of rats.[6] All the GA-containing polymers degrade fast and are replaced by bony tissue at six month post-implantation. Amorphous PLA stereocopolymers behaved almost similarly except highly crystalline PLA 100 which appeared much more stable, as no significant degradation was detected after one year on the basis of X-rays pictures and molecular weight determination. The same trends were observed for implants located in the lateral cortex of sheep femurs but the time-scale was larger and degradation rates of the various stereocopolymers were found different (Table 3).

Table 3

The fate of poly[α-hydroxy acids] after implantation in vivo
(5 x5 mm cylinders implanted into lateral cortex of femur of sheep)

Polymer	Composition of the monomer feed %			Fate of the polymeric implant	
	LA		GA	six months	one year
	L	DL			
PLA 100	100	-	-		no degradation
PLA X	100-2X	X	-	mild foreign body reaction, partial degradation increasing with the content in D-repeating units.	
PLA 50	-	50	-	withish and deformed	no implant visible cortical brindging
PLA 25 GA 75	25	-	75	degraded, replaced by woven bone	resorbed + woven bone
PLA 50 GA 50	50	-	50	degraded, replaced by woven bone	resorbed + woven bone
PLA 75 GA 25	75	-	25	degraded, replaced by woven bone	resorbed + woven bone

In any case, one can consider that poly[α-hydroxy acids] are bioresorbable polymers with excellent biocompatibility and adjustable bioresor-
bability through the amounts of L- and D-lactic acid units and the amounts
of GA and LA repeating units. Furthermore, the fast degrading polymers are
replaced by bony tissue within one year at most when they fill holes
artificially created in cortical bone in animals. For the same time,
long-lasting PLA 100 has been shown to remain stable.[6]

THE LONG TERM FATE OF LONG LASTING PLA 100 IN VIVO

For the study of the long-term fate of long-lasting PLA 100,
experimentations were initiated several years ago. These experimentations
were based on in vivo implantation of six-hole bone plates fixed by metallic
screws on tibiae of sheep. The plates were processed by compression moulding
of highly crystalline isotactic PLA 100 obtained by ring opening polymerization of L-lactide (Mp=96°C; $[\alpha]_D^{25}$=-306 for c=1.25 g.100cm^{-3} in benzene)
at 150°C for three weeks using Zn powder as the initiator.[3] Low molecular
weight by-products present in the crude polymer were removed by dissolution
of the crude polymer in chloroform and reprecipitation with methanol. After
drying, the purified polymer was compression moulded (T=210°C, P=20 MPa) to
yield the six hole bone plates. These plates where then sterilized using
ethylene oxide and allowed to stand for degasing under vacuum for a week.
The plates were implanted on sheep tibiae without osteotomy for 5, 13, 24,
36 and 48 months. Data corresponding to the first three years have already
been reported.[4] After removal, samples were taken from the plates and conditioned as usual for histologic examination and GPC measurements in dioxane.
For the sake of clarity, these data are recalled herein to be compared with
the data obtained for the fourth year post-implantation. GPC molecular
weights are given with respect to polystyrene standards.

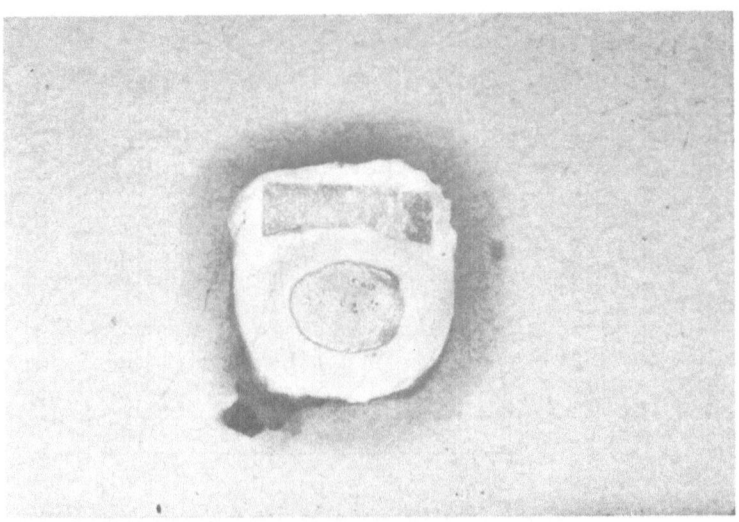

Fig. 1. Histology of a bone plate in long-lasting PLA 100 showing the embedding
of the plate within exostositic bone and partial degradation after 4 year
implantation

From 1 to 4 years, histology showed that plates were surrounded by a
fibrous capsule with increasing thickness (100 μm at one year, 800 μm at
3 years) and some moderate chronic inflammation. Fibrous tissue was always

interposed between cortical bone and plate. Giant cells and macrophages containing PLA debris were found in direct contact with the polymer. The capsule showed a high degree of vascularisation after one year which decreased later on. Furthermore, the bone plates showed cracks which were not related to degradation since molecular weights remainded almost the same as shown by GPC. These cracks seemed to be due to stress concentration at the level of the contacts with metallic screws. They were invaded by fibrous tissue that became ossified with time. It was only after 3 years that the plates started to exhibit histological features of degradation at their periphery. After four years, the core was found to be fragmented and surrounded by exostositic bony tissue (Fig. 1). Microradiogramms showed a thin remodeled bony layer beneath the plate after one year only. Aortic lymph nodes, spleen, liver, lungs, and kidneys were examined after 2 years and did not show any abnormality or polymer debris.

Fig. 2 shows the GPC chromatograms corresponding to various implantation times. Initially, M_{GPC} was 210,000. A slight decrease of molecular weight became detectable after 13 months implantation. At the end of the second year, the GPC chromatograms showed a typical bimodal shape already observed in other poly[α-hydroxy acids], in particular PGA (7). This bimodal shape agrees with a degradation occuring preferentially in the amorphous part of the semi-crystalline structure as already suggested by in vitro experiments (7) and alkaline hydrolysis (8). The formation of long lasting low molecular weight particles agrees with the well-defined GPC peaks and the decrease in degradation rate observed in the 36-48 month range. The latter point is well demonstrated by the S-shaped curve in Fig. 3 found when one plots the variation of molecular weight (maximum of the GPC peak) vs. time. In the fourth year, the plate was dramatically affected in agreement with the drastic decrease of M_{GPC}. The plate turned to powder and was invaded by bony tissue.

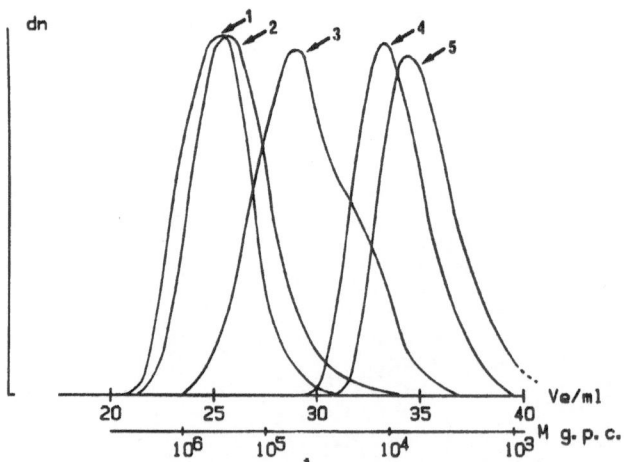

Fig. 2. GPC chromatograms of long-lasting PLA 100 implanted as six hole bone plates on femurs of sheep for various period of time : 1) 0 to 5 months; 2) 13 months; 3) 2 years; 4) 3 years; 5) 4 years.

Therefore, one can conclude from this long-term experimentation that the degradation of long-lasting PLA 100 bone plates shows characteristics similar to those of more rapidly degrading poly[α-hydroxy acids]. The time-scale, none the less appears to be much longer in comparison with the time required for bone healing in usual forms of fracture reduction. The total replacement of the plate by bone has not yet been observed. PLA X stereocopolymers with 2 to 4 % D-units appear to be more adapted than PLA 100 due to both relatively faster degradation and lower melting points. The decrease of the melting temperature of PLA X stereocopolymers with the content in D-units is of special interest for the fabrication of the composite combination as it enlarges the difference between the melting temperatures of matrix and reinforcing fibers.

DYNAMIC MECHANICAL PROPERTIES OF THE COMPOSITE

Bioresorbable composite test bars are usually fabricated so far by combining reinforcing plies made of PGA ERCEDEX[R] sutures n°1 from Robert et Carrière (France) woven to two dimensional fabric and embedded in a PLA 100 matrix by compression moulding. Before moulding, the plies are impregnated with a 5% solution of PLA 100 in chloroform in order to insure good adhesion between fibers and matrix during the moulding.

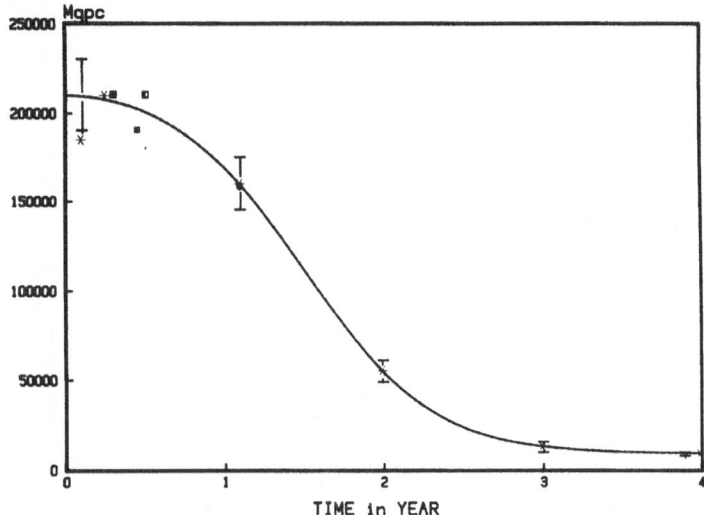

Fig. 3. Variation of the maximum of the GPC peak with regard to implantation time of long lasting PLA 100 bone plate.

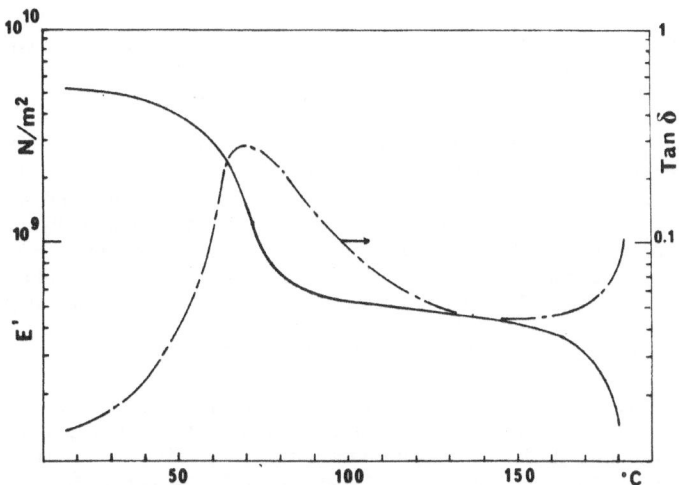

Fig. 4. Variations of E' and tan δ of a 60/40 W/W PLA 100/PGA threads composite with regard to temperature at 7.8 Hz.

The dynamic mechanical properties of composite materials thus obtained are currently evaluated by using a Metravib viscoelasticimeter equipped for traction-compression measurements. In particular, E', E" and tan δ (storage modulus, loss modulus ans loss factor respectively) were measured for a 60-40 W/W PLA 100/PGA composite with 50% of the reinforcing threads aligned parallel to the stress direction, the rest being perpendicular. Data were collected at various temperature and frequencies (7.8-1000 Hz) as usual. Comparison between the composite (Fig. 4) and PLA 100 (Fig. 5) shows that the storage modulus at 7.8 Hz is increased by c.a. 20% (3.8 to 4.5 GPa) and that the width of the viscoelastic transition zone is narrowed.

The use of the time-temperature equivalence principle led to master curves for the matrix and the composite (Fig. 6). The difference between the two curves shows, in the zone of body temperature corresponding to high frequencies, that the reinforcing plies do improve the mechanical properties of PLA 100. The 20% increase of the storage modulus is in line with the expected contribution of the reinforcing plies according to their volumic fraction Φ_2 =0.33 and the fact that only 50% of the fibers are aligned in the traction-compression direction and are effectively active as reinforcement in traction-compression testings.

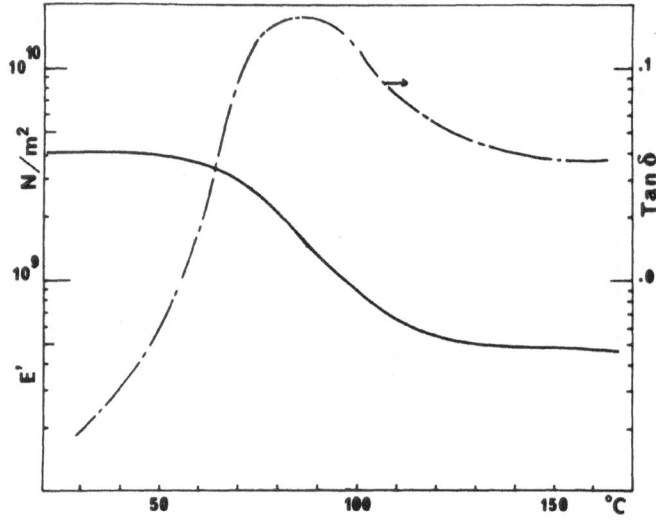

Fig. 5. Variations of E' and tan δ of PLA 100 with regard to temperature at 7.8 Hz.

These findings reflect a good adhesion between the reinforcing plies and the matrix. Experimental home-made PGA fibers have been obtained whose characteristics are: diameter 16 μm, σ_B=520 MPa, $\Delta L/L$=40% and E=17 GPa. When the spinning of such fibers will be under control, semi-crystalline PLA X stereocopolymers should allow one to rise the longitudinal Young modulus of properly designed composite bone plates or nails up to the lower values admitted for bone (10 GPa). Work in this direction is underway.

BONE FRACTURE INTERNAL FIXATION IN CLINICAL TRIALS

Our first attempts to use the totally bioresorbable composite material for the fixation of bone osteotomy in animals was carried out in sheep as most of the characterization work was performed using this animals. However, we soon realized that the experimentation will be worthless because of the behavior of the animals which poorly supported the post-operative complications as recalled in the introduction.

Therefore, three years ago, it was decided to change the strategy and to turn to clinical experimentation with the condition that the trials should be limited to easily accessible bone sites and mildly loaded fractures rather than of long bone fractures in animals right away.

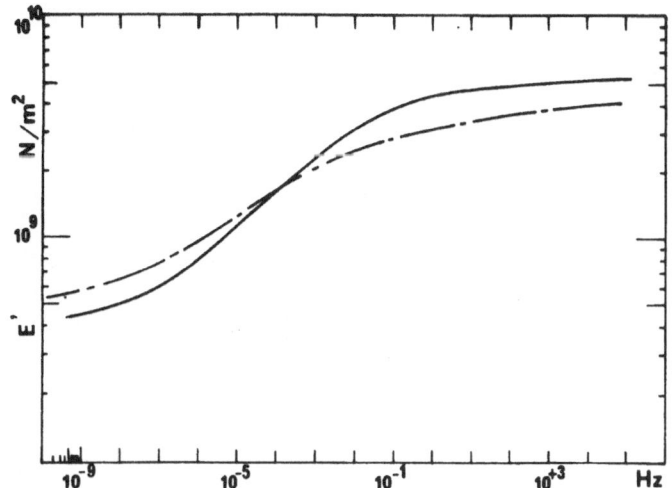

Fig. 6. Comparison between the master curves of PLA 100 (— - —) and of a 60/40 W/W PLA 100/PGA threads composite (———) obtained by using the time-temperature equivalence principle with reference temperature of 37°.

For the sake of security, the first clinical investigations have been limited to the maxillo facial sphere and especially to mandibular and skull fractures. Small size composite bone plates adapted to maxillo facial surgery were cut in circular composite plates (diameter 8 cm, thickness 2 mm) reinforced with 6 plies of the PGA mesh embedded by compression moulding as in the case of six holes bone plates described previously. It is of great interest to point out here some of the advantages of our composite with respect to composites based on mineral fibers. Indeed, the PLA/PGA composite is entirely made of organic polymers. Therefore, it is possible to cut and shape devices by various methods. This allows easier design of the plates and better adjustement to the fracture sites. Of particular interest are the possibility of cutting and shaping the device by using a surgical CO_2 LASER equipment (Fig. 7) and some very simple machines like an oven or better a heating gun with an air jet hot enough to bring the matrix to $T > T_G$ and thus in a physical state allowing plastic deformation. Finally, the definite shape can be fixed when the device is cooled to T below T_G. Some of the bone plates have been used for the fixation of mandibular fractures as exemplified in Fig. 8. More than thirty patients have been operated. These clinical experiments have led so far to succesful healing. In most cases, plates were not retrieved and thus informations on long term post operative follow-ups are not available. Anyhow, post implantation times are still too short to observe a total resorption of the plates based on long lasting PLA 100.

REPLACEMENT OF PLA 50 BY BONE WHEN USED TO FILL BONE DEFECTS IN HUMANS

Poly[α-hydroxy acids], PLA X and PLA X GA Y, have been proposed as candidates for the replacement of bone defects because of their very good biocompatibility, their adjustable biodegradability and their resorption via by-products which are eliminated through natural pathways.[9]   PLA X and PLA X GA Y have been used in the repair of experimental fractures of the orbital floor in monkeys[10] and as intraosseous appliances in the treatment of mandibular fractures in dogs.[11]   Although this was not precised by authors, what is believed to be PLA 25 GA 50 has been evaluated comparatively with $CA_3(PO_4)_2$ regarding its osteogenic potential in experimentally created defects in rat tibiae.[6]   It was shown that copolymer implants resulted in a nice gradual bone formation, processing slowly from the wound peripheries. Injection moulded PLA X and PLA X GA Y have been shown to degrade at different rates depending on composition when implanted in the cortex of rat tibiae or sheep femur as recalled above.[12]   In all cases bony tissue was observed after complete degradation at least for fast degrading poly[α-hydroxy acids].

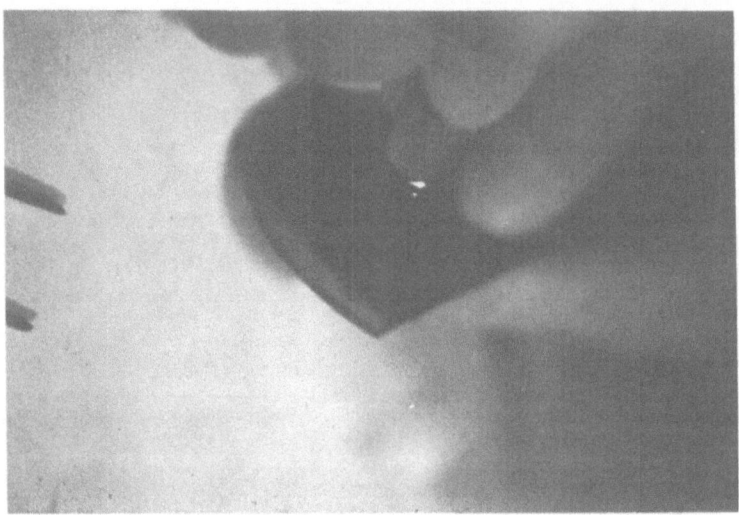

Fig. 7. Cutting of a small bone plate from a 2 mm thick circular composite plate by using a surgical CO_2 LASER.

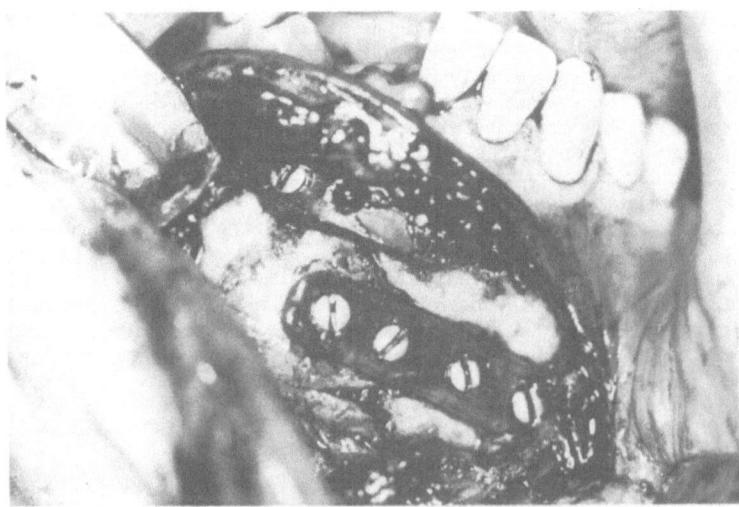

Fig. 8. Osteosynthesis of a mandibular fracture in humans by using PLA 100/PGA threads composite plates fixed by metallic screwing in this particular case.

More recently, osteogenic potential of PLA 50 GA 50 was evaluated up to 42 weeks comparatively with natural healing of experimental bone defects.[13] Bone genesis was shown to be increased by the copolymer at short implantation times but no significant difference was observed at 42 weeks.

Whether poly[α-hydroxy acids] can have osteogenic activity is of special interest for internal fixation of bone fracture. Indeed, one can expect that the healing will be faster and better if the bone tissue is regenerated rapidly beneath the plate, around the plate and wherever there is a lack of hard tissue.

We have recently experimentally used PLA 50 to fill up the hole obtained after curretage of a large maxilla cyst with nasal and sinusal communication. The implant was carefully shaped to fit the cavity as shown in Fig. 9.

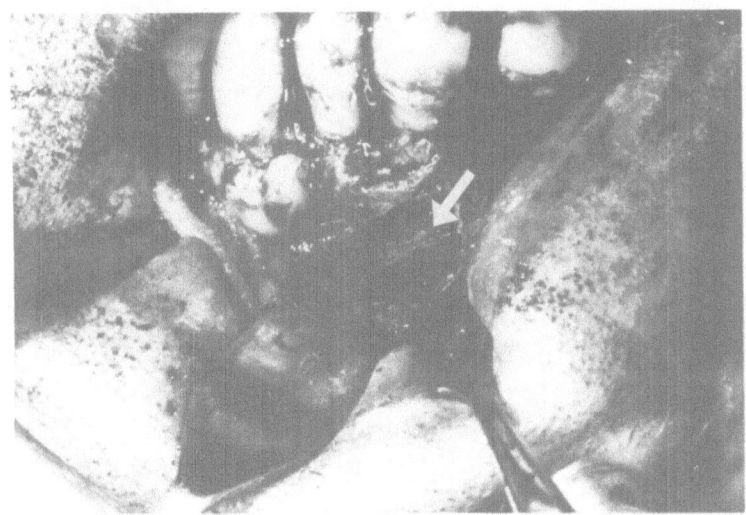

Fig. 9. Upper maxilla cyst of large size filled with a PLA 50 implant shaped to fit the site as much as possible for self-locking. (arrow: implant)

Fig. 10 shows the fate of the implantation site after various periods of time as monitored by X-rays. In spite of the sinusal communication, no infection was detected and thus no antibiotics was necessary. After 3 months, the cavity was no longer detectable on X-rays pictures.

A biopsy, whose trace is visible on the corresponding X-rays picture (Fig. 10), revealed a significant osteogenic activity (Fig. 11). The implant was almost totally resorbed and ossifying fibroblastic tissue was observed. A number of young and nature osteocytes surrounded by layer of non-calcified osteoids were observed. In two sites, the osteoids had undergone a more advanced calcification. Macrophages and giant cells were also present which were somewhat in the same line as reported in the case of animals. It is difficult to conclude from these qualitative experiments that PLA 50 increased the rate of restoration of the bone tissue previously destroyed by the cyst. However, our experience of pure surgical treatment of vast maxillary cyst does not allow us to visualise such rapid reconstruction of the bone.[14] The average time laps required for a satisfactory partial filling of the cavity is never less than a year. If similar findings were born out in long bones, it is likely that they would contribute to improve the stability of the osteosynthesis early in the healing process.

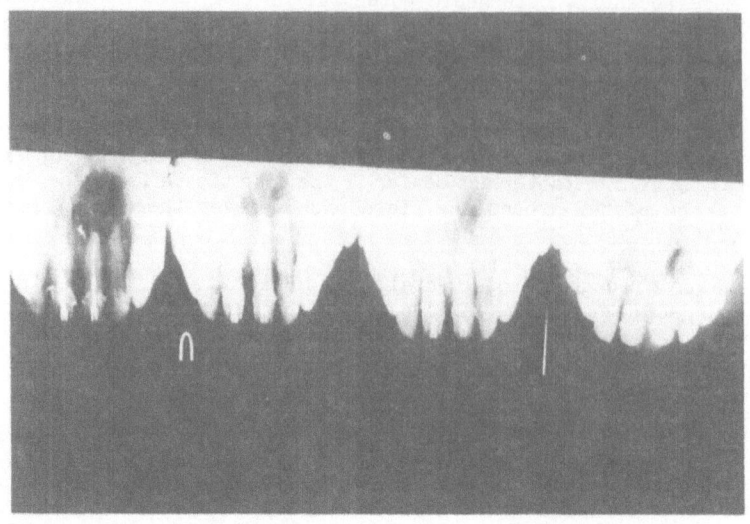

Fig. 10. Fate of the cyst cavity filled with a PLA 50 implant after 0, 1, 2, and 3 months post-operation (from left to right) as monitored by X-rays.

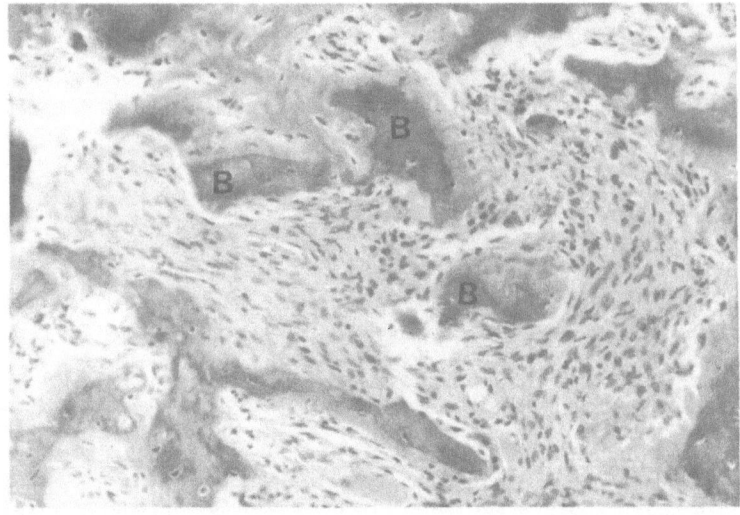

Fig. 11. Histology from biopsy at the implantation site of the PLA 50 implant at the end of the third month of implantation. (B for osteocytes and new bony tissue)

CONCLUSIONS

After a stepwise investigation of the possibility of effective use of poly[£a§-hydroxy acids] for bone surgery and more precisely for internal fixation of bone fracture in human, there is no data so far that might prohibit further efforts towards optimization of internal fixation devices (plates, nails, screws, etc...) by taking advantage of the characteristics of the totally bioresorbable composite. However, it is important for surgeons to become familiar with this new and promising materials. It is likely that they will have to adapt the current surgical procedures set up for metallic appliances to the particulars of the composite. Developments of new procedures is predictable too

ACKNOWLEDGEMENTS

Part of this work has been supported by the DGRST, research grant n°81 M 0078, and by the CNRS, ATP "MAT 3" grant n°9 83 88.

REFERENCES

1) G.W. Hastings, Carbon fibre composite for orthopaedic implants, Composites, july:193, 1978.
2) M. Vert, F. Chabot, J. Leray, and P. Christel, Bioresorbable polyesters for bone surgery, Makromol. Chem., Suppl. 5:30, 1981.
3) M. Vert, F. Chabot, J. Leray, and P. Christel, Nouvelles pièces d'ostéosynthèse, leur préparation et leur application" Fr. Pat. n° 78 28,183.
4) P. Christel, F. Chabot, J. Leray, C. Morin, and M. Vert, Biodegradable composite for internal fixation, in "Biomaterials 1980", G.D. Winter, D.F. Gibbons, and H. Plenck Jr., Eds., John Wiley & Sons Ltd, 1982, 271.
5) F. Chabot, M. Vert, S. Chapelle, and P. Granger, Configurational structure of lactic acid stereocopolymers as determined by $^{13}C-(^1H)$n.m.r., Polymer, 24:53, 1983.
6) M. Vert, P. Christel, F. Chabot and J. Leray, Bioresorbable plastic materials for bone surgery, in "Macromolecular Biomaterials", G.W. Hastings and P. Ducheyne Eds, CRC Press, (1984), 119.
7) A.M. Reed, Ph. D. Thesis, University of Liverpool (1978)
8) E.W. Fisher, M.J. Sterzel, and G. Wegner, Investigation of the structure of the solution groven crystals of lactide copolymers by means of chemical reactions, Polym., 251:980, (1973)
9) J.F. Nelson, H.G. Stanford and D.E. Cutright, Evaluation and comparisons of biodegradable substances as osteogenic agents, Oral Surg., 43:836, (1977)
10) D.E. Cutright and E.E. Hunsuck, The repair of fractures of the orbital floor using biodegradable polylactic acid, Oral Surg., 33:28, (1972)
11) L. Getter, D.E. Cutright, S.N. Bashar and J.K. Ausburg, A biodegradable intraosseous appliance in the treatment of mandibular fracture, J. Oral Surg., 30:344, (1972)
12) L. Sedel, F. Chabot, P. Christel, X. De Charentenay, J. Leray and M. Vert, Les implants biodégradables en chirurgie orthopédique, Rev. Chir. Orthop., Suppl. II, 64:92, (1978)
13) J.O. Hollinger, Preliminary report on the osteogenic potential of a biodegradable copolymer of polylactide (PLA) and polyglycolide (PGA), J. Biomed. Mat. Res., 17:71, (1983)
14) M. Chanavaz, F. Chabot, M. Donazzan and M. Vert, Clinical applications of a bioresorbable lactic acid/glycolic acid copolymer for limited bone augmentation and bone replacement, Excerpta Medica, in press.

RESIN BASED DENTAL COMPOSITES - AN OVERVIEW

Joseph M. Antonucci
National Bureau of Standards
Gaithersburg, MD 20899

INTRODUCTION

The quest for a durable, esthetic, adhesive and biocompatible material with both good handling and setting qualities suitable for the restoration of lost or damaged tooth structure has long challenged dental material researchers. A significant step toward the realization of this goal was the development of resin based dental composites which overcame many of the deficiencies of silicate cements (purely inorganic composites) and the so-called unfilled acrylic resin restoratives (purely organic composites).

Dental silicate cements, which set by an acid-base process that occurs between concentrated aqueous solutions of phosphoric acid and ion-leachable aluminosilicate glass powders, are noted for their translucency, excellent thermal properties, and an anticariogenic fluoride release that is beneficial to contiguous tooth structure (1). The shortcomings of this material are many. It is highly vulnerable to erosion, especially under acidic oral conditions (e.g. in plaque covered areas); it has a brittle nature and is weak in tensile and flexural strength; and it tends to be optically unstable, technique sensitive and irritating to dental tissues (2-5). As an esthetic restorative material the silicate cement is rarely used today.

Unfilled acrylic restoratives are formed by the ambient redox polymerization of a dough derived from mixing methyl methacrylate (MMA) with fine beads of poly(methyl methacrylate), PMMA, using an initiator

277

system consisting of benzoyl peroxide and an activator such as a tertiary aromatic amine, e.g. N,N-dimethyl-p-toluidine. This type of direct filling material is noted for ease of manipulation and application, low solubility in oral fluids, excellent optical properties, a smooth surface texture and a mechanical nature that is ductile rather than brittle (5-7). On the debit side are such unfavorable properties as a coefficient of thermal expansion many times greater than that of enamel or dentin, a high setting exotherm and polymerization shrinkage (conducive to microleakage, pulp irritation and secondary caries), a low modulus of elasticity, high water uptake and questionable biocompatibility due to the volatility of MMA and its potential for infiltration into dental tissues (5-7). Various attempts to remedy these deficiencies by the addition of crosslinking monomers to MMA and/or by the addition of glass or ceramic fillers to the PMMA powder were only modestly successful (8). Neither the silicate cements nor the unfilled acrylics exhibit any adhesion to tooth structure.

The modern development of dental composites owes much to the pioneering studies of Dr. R. L. Bowen at the National Bureau of Standards. His recognition of the excellent matrix forming potential of epoxy resins as well as their poor ambient polymerization characteristics under clinical conditions led him to the discovery of a unique hybrid monomer which combined the low polymerization contraction of epoxy resins with the excellent setting behavior of acrylic monomers (9,10). His classical syntheses (Figure 1) of the bulky, thermosetting dimethacrylate, BIS-GMA, 2,2-bis[p(2'-hydroxy-3'-methacryloxypropoxyphenyl)]propane, coupled with the preparation of silanized silica that combined translucency and radiopacity while matching the refractive indices of the resin matrix ushered in the modern era of esthetic dental composites (9-12). Compared to both the unfilled and early acrylic composites based on the MMA/PMMA system, BIS-GMA based composites have low polymerization exotherms and shrinkages, high moduli, relatively low water sorption and solubilities, improved thermal properties, esthetics and biocompatibility.

278

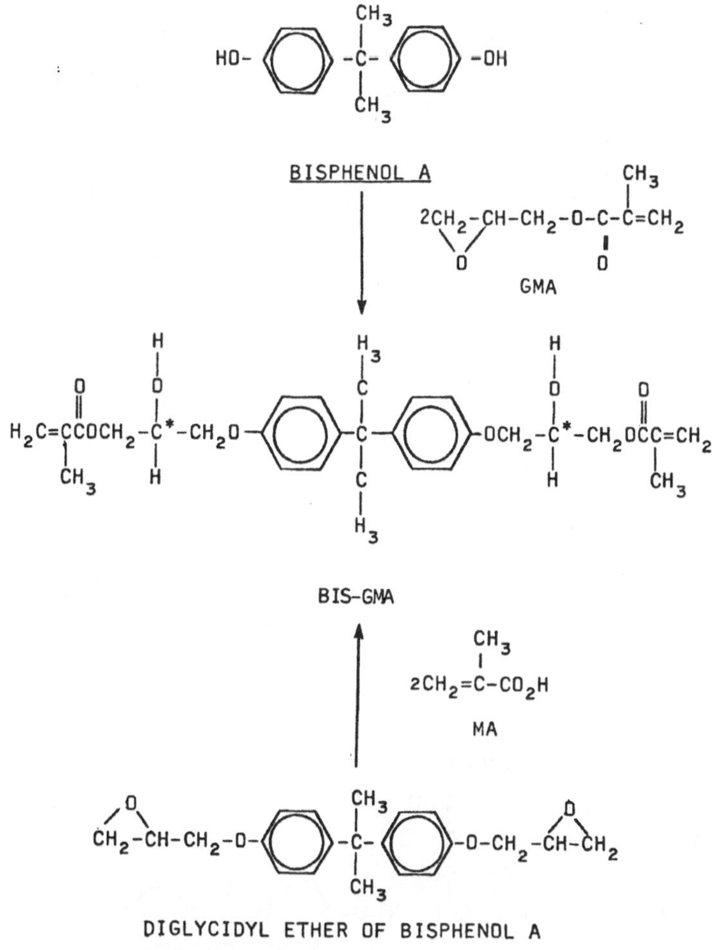

DIGLYCIDYL ETHER OF BISPHENOL A

Figure 1. Synthesis of BIS-GMA.

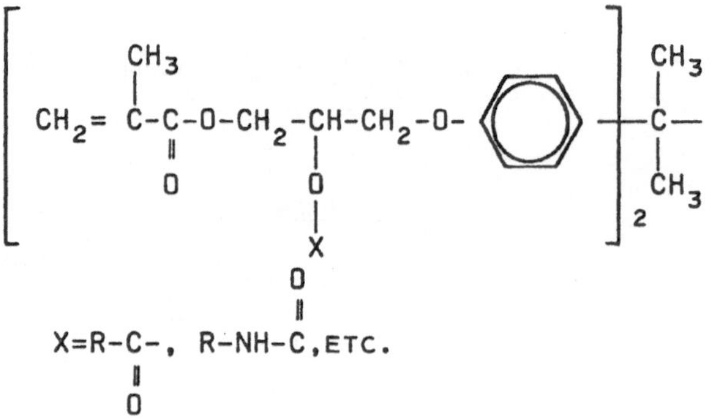

Figure 2. Derivatives of BIS-GMA.

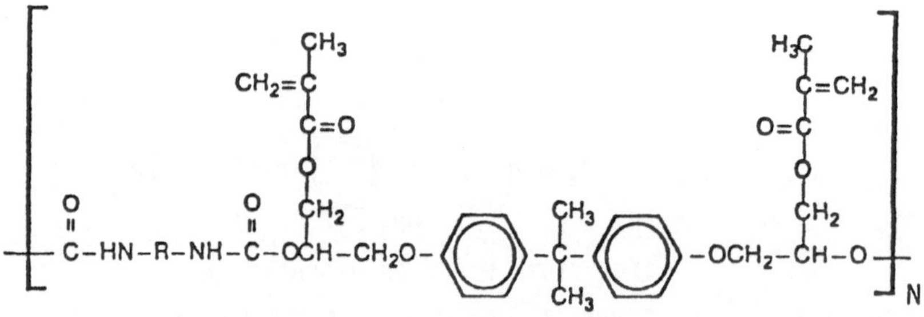

Figure 3. Oligomeric Urethane Derivative of BIS-GMA.

A. Chemistry and Composition of Dental Composites

The essential constituents of resin based dental composites are: (1) a vinyl monomer system which on free radical polymerization provides the matrix or continuous phase, (2) a free radical initiator system effective under clinical conditions, (3) glass, ceramic, organic, metallic and hybrid organic-inorganic fillers dispersed in the organic matrix constitute the reinforcing or discontinuous phase, (4) an interfacial bonding phase derived from coupling or surface active agents capable of reacting or interacting with the two disparate phases, and (5) various types of inhibitors and stabilizers for optimizing the storage stability of the uncured material and the color and general chemical stability of the cured composite. Other minor components include pigments and fluorescent agents which aid in harmonizing the appearance of the restoration to that of the tooth. Ancillary materials such as cleansers, etchants, bonding agents, protective liners and fluoride release agents are used in conjunction with the placement of dental composites to aid in their adhesion to contiguous enamel and dentin and for the protection and preservation of sound tooth structure. Dental sealants have similar compositions but are unfilled or not as highly filled as composites and usually contain a higher proportion of diluent to decrease the viscosity.

A-1. Continuous or Resin Matrix Phase

Although BIS-GMA is widely used in commercial dental resin formulations as the major or base monomer, alternative base resins as well as diluent monomer systems have been investigated in an effort to further enhance the quality of the polymeric matrix (13-19). Some are simple derivatives of BIS-GMA, e.g. esters (20) and urethanes (21) obtained by reactions involving its secondary hydroxyl groups (Figure 2). Others are oligomeric derivatives such as the urethane prepolymer multifunctional methacrylates derived from the chain extension reaction of BIS-GMA with organodiisocyanates such as 1,6-hexamethylene diisocyanate, Figure 3 (21,22). Bulky, urethane linked di-and polymethacrylates have been

synthesized from hydroxylalkyl methacrylates and multifunctional
organoisocyanates as shown in Figure 4 (23,24).

Recently a unique hexafunctional methacrylate, PNC-HEMA, was
synthesized from hexachlorocyclotriphosphazene (PNC) and 2-hydroxyethyl
methacrylate (HEMA) as outlined in Figure 5 (25). The presence of the
phosphazene aromatic ring, with its strong heteroatomic type of covalent
bonding, confers on homopolymers and copolymers of PNC-HEMA some
interesting semiorganic properties such as relatively low coefficients of
thermal expansion and very high values for surface hardness and
compressive strength. Shrinkage on polymerization also is relatively low
for this type of resin system (25). Composites based on PNC-HEMA resins,
especially those using Si_3N_4 as a filler, exhibited extremely high
compressive strengths and Knoop hardness values and low thermal expansion,
but had diametral tensile strengths somewhat lower and water sorption
values somewhat higher than conventional composites (26).

Since the water-related properties of dental composites may have a
significant influence on their performance, various homologs and analogs
of BIS-GMA without hydroxyl groups have been synthesized (Figure 6).
Some of these have such low viscosities that they do not require diluent
monomers. Their composites are characterized by their significantly lower
water sorption (27,28). In contrast to BIS-GMA and many
urethane-methacrylate based composites which have higher water uptake and
exhibit hygroscopic expansion, composites with hydrophobic properties
exhibit better dimensional stability (28). It has been argued that the
hygroscopic expansion arising from relatively high water uptake may be
beneficial in closing margin gaps existing at the restorative-tooth
interface and reducing stresses in the composite (29-32). Others contend
that to prevent microleakage due to gap formation it is necessary to
prevent capillary penetration of oral fluids by use of composites having
contact angles versus saliva of greater than 90° (33-37). A hydrophobic
composite of extremely low water uptake and high contact angle was

282

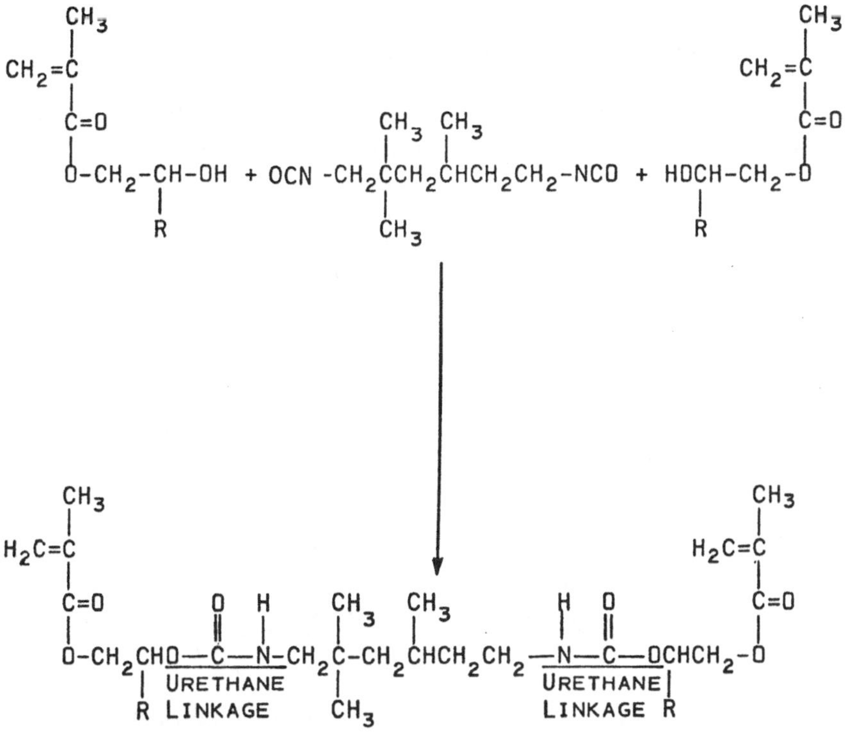

R = H,CH₃, ETC.

Figure 4. Synthesis of Typical Urethane Methacrylates.

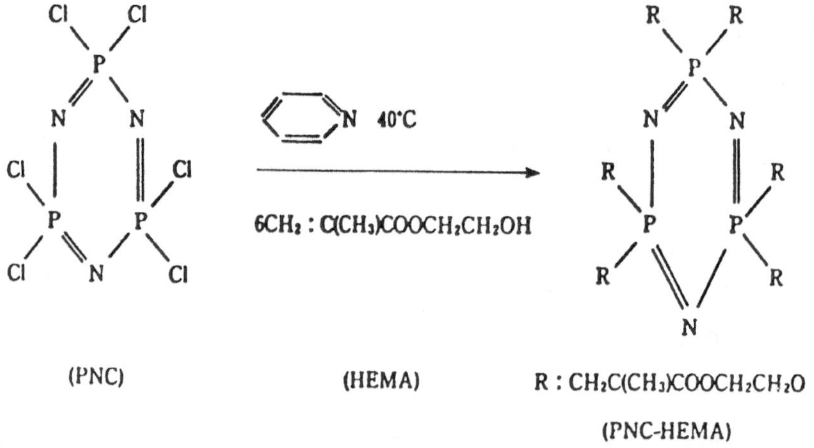
Cl
Cl
P
N
N
Cl
Cl
P
P
Cl
N
Cl
N 40°C
6CH₂ : C(CH₃)COOCH₂CH₂OH
R
R
P
N
N
R
R
P
P
R
R
N
(PNC)
(HEMA)
R : CH₂C(CH₃)COOCH₂CH₂O
(PNC-HEMA)

Figure 5. Synthesis of PNC-HEMA.

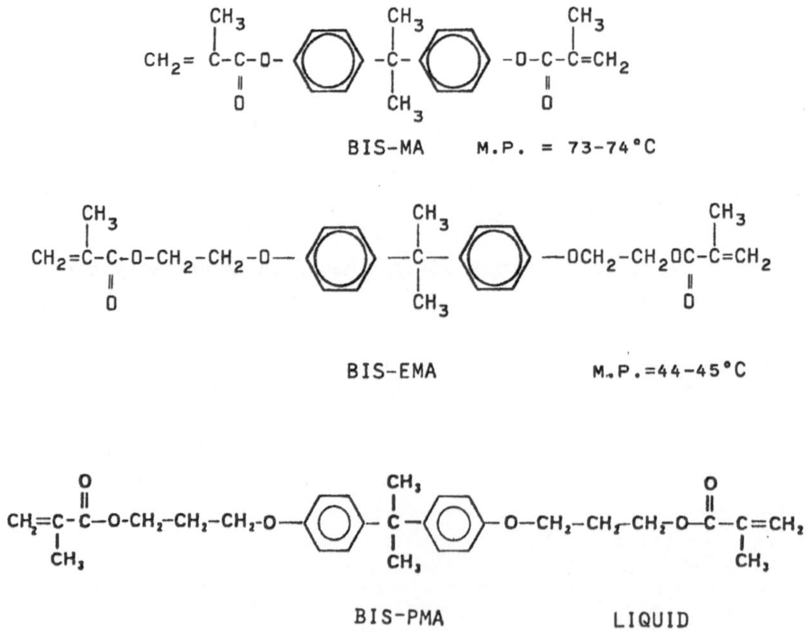
BIS-MA
M.P. = 73-74°C
BIS-EMA
M.P.=44-45°C
BIS-PMA
LIQUID

Figure 6. Non-hydroxylated Homologs of BIS-GMA.

obtained using a highly fluorinated resin system (Figure 7) comprising 3 parts by weight of octafluoro-1,1,5-trihydropentyl methacrylate (OFPMA) and 1 part by weight of 2,2-bis[p-(2'methacryloxyethoxyphenyl)]-propane, BIS-EMA (36). This composite did show reduced microleakage and improved stain resistance but had relatively low mechanical strength and a high contraction on polymerization (36-38).

Much stronger, hydrophobic composites were prepared from oligomeric highly fluorinated methacrylates such as those shown in Figure 8 (39-41). These low surface energy prepolymer multifunctional monomers are viscous liquids that are miscible with a variety of diluent monomers. Thermoset composites of low water uptake, acceptable mechanical strength, low residual vinyl unsaturation and polymerization shrinkage have been prepared (39-44). These dental composites have solubility parameters far less than those found in the oral cavity and, therefore, their polymeric matrices are less susceptible to plasticization and the possible degradative effects induced by constant exposure to oral fluids. In a related study it was demonstrated that siloxane-containing resin systems also yield composites of low water sorption, acceptable mechanical properties and enhanced oral environmental resistance (45).

Monomers that expand or show no volume change on polymerization have been synthesized and may offer a method of counteracting margin gap formation which contributes to the microleakage of composites. In addition, a non-shrinking or slightly expanding dental resin system may relieve stresses that inevitably develop in the composite and at the composite-tooth interface due to polymerization shrinkage (46-49). An experimental resin based on BIS-GMA and the solid spiro orthocarbonate, 3,9-dimethylene-1,5,7,11-tetraoxaspiro-5,5-undecane, which can polymerize by a free radical ring opening mechanism (Figure 9), yielded low shrinking composites with improved bonding to enamel (49).

A-2 Mode of Polymerization

Free radical polymerization of resin based dental materials is

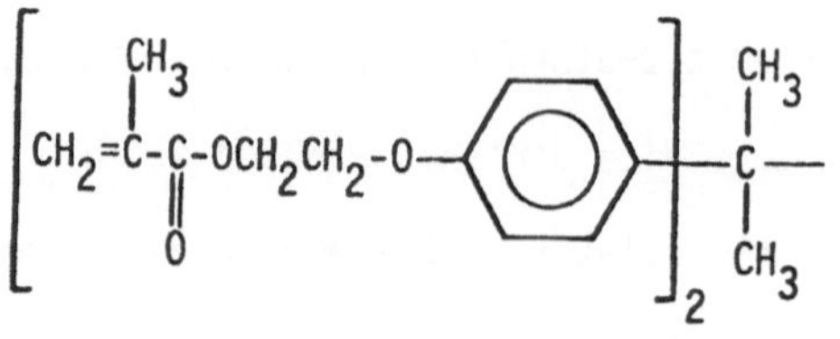

2,2-BIS[p-(β-methacryloxyethoxy)phenyl]propane

BIS-EMA

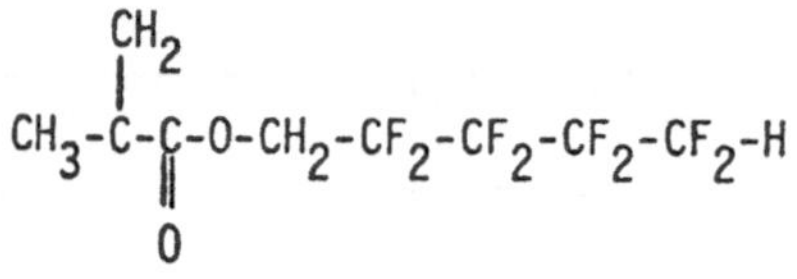

Percent F = 50.7, MW = 300

1,1,5-trihydrooctafluoropentyl
Methacrylate

OFPMA

Figure 7. Hydrophobic Dental Rein System Based on OFPMA.

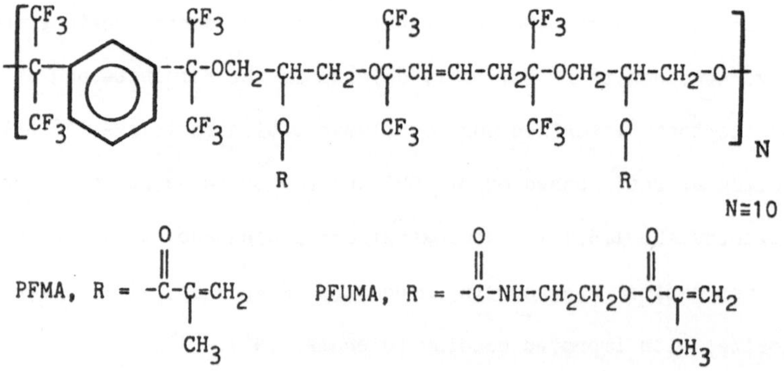

N≈10

Figure 8. Chemical Structures of PFMA and PFUMA

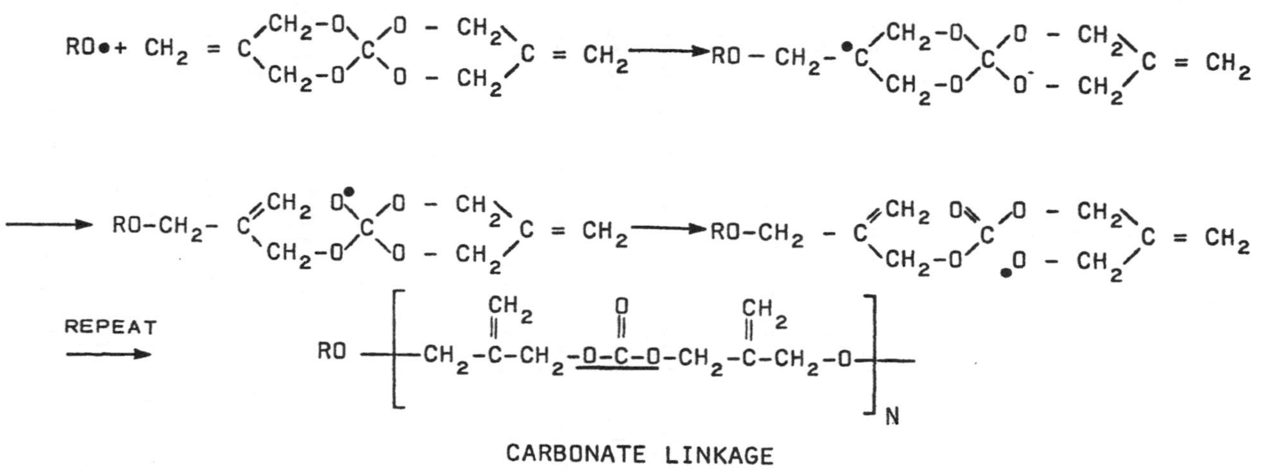

CARBONATE LINKAGE

Figure 9. Radical Ring Opening Polymerization of Spiro Orhto Carbonate Vinyl Monomers with Expansion in Volume.

initiated chemically by redox systems or photochemically by either long wave ultraviolet (320-400nm) or longer, visible irradition (400-500nm) using photoinitiators or photosensitizers.

The classical initiator system of benzoyl peroxide plus a tertiary aromatic amine such as N,N-bis(2-hydroxyethyl)p-toluidine is used in many dental composites (50,51). Studies have indicated that the nature of the aryl amine polymerization accelerator and the peroxide/amine ratio are important factors in determining the setting characteristics, physical and mechanical properties, esthetics and color stability of composites (50-55).

Hydroperoxides (e.g. cumene hydroperoxide) and thioureas (e.g. acetylthiourea) form an alternate redox initiator system that exhibits improved storage and color stability (56). Less color prone benzoyl peroxide/amine initiator systems were developed using ascorbic acid (or derivatives) and/or polythiols as synergistic activators (55,57,58). Ascorbic acid and its derivatives were shown to be versatile promoters for the ambient decomposition of not only acyl peroxides such as benzoyl peroxide but also for peresters and hydroperoxides (57). Chemically cured composites of necessity require a two component formulation (e.g. paste/paste, paste/liquid, powder/liquid).

Benzoin ethers (e.g. benzoin methyl ether) provide initiating radicals via a photofragmentation mechanism upon exposure to UV irradiation at ca 360nm (59). In visible light activated composites α-diketones such as camphorquinone are used as the photoxidant and tertiary amines (e.g. N,N-dimethylaminoethyl methacrylate) as the photoreductant. Exposure to visible light (ca 470nm) promotes the α-diketone to an excited triplet state (acceptor) where it combines with the reductant (donor) to form an exiplex which breaks down to initiating radicals (51,59-62).

Both the UV and visible light initiators provide a "command set", single component, restorative material that yields composites of lower

porosity and less surface tackiness (due to air inhibition) than the usual
chemically activated composites. The degree of cure varies with depth for
these materials and a layering technique is employed for restorations of
great bulk. The depth of cure is greater for visible light composites
than for similar UV activated materials (59, 63, 64). Chemically
activated composites are more uniformly polymerized except at the surface
unless measures are taken to prevent oxygen inhibition.

A-3. Discontinuous or Reinforcing Filler Phase

On a weight basis the major component of dental composites is usually
the reinforcing filler (e.g. 50-85%). For many properties of the
composite the volume percent of the dispersed phase is a more significant
parameter. The reinforcing filler performs many functions in a composite
such as stiffening the lower modulus resin binder thereby increasing
mechanical properties, enhancing dimensional stability, moderating the
exotherm of polymerization and the mismatch between the thermal expansion
of the organic matrix and tooth structure, reducing water sorption and
polymerization shrinkage and aiding in matching tooth appearance. By
using glass or ceramic fillers that have refractive indices approximating
those of the matrix they can be used to form translucent fillings that
match the translucency of tooth structure. The selective inclusion of
compounds with elements of high atomic number (e.g. barium, strontium,
lanthanum, zinc, zirconium, titanium, etc.) in the preparation of glass
fillers yields esthetic composites with a degree of radiopacity
(12,19,65-69).

A variety of types, shapes and sizes of fillers have been used in
dental composites, e.g. quartz, fused silica, borosilicate and
aluminosilicate glasses, silicon nitride, calcium silicate, calcium
phosphates, aluminum oxide, metals, etc. (19). In addition, submicron
fillers such as precipitated or pyrogenic silicas (0.14-0.007um) averaging
0.04um in size have been used in microfilled composites (70-71). The high
surface area of this type of filler makes it difficult to achieve high

filler loadings in this type of composite, e.g. 50 weight percent is usually the maximum. To enhance their miscibililty and dispersion in resin systems, small, organic-inorganic, macrofillers are made from pulverized, prepolymerized composites derived from the silanized, microfine fillers and the same or similar monomer systems. Composites formulated with these prepolymerized composite fillers also are termed microfilled composites (71).

Compared to conventional composites with their larger filler sizes (0.7 to 100um, but usually 3-50um), microfilled composites have a smoother, more easily polishable surface texture which may reduce the adherence of plaque and stains. On the other hand, they have lower moduli and tensile strength, exhibit more creep and have higher water uptake, thermal expansion and polymerization shrinkage than conventional or hybrid composites.

Hybrid composites, which incorporate major quantities of the smaller sized macrofillers along with small amounts of microfillers, achieve almost as smooth a surface texture as the microfilled composites without compromising (and actually improving) other properties. Some of the newer hybrid composites have a multimodal dispersed phase consisting of different types, shapes and sizes of fillers.

A novel approach at developing stronger interaction between the continuous and the dispersed phases involved the use of porous, three dimensional glass network particles derived from extremely fine sintered glass fibers of small diameter (72). The objective was to develop composites with a condensable filler phase which can strongly interlock with a minimum of the resin binder.

Another innovative approach to enhance the interfacial bonding of the inorganic and organic phases of the composite is through the use of "semiporous" glass fillers obtained by selectively acid etching the more soluble phase of glass particles having two interconnected vitreous phases (73-74). Properly done this results in a glass filler having superficial

290

surface porosity into which the resin can flow and mechanically interlock on polymerization, thus complementing the usual bonding through silane coupling agents.

A new type of vitreous filler having a small uniform size (0.2-0.3um) and spherical shape has recently been developed. The fillers are intermediate in size between microfine silica and small macrofillers. Composites prepared with these fillers compare favorably in mechanical and physical properties with conventional or hybrid composites and in polishability and surface texture with microfilled composites (75,76).

The search for dental composites of superior wear resistance for use in stress-bearing applications has spurred research into new types of stable fillers of sizes and shapes conducive to optimal packing efficiency (75-79).

A-4. Interfacial Phase

Although only a minor component of resin based dental restorative materials, interfacial bonding agents exert a profound effect on the durability of these composites. The quality of the interfacial bonding phase existing between the polymeric matrix and the dispersed phase has a significant effect on the ultimate properties and the clinical performance of dental composites. Even composites prepared from the best of resin binders and reinforcing fillers will be deficient in durability if water and other contaminants penetrate and disrupt the interfacial bonding phase.

Bifunctional coupling agents such as organofunctional silanes, titanates, zirconates, etc. are used in composites to promote adhesion between mineral fillers and organic resin binders (80). Alkoxysilanes having terminal vinyl groups have been the most widely used type of coupling agent for dental composites. Initially, a vinyltrialkoxysilane was used but it was later found that 3-methacryloxypropyltrimethoxysilane was more effective (12,81).

Alkoxysilanes can react with surface moisture, usually present at least as a monolayer on mineral surfaces, to generate silanol groups which can strongly hydrogen bond to hydroxylated surfaces. In addition, silanol groups can react chemically with surface hydroxyl groups of the filler via covalent bond formation. There is some direct (e.g. spectroscopic) evidence that suggests that these kinds of reactions do occur between silane coupling agents and many types of mineral fillers used to reinforce composites (80,82). Organofunctional silanes then can be visualized as reacting by both hydrogen bonding and/or covalent attachment to mineral fillers by virtue of their silanol or derivative groups and by copolymerization with the resin system via their terminal vinyl groups. Indirect evidence for this interfacial bonding is provided by the observed enhancement in mechanical strength and resistance to wear and attack by water and other chemicals of silanized composites (80,83).

The effectiveness of coupling agents in a composite depends on a number of factors: (1) the nature of the resin binder and filler, (2) the structure and chemical reactivity of the coupling agent (3) the amount used and (4) the mode of application.

A-5. Inhibitors and Stabilizers

For reasonable storage stability unpolymerized composite restorative materials contain small quantities of inhibitors and antioxidants. The presence of at least trace amounts of oxygen also is necessary for the phenolic type of inhibitors to be effective. Formerly hydroquinone was used in many dental resin systems but it caused serious discoloration in the polymeric materials and has been replaced by the monomethyl ether of hydroquinone (MEHQ) or by the sterically hindered phenol, 2,6-di-tert-butyl-p-cresol or butylated hydroxytoluene (BHT). Of the three inhibitors, BHT contributes the least discoloration to composites and may have superior biocompatibility as well, but it is not as effective in inhibiting polymerization as MEHQ (81). Another effective, potentially biocompatible inhibitor, is the phenolic α-amino acid, tyrosine (83).

In addition to their function as polymerization inhibitors the

292

hindered type of phenols, e.g. BHT, also can serve as an antioxidant. An effective approach is to use a copolymerizable inhibitor-antioxidant as part of the stabilizing system (84).

Composite restorations have a tendency to darken somewhat with age and at least part of this color instability may be indicative of chemical changes occurring in the organic matrix and/or by-products from the initiator or inhibitor systems. It has been shown that significant amounts of residual vinyl groups reside in the polymer matrix (85). These and other labile structures present in the organic phase may serve as sites for oxidative degradation.

In addition to antioxidants, UV stabilizers such as 2-hydroxy-4-methoxybenzophenone, 2-(2'-hydroxy-5'-methylphenyl)benzotriazole and phenyl salicylate are added to composites to preserve their color stability. Since they are absorbers of UV irradiation, UV stabilizers are not compatible with UV activated composites.

Adhesion to Tooth Structure

Currently used resin based dental composites are not adhesive to tooth structure. However, the discovery of the acid etch technique by Dr. M. G. Buonocore of the Eastman Dental Center at the University of Rochester School of Medicine and Dentistry made it possible to bond resin based dental materials to enamel by a micromechanical interlocking mechanism (86,87). Surface microporosity is generated on enamel by a brief pretreatment with aqueous phosphoric acid or certain types of organic acid e.g. pyruvic, citric, etc. (88). The acid etch technique is ineffective and contraindicated for use with dentin.

Adhesion to dentin has presented a more challenging problem. On a weight basis dentin consists of 69% hydroxyapatite, 18% organic matter (mainly collagen) and about 13% water. An adhesive bonding or coupling agent for this substrate would mean less invasive cavity preparations with decreased loss of sound tooth structure and a reduction in microleakage with its potential for secondary caries formation. Considerable effort has been devoted to the development of coupling agents that can mediate bonding between dental resins and apatitic substrates (88).

A surface active comonomer that can bond to apatitic substrates by chelation of surface Ca^{+2} and other multivalent cations is the adduct of glycidyl methacrylate (GMA) and N-phenylglycine (NPG), NPG-GMA (89). A similar coupling agent (NTG-GMA) has been made from N-p-tolylglycine and GMA (90). Other types of coupling agents are functional vinyl monomers that have groups capable of reacting with collagen by specific chemical reactions, e.g. esterification, urethane or schiff base formation (90-96). Another approach for chemically bonding to collagenous substrates involved graft polymerization techniques using free radical initiation (88).

Recently three types of adhesion-promoting systems have demonstrated in vitro rather strong adhesion to dentin. One system involves pretreatment of dentin with an aqueous solution of 2-hydroxyethyl methacrylate (HEMA) and glutaraldehyde (96). The critical component of the second bonding system is the functional monomer 4-methacryloxyethyl trimellitic anhydride, 4-META (91,92). The third tooth-structure conserving bonding procedure utilizes a brief application of a cleanser mordant aqueous solution of ferric oxalate, a surface active comonomer such as NPG-GMA or NTG-GMA and PMDA, the diadduct of HEMA and pyromellitic anyhdride (97,98). The latter two bonding systems display significant adhesion to enamel as well.

Properties of Dental Composites

Table 1 compares some properties of the different types of dental composites, unfilled acrylic resin restoratives, glass ionomer cements and amalgams. In general, microfilled composites are somewhat inferior to conventional and the newer hybrid composites in physical and mechanical properties but have superior surface finish. Because of their higher filler content, the newer hybrid composites are stronger, have less water sorption, lower thermal expansion and polymerization shrinkage, and show less wear than both the microfilled and macrofilled composites.

294

Table 1

Some Typical Properties of Dental Restoratives

Property	Unfilled MMA/PMMA	Microfilled Composite	Conventional & Hybrid Filled Composite	Glass Ionomer Cements	Amalgam
Inorganic Content (Wt %)	—	27-51	60-86	—	—
Compressive Strength (MPa)	55-76	221-330	127-473	140-229	320-500
Tensile Strength (MPa)	14-28	28-56	28-63	7-16	45-70
Modulus of Elasticity (GPa)	2.4	3.2-5.4	6-17	9-20	20-45
Transverse Strength (MPa)	55	80-100	90-139	7-40	124
Knoop Hardness (KHN)	10-16	25-52	60-90	60	110
Linear Coefficient of Thermal Expansion (10^{-6} °C^{-1})	92	46-70	25-40	10-15	22-28
Water Sorption (mg/cm^2)	2	0.94-2.20	0.11-0.74	—	0
Polymerization Shrinkage (Vol %)	5.2-8.0	1.9-5.8	1.2-5.3	—	0
Thermal Conductivity	5.7	15-20	25-33		550

Polyelectrolyte Based Restoratives

The glass ionomer cement, a polyelectrolyte based composite developed by Dr. A. D. Wilson, has its roots in both the traditional dental silicate cement and the modern zinc polyacrylate or polycarboxylate cement discovered by Dr. D. C. Smith (1,99-101). Translucent, brittle cements, strong in compression but weak in tension, are formed by the admixture of ion-leachable calcium aluminofluorosilicate glass powders and aqueous solutions of a poly(alkenoic acid), e.g. poly(acrylic acid). The formation of the matrix occurs by a series of acid-base reactions involving multivalent cations (e.g. Ca^{+2}, $A\ell^{+3}$) that bind the pendant carboxylate anions primarily into hydrated, ionic crosslinks thereby transforming the polyacid into a stiff hydrogel into which are tightly imbedded partially reacted glass particles. In contrast to resin based composites, the glass ionomer cements are not subject to oxygen inhibition during hardening but are sensitive to moisture during the early stages of the setting reaction. The introduction of water-setting glass ionomer cements in which a powder blend of freeze dried polyacid and leachable glass are activated by the addition of water has made this material less technique sensitive and easier to manipulate. Their working and setting behavior also has recently been improved by the use of chelating agents and pretreatments of the glass powder with acids (102).

The most noteworthy property of the polyelectrolyte cements is their adhesion to enamel, dentin and base metals. A recent study has shown that the polyelectrolyte ions react with the apatite structure by displacement of both calcium and phosphate ions while presumably forming an intermediate layer between the composite and enamel that consists of polyacrylate, phosphate and calcium ions (103). The mechanism for bonding to dentin is less clear but may involve adhesion through hydrogen bonding. The glass ionomer cement is used as a luting agent and as a pit and fissure sealant but its main use is as a cosmetic filling, especially for the repair of cervical lesions.

An interesting new application of glass ionomer cements is as an intermediary bonding layer for the non-adhesive resin based composites by first acid-etching the surface of the hardened cement prior to placement of the composite. The composite then bonds micromechanically to the etched cement in a manner similar to its attachment to acid-etched enamel (103).

Metallic fillers have been added to glass ionomer cements in an effort to moderate their excessively brittle nature. The flexural strength was increased, but the matrix-metal bond failed in fracture and the composites were prone to wear and had poor esthetics (103). A more successful approach at enhancing the toughness of this brittle material involved fusing fine malleable metal powders (silver and gold) onto ion-leachable glasses. This new type of polyelectrolyte composite, the cermets, retain the adhesiveness, cariostatic and biocompatible features of glass ionomer cements but their esthetics are compromised. The ductile, metallized filler imparts burnishability, toughness, radiopacity and improved wear resistance to the cement, which may have potential as a posterior restorative material. The enhanced wear may be attributable to the low coefficient of friction of the polished, metallized surface, the energy absorbing properties of the filler which acts as a toughening agent for the brittle matrix, and the absence of the porosity normally found in glass ionomer cements (103,104).

Future Trends in Restorative Material Research

In spite of the steady improvement in the properties of resin - and polyelectrolyte based composites, amalgams remain the primary posterior restorative material. It is instructive to compare the properties of amalgam with those of composites and the glass ionomer cements previously cited. The deficiences of amalgam beside the esthetic factor are: (1) a slow setting behavior, (2) relatively high thermal conductivity which requires thermal insulation to protect the pulp, (3) susceptibility to corrosion and creep (especially in the early stages of the hardening process) and (4) poor margin adaptability traceable to its inability to bond to tooth structure, i.e., it acts only as a mechanical plug.

Among the many advantages of amalgam are: (1) ease of placement and finishing, (2) smooth surface texture and low friction, (3) excellent mechanical properties, particularly its low wear, (4) zero water sorption and excellent resistance to softening by the oral environment (e.g. plasticization), (5) virtually no change in volume on hardening, which while it does not seem to enhance margin sealing should reduce or eliminate internal contractile stresses in the restorative, and (6) acceptable biocompatibility. Also, compared to other posterior restorative materials, amalgam causes the least wear of opposing tooth structure (105,106).

For composites and glass ionomer cements to be considered as alternative stress-bearing restorative materials they will need to emulate many, if not all, the desirable properties of amalgam while eliminating its shortcomings.

This means that material restorative research should not only aim at toughening and strengthening polymer based composites by optimizing the volume fraction of the dispersed phase with a, tough, stable filler system, but also should develop more durable resin binders and interfacial coupling agents since it is the organic phase of the composite or the polyelectrolyte matrix of the glass ionomer cement that is the first line of defense against the constant assaults of the oral environment.

Future resin systems must achieve the seemingly contradictory goals of achieving a high degree of thermoset polymerization with only minimal, preferably zero, polymerization shrinkage. Ideally resins with a slight expansion on setting may be desirable as this can improve margin sealing and augment adhesion to tooth structure by a strong micromechanical interlocking mechanism. In addition, for maximum chemical and mechanical stability, the chemical nature of the polymeric matrix and interfacial bonding phase must be such that they yield strong virtually frictionless

298

composites that are impermeable to the softening and potentially degradative effects of oral fluids.

One promising approach that we are pursuing at the National Bureau of Standards is to develop low surface energy resin systems, e.g. thermosetting, highly fluorinated bulky resins and similar types of polysiloxane-containing resins, that are capable of a high degree of polymerization leading to networks of high crosslink density with only minimal polymerization contraction. In their tough, ductile nature, low water uptake and volume change, and low solubility parameters far removed from that of oral fluids, these types of polymeric binders are expected to enhance the wear resistance of composites. With the concomitant development of more stable, hydrophobic coupling agents having better compatibility with these low surface energy resins plus the ability to bond more effectively to fillers, the outlook for the development of more durable dental composites looks bright.

Summary

Vinyl resin based composites have become the primary esthetic anterior filling material used in dentistry and also are widely used in a variety of other dental applications, e.g. pit and fissure sealants, glazes, orthodontic bonding agents, core build-up, crown and bridge materials, laminating veneers, and, at least in their early developmental stage, a potential posterior restorative material. Significant advances in the art of conservative and preventive dentistry have accompanied the development of resin and polyelectrolyte based dental composites and the increasingly effective bonding systems for enamel and dentin.

This work was supported by NIDR/NBS Interagency Agreement Y01-DE-30001.

References

1. Wilson, A.D., Chem. Soc. Rev. 7 265, 1978.

2. Paffenbarger, G. C. Schoonover, I.C. and Sounder, W., JADA 25 32, 1938.

3. Henschel, C.J., J. Den. Res. 28, 528, 1949.

4. Wilson, A.D., in Scientific Aspects of Dental Materials, Von Fraunhofer, J.A. ed., Buttersworths, London, Chap.6, 1975.

5. Bowen, R. L., Paffenbarber, G. C., and Mullineaux, A. L., J. Pros. Dent. 20 426, 1968.

6. Coy, A.D., JADA 47, 532, 1953.

7. Paffenbarger, G. C., Nelsen, R.J., and Sweeney, W.T., JADA 47 16, 1953.

8. Brauer, G.M., JADA 72 1151, 1966.

9. Bowen, R.L., J. Dent. Res. 35 360, 1956.

10. Bowen, R.L., U.S. Patent 3,066,112, 1962.

11. Bowen, R.L., JADA 66 57, 1963.

12. Bowen, R.L., JADA 69 481, 1964.

13. Glenn, J.F., in Biomedical and Dental Applications of Polymers, Gebelin, C.G. and Koblitz, F.F., eds. Plenum Press, New York, pp. 317-345, 1981.

14. Halpern, B.D. and Karo, D.W., ibid, pp 337-345.

15. Antonucci, J.M., ibid, pp 357-371.

16. Cowperthwaite, G.F., Foy, J.J., and Malloy, M.A., ibid, pp 379-385.

17. Dulik, D.M., Bernier, R., and Brauer, G.M., J. Dent. Res. 60 983, 1981.

18. Dermann, K., Rupp, N.W., and Brauer, G.M., J. Dent. Res. 61 1250, 1982.

19. Glenn, J.F., in Biocompatibility of Dental Materials, Vol. 3, Smith, D.C. and Williams, D.F., eds, CRC Press Inc., Boca Raton, FL, Chap. 5, 1982.

20. Stoffey, D.G. and Lee, H.L., U.S. Patent 3,755,420, 1973.

21. Waller, D.E., U.S. Patent 3,629, 87, 1971.

22. Waller, D.E., U.S. Patent 3,709,866, 1973.

23. Foster, J. and Walker, R.J., U.S. Patent 3,825,518, 1974.

24. Foster, J. and Walker, R.J., U. S. Patent 3,862,920, 1975.

25. Anzai, M. and Ohashi, M., J., Nihon Sch. Dent. 26 109, 1984.

26. Anzai, M. and Ohashi, M., J., Nihon Univ. Sch. Dent. 26, 238, 1984.

27. Atsuta, M., Nakabayashi, N. and Masuhara, E., J. Biomed. Mater. Res. 5 183, 1971.

28. Schmitt, W. and Purrman, R., U.S. Patent 3,810,938, 1974.

29. Asmussen, E. and Joergensen, K.D., ACTA ODONT Scand. 30 3, 1972.

30. Asmussen, E., ACTA ODONT Scand 33 337, 1975.

300

31. Brauer, G.M., Dulik, D.M., Hughes, H.N., Dermann, K. and Rupp, N.W., J. Dent. Res. 60 1966, 1981.

32. Bowen, R. L., Rapson, J.E. and Dickson, G., J. Dent. Res. 61 654, 1982.

33. O'Brien, W., Craig, R.G. and Peyton, F., J. Prosthet Dent. 19 399, 1968.

34. Craig, R.G., J. Dent. Res. 58 1544, 1979.

35. Craig, R. G. and Wang, T.K., J. Oral Rehabil. 7 361, 1980.

36. Douglas, W.H., Craig, R.G. and Chen, C.J., J. Dent. Res. 58 1981, 1979.

37. Douglas, W.H., Chen, C.J. and Craig, R.G., J. Dent. Res. 59 1507, 1980.

38. Douglas, W. H. and Craig, R.G., J. Dent. Res. 61 41, 1982.

39. Antonucci, J.M., Griffith, J.R., Peckoo, R.J. and Termini, D.J., J. Dent. Res. 58 242, 1979.

40. Antonucci, J.M., Venz, S., Stansbury, J.W. and Dudderar, D.J., J. Dent. Res. 62 285, 1983.

41. Antonucci, J.M., Stansbury, J.W., and Venz, S., J. Dent. Res. 64 209, 1985.

42. Venz, S., Rupp, N.W. and Antonucci, J.M. J. Dent. Res. 62 254, 1983.

43. Antonucci, J.M., Venz, S., Stansbury, J.W. and Dudderar, D.J., Proc. of The First Medical Plastic Conf. of the Society of the Plastics Industry, Inc., New Brunswick, NJ, 1983.

44. Venz, S. and Antonucci, J.M., J. Dent. Res. 64 229, 1985.

45. Kuo, J.S., Antonucci, J.M. and Wu, W., J. Dent. Res. 64 178, 1985.

46. Endo, T. and Bailey, W.J., J. Poly. Sci, Poly. Chem. Ed. 13 2525, 1975.

47. Endo, T. and Bailey, W.J., J. Poly. Sci, Polym. Lett. Ed. 18 25, 1980.

48. Han, Y.K. and Choi, S.K., J. Poly. Sci., Poly. Chem. Ed. 21 353, 1984.

49. Thompson, V.P., Williams, E.F. and Bailey, W.J., J. Dent. Res. 58 1522, 1979.

50. Brauer, G.M., Ref. 13, pp. 395-409.

51. Brauer, G.M. and Argentar, H., ACS Symposium Series No. 212, Bailery, F.E. Jr., Ed. ACS, 1983.

52. Bowen, R. L. and Argentar, H., J. Appl. Poly. Sci. 17 2213, 1973.

53. Brauer, G.M. Dulik, D.M., Antonucci, J.M., Termini, D.J. and Argentar, H., J. Dent. Res. 58 1994, 1979.

54. Brauer, G.M., Stansbury, J.W. and Antonucci, J.M., J. Dent. Res. 60 1343, 1984.

55. Antonucci, J.M., Peckoo, R.J., Schruhl, C. and Toth, E.E., J. Dent. Res. 60 1325, 1981.

56. Temin, S.C. and Richards, C.L., U.S. Patent 3,991,008, 1976.

57. Antonucci, J.M., Grams, C.L. and Termini, D.J., J. Dent. Res. 58 1887, 1979.

58. Antonucci, J.M., Stansbury, J.W. and Dudderar, D.J., J. Dent. Res. 61 270, 1982.

59. Kilian, R.J., Ref.13, pp 411-417.

60. Buonocore, M.G. and Davila, J., JADA 86 1349, 1973.

61. Dart, E. C. and Nemcek, J., U.S. Patent 4,071,424, 1978.

62. Ledwith, A., Pure Appl. Chem 49 431, 1977.

63. Cook, W.D., J. Dent. Res. 59 800, 1980.

64. Salako, N.O. and Cruickshanks-Boyd, D.W., Br. Dent. J., 146, 375, 1979.

65. Bowen, R. L. and Cleek, G.W., J. Dent. Res. 48 79, 1969.

66. Bowen, R. L. and Cleek, G.W., J. Dent. Res. 51 177, 1972.

67. Barton, J.A., Burns, C.L., Chandler, H.H. and Bowen, R.L., J. Dent. Res. 52 731, 1973.

68. Juracic, A., U.S. Patent 3,971,754, 1976.

69. Müller, G., J. Dent. Res. 53 1342, 1974.

70. Craig, R. G., Symposium on Composite Resins in Dentistry, Dental Clinics of North America, 25 219, 1981.

71. Lutz, F. and Phillips, R.W., J. Prosthet. Dent. 50 480, 1983.

72. Ehrnford, L., J. Dent. Res. 60 1759, 1981.

73. Bowen, R. L. and Reed, L.E., J. Dent. Res. 55 738, 1976.

74. Bowen, R. L. and Reed, L.E., J. Dent. Res. 55 748, 1976.

75. Tani, Y., Dental Outlook 2 1, 1983.

76. Tani, Y., Suzuki, K., Hanada, T. and Yuasa, S., J. Dent. Res., 64 179, 1985.

77. Soderholm, K.J.M., J. Dent. Res. 60 1867, 1981.

78. Soderholm, K.J.M., J. Dent. Res. 62 126, 1983.

79. Cross, M., Douglas, W.H. and Fields, R.P., J. Dent. Res. 62 850, 1983.

80. Pludderman, E.F., in Silane Coupling Agents, Plenum Press, NY, 1982.

81. Bowen, R. L., J. Dent. Res. 58 1493, 1979.

82. Chen, T.M. and Brauer, G.M., J. Dent. Res. 61 1439, 1982.

83. Antonucci, J.M., Misra, D.N., and Peckoo, R.J., J. Dent. Res. 60 1332, 1981.

84. Bowen, R. L. and Argentar, H., J. Dent. Res. 51, 1071, 1972.

85. Ruyter, I.E. and Svendsen, S.A., ACTA ODONTOL Scand. 36 75, 1977.

86. Buonocore, M.G., J. Dent. Res. 34 849, 1965.

87. Silverstone, L.M., Ref.19, Vol.I, pp 61-74, 1982.

88. Brauer, G.M., Ref.4, Chap. 2.

89. Bowen, R. L., J. Dent. Res. 44 600, 895, 903, 906, 1369, 1965.

90. Bowen, R.L., Cobb, E.N. and Rapson, J.E., J. Dent. Res. 61 1070, 1982.

91. Yamauchi, J., Nakabayashi, N. and Masuhara, E., ACS Poly. Prepr. 20 594, 1979.

92. Atsuta, M., Abell, A.K., Turner, D.T., Nakabayashi, N. and Takeyama, M., J. Biomed Mater. Res. 16 619, 1982.

93. Antonucci, J.M., J. Dent. Res. 57 500, 1978.

94. Antonucci, J.M., Brauer, G.M., and Termini, D.J., J. Dent. Res. 59 35, 1980.

95. Asmussen, E. and Munksgaard, E.C., Scand. J. Dent. Res. 91 153, 1983.

96. Asmussen, E. and Munksgaard, E.C., J. Dent. Res. 63 1087, 1984.

97. Bowen, R. L. and Cobb, E.N., JADA 107 734, 1983.

98. Bowen, R.L., Cobb, E.N. and Misra, D.N., Ind. Eng. Chem. Prod. Res. Dev. 23 78, 1984.

99. Smith, D.C., Br. Dent. J. 125 381, 1968.

100. Wilson, A.D. and Kent, B.E., Br. Dent. J. 132 133, 1972.

101. Wilson, A.D., Ref. 19, Chap.3.

102. Wilson, A.D. and Prosser, H.J. Br. Dent. J. 157 449, 1984.

103. Wilson, A.D., Prosser, H.J. and Powls, D.R., J. Dent. Res. 62 590, 1983.

104. McLean, J.W., Br. Dent. J. 157 4322, 1984.

105. Sakaguchi, R., Delong, R., and Douglas, W.H., J. Dent. Res. 64 370, 1985.

106. Douglas, W.H. Delong, R. and Sakaguchi, R., J. Dent. Res. 64 370, 1985.

THE RELEVANCE OF THE IN SITU MUCIN ADSORPTION ON SOLID SURFACES IN DENTISTRY AND OPHTHALMOLOGY

Adam Baszkin, Jacques Emile Proust, Eric Perez and
Marie Martine Boissonnade

Vectorisation et Biodisponibilité des Médicaments, CNRS
Physico-Chimie des Surfaces, UER Biomédicale, 45 rue des
Saints-Pères, 75270 Paris cedex 06, France

INTRODUCTION

Mucosal surfaces in contact with the external biological environment are generally covered with a mucus layer. The primary function of this layer in the oral cavity and in the respiratory, gastrointestinal and reproductive tracts is to lubricate ephithelial cells and protect them from attack by microorganisms, the toxins they produce, and other antigens[1-3].

In the eye the corneal mucus layer is necessary to ensure the tear film stability. The rapidly moving eyelid removes lipid-contaminated mucus in the form of fibrils and threads and distributes fresh mucus over the epithelial surface. The presence of the mucus layer maintains a low interfacial tension at the epithelium-aqueous tear layer interface which is an important condition of tear film stability. When a contact lens is placed on the eye, the presence of the pre-lens and post-lens tear mucus layers is imperative for ensuring its good biocompatibiliy. The adsorption process of different lachrymal liquid constituents, the composition and molecular structure of the adsorbed layers are essential therefore in evaluating the contact lens compatibility in the eye[4-8].

In the oral cavity, the selective adsorption of salivary proteins maily glycoproteins is an initial formation step of the 0.1-0.5 μm thick acquired enamel pellicle on tooth surfaces[9-11]. The acquired enamel pellicle is a protein layer attached to the surface and can be considered as intermediate between enamel and dental plaque. The enamel surface, the hydroxyapatite, is mainly covered with HPO_4^{2-} and $H_2PO_4^-$, in addition to an adherent liquid layer which is rich in calcium and phosphate ions. The acidic or basic peptides of the salivary glycoprotein may interact therefore with the tooth surface through several modes including calcium bridging, water bridges, direct protein carboxyl phosphate group interaction via hydrogen bridging or directly between amino and phosphate groups in the case of basic peptides[12]. The enamel pellicle is then colonized with different bacteries (streptococcus mutans and streptococcus sanguis) giving rise to a definite formation of the dental plaque[13-14].

305

Glycoproteins and mucins in particular are considered therefore to play a major role in the oral cavity and in the eye.

For all the above mentioned reasons the direct measurement of adsorbed mucin layers onto different surfaces is of extreme importance in the science and technology of biomaterials. Such different surfaces as dentures and contact lenses in spite of different functions have one commun feature, both are placed on a mucosal surface. To perform these experiments we have extracted bovine submaxillary mucin (BSM) from salivary glands and developed original in situ adsorption techniques based on the use of ^{14}C-labelled mucin prepared also in the laboratory.

Bovine submaxillary mucin (BSM) the sialic acid-containing glyco-protein is a large, highly asymmetric, rod-like molecule, with a mole-cular weight of 4×10^6. It consists of a long protein chain with numerous disaccharide and oligosaccharide side chains[15]. Light scatte-ring measurements show that at low ionic strength the molecule is a rigid rod and when the ionic strength is increased, the molecule becomes a stiff compact coil[16].

EXPERIMENTAL

Solid Surfaces

Adsorption of BSM was studied on silicone and poly(vinyl pyrroli-done) grafted silicone contact lenses manufactured by Essilor Interna-tional, France. The grafting procedure is described in [17]. Contact lenses made of poly(dimethyl siloxane) were circular samples, 15 mm in diameter and 50 µ thick. They were irradiated with a mercury vapour lamp (Philips HPK 125) in the presence of air. Then they were immersed in N-vinyl pyrrolidone for 30 minutes at 95°C which was made free of air by bubbling nitrogen through it. The samples were washed three times in boiling water and dried. Mica surfaces which represent a surface of certain bioceramics applied as dental implant materials were used for adsorption experiments[18].

BSM Isolation and radiolabelling

BSM was isolated from fresh salivary glands obtained at a slaughter house by a procedure of successive dissolutions and precipitations described by Tettamanti and Pigman[15]. Each step of the preparation as well as the final lyophilized mucin were analyzed by colorimetric methods to determine the total protein and sialic acid contents[17,19].

Radiolabelling of the lyophilized mucin was achieved by acetylation with (1 - ^{14}C) acetic anhydride in dimethylsulphoxide solution. The details of this preparation are described in [19].

Adsorption measurement techniques

Two distinct techniques were used to measure mucin adsorption on studied surfaces. They are shown in Fig.1. The densities of silicone contact lenses are lower than that of water and the polymers float when placed on top of mucin aqueous solutions. Silicone contact lenses are gently placed at the surface of BSM solutions, care being taken to ensure that no air bubbles (Fig.1a) are trapped between polymer and

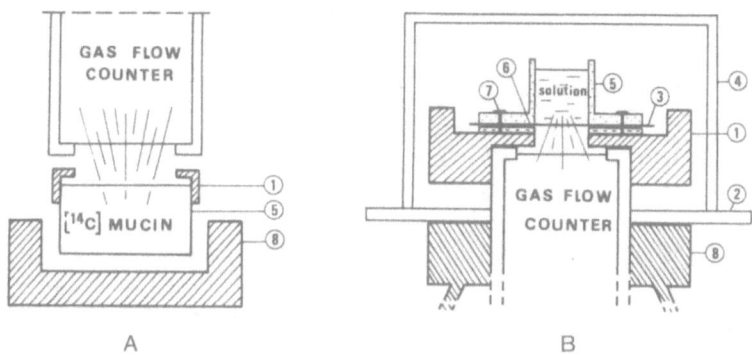

Fig.1. Adsorption measuring devices. (A) for adsorption on contact
lenses. (B) for adsorption on mica. (1), (2), (8) supports
ensuring reproductibility of geometry ; (3) mica window ;
(4) cover ; (5) glass container ; (6) "0" ring ; (7) cell
assembling screws.

solution. The floating silicone sample is covered with a Teflon window
of a known area. For mica surfaces a specially-constructed circular
container was used (Fig.1b). The flat, ground part of the glass cell was
covered with molten paraffin. Three fixing screws and a Viton "0" ring
ensured a tight seal between glass and plastic parts of the container
with a mica window inserted between them. The radioactivity-counting
device (gas-flow chamber) placed at a fixed position below a sample
measured the radioactivity and displayed it on a recorder as a function
of time. Counting corrections are necessary to obtain the net adsorption
values : allowance was made for the absorption of γ-radiation by a
polymer or a mica surface and the radioactivity originating from the
bulk of the BSM solutions was subtracted. A calibration graph is also
needed, to convert the adsorption values from counts min^{-1} into µg of
BSM cm^{-2}. Details of the method may be found in [18].

RESULTS AND DISCUSSION

The kinetics of mucin adsorption on contact lenses and mica surfaces is shown in Fig.2. The initial rate of adsorption is high, but no
stable adsorbed values are obtained after 20 hrs of adsorption.

Thick mucin layer are obtained on all surfaces with increasing bulk
protein concentration (Fig.3).

After 20 hrs of adsorption and for each studied bulk concentration
of mucin, the desorption experiments were performed by replacing mucin
solutions with zero protein concentration aqueous substrates. The
desorbed amounts as a function of BSM bulk concentration are indicated
in Fig.4. These quantities correspond to reversibly adsorbed (loosely
bound) mucin and they are identical for all studied surfaces.

Several observations can be made about the adsorption/desorption
results on studied surfaces and their relevance to dental and
ophtalmological applications.

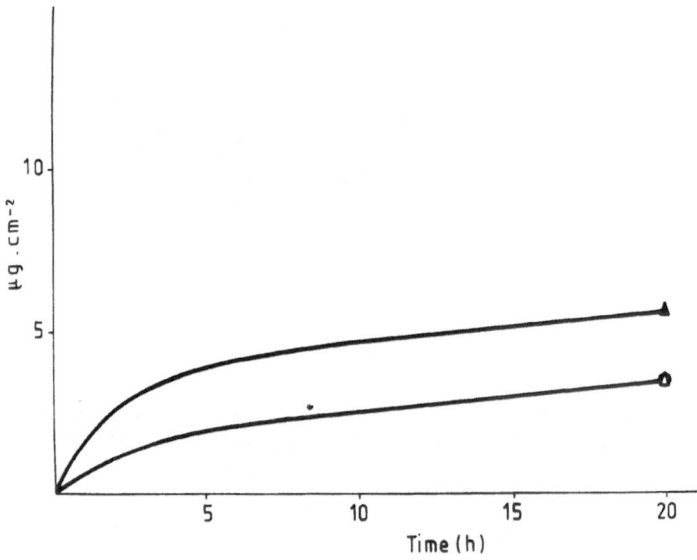

Fig.2. Kinetics of BSM adsorption on contact lenses and mica. (▲) PVP grafted silicone ; (Δ) silicone ; (O) mica. BSM concentration = 5.10^{-2} mg ml^{-1} ; 0.15 M NaCl ; pH = 7.2.

1) Grafting of silicones with poly(vinyl pyrrolidone) leads to significant difference in adsorption of mucin. The increase of adsorption due to the grafting is about 25% for the highest concentration studied (0.2 mg/ml). In spite of the relatively low degree of grafting (3-5% as measured by weight increase) and its insignificant influence upon the wettability of silicone lenses (contact angles with water on ungrafted and grafted silicone surfaces are respectively equal to 106° and 103°), the stability of tear film solutions is much higher on grafted samples[20]. It seems therefore that some functional groups are generated during the grafting procedure on silicone surfaces capable to anchor mucin molecules. The irreversibly bound mucin fraction which is much higher on grafted samples would also explain their better wettability by tear solutions, a necessary condition for their practical application.

2) Identical behavior of mica and ungrafted silicone surfaces in adsorption experiments (Figs.2 & 3) may be explained by the fact that both surfaces represent extreme physicochemical properties. Mica is a highly hydrophilic surface while the ungrafted silicone is very hydrophobic. In fact the protein coverage, the selectivity and the amounts of adsorbed proteins as well as the platlet and cell adhesion depend to a large extent on the hydrophilic/hydrophobic force balance at the polymer/solution interface. Moderate wettability and an equilibrated hydrophilic/hydrophobic force balance of solid surfaces seems to favor protein adsorption[21-23].

3) A mucin coat on the tooth enamel and on other solid surfaces in the oral cavity could promote the initial bacterial colonization of these surfaces and subsequently lead to dental caries, periodontal

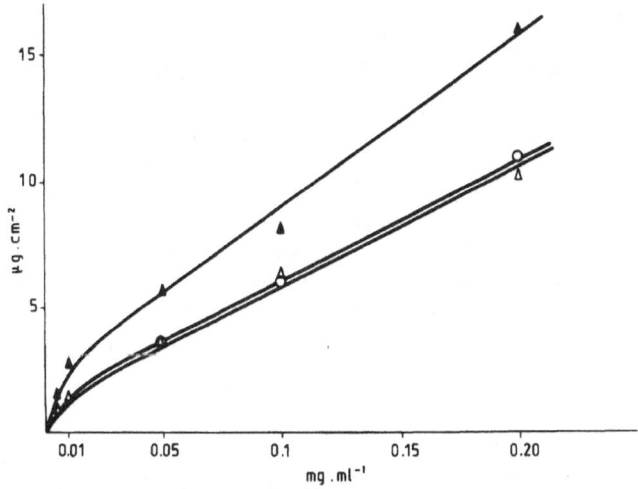

Fig.3. Adsorption vs BSM concentration in solution. Symbols as in
Fig.2. Adsorption time : 20 hrs ; 0.15 M NaCl ; pH = 7.2.

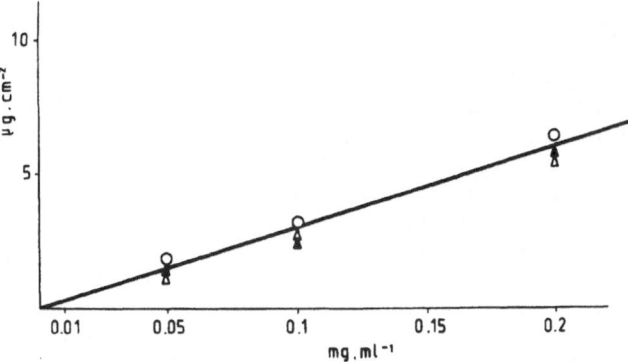

Fig.4 Loosely bound (easily desorbing) BSM vs protein
concentration in solution. Symbols as in Fig.2.

diseases and dental plaque formation. The salivary clearence as assured by the saliva flow in the mouth plays therefore an essential role. High percentage of the loosely bound (easely desorbing) mucin fraction of the totally adsorbed layers would, therefore, be a positive factor in the immunological protection of dental surfaces in the oral cavity.

The contamination of contact lenses is the result of the exchange which takes place between the loosely adsorbed mucin molecules and the fresh molecules coming from the tear/liquid interface. Here again a high fraction of loosely bound mucin molecules, as well to grafted as to ungrafted polymer would play an important role in functioning of the silicone contact lenses.

REFERENCES

1. G.H. Bell, D. Emslie-Smith & C.R. Paterson, "Textbook of Physiology and Biochemistry", Churchill Livingstone, Edinburgh, London, N.Y. (1976).

2. A. Gottschalk, "Glycoproteins. Their Composition, Structure and Functions", Elsevier, N.Y. (1972).

3. P.C. McNabb & T.B. Tomasi, "Host defense mechanism at mucosal surfaces", Ann. Rev. Microbiol. 35:477 (1981).

4. F.J. Holly, "Tear film physiology and contact lens wear. I. Pertinent aspects of tear film physiology", Am. J. Optom. Physiol. Opt. 58:324 (1981).

5. F.J. Holly, "Basic aspects of contact lens biocompatibility", Colloids & Surfaces 10:343 (1984).

6. J.C. Moore & J.M. Tiffany, "Human ocular mucus. Chemical studies", Exp. Eye Res. 33:203 (1981).

7. C.C.W. Chao, J.P. Vergnes & S.I. Brown, "Fractionation and partial characterisation of macromolecular components from human ocular mucus", Exp. Eye Res. 36:139 (1983).

8. A.D. Adams, "The morphology of human conjunctival mucus", Arch. Ophthalmol. 22:69 (1979).

9. R.M. Frank & G. Hoover, "An ultrastructural study of human supragingival dental plaque formation", in "Dental Plaque", W.D. McHugh ed., E. & S. Livingstone, Edinburgh (1970).

10. G. Embery, G. Rolla & J.B. Stanbury, "Interaction of acid glucosa-minoglycans (mucopolysaccharides) with hydroxyapatite", Scand. J. Dent. Res. 87:318 (1979).

11. W.B. Clark, L. Bammann & R.J. Gibbons, "Comparative estimates of bacterial affinities and adsorption sites on hydroxyapatite surfaces", Infect. Immun. 18:514 (1977).

12. A.C. Juriaanse, J. Arends & J.J. Ten Bosch, "The adsorption of acidic and basic homopolypeptides to whole bovine dental enamel" and " The role of surface ions in a model for peptide adsorption to whole bovine dental enamel", J. Colloid & Interface Sci. 76:212 & 220 (1980).

13. W.F. Liljemark & S.V. Schauer, "Studies on the bacterial components which bind streptocòccus sanguis and streptococcus mutans to hydroxyapatite", Arch. Oral. Biol. 20:609 (1975).

14. M. Carraz & Y. Frobert, "Adhérence microbienne aux muqueuses", Revue Inst. Pasteur Lyon 14:25, suppl.I (1981).

15. G. Tettamanti & W. Pigman, "Purification and characterization of bovine and ovine submaxillary mucin", Arch. Biochem. Biophys. 124:41 (1968).

16. F.A. Bettelheim & S.K. Dey, "Molecular parameters of submaxillary mucins", Arch. Biochem. Biophys. 109:259 (1965).

17. A. Baszkin, J.E. Proust & M.M. Boissonnade, "Adsorption of bovine submaxillary mucin on silicone contact lenses grafted with poly(vinyl pyrrolidone)", Biomaterials 5:175 (1984).

18. E. Perez, J.E. Proust, A. Baszkin & M.M. Boissonnade, "In situ adsorption of bovine submaxillary mucin at the mica/aqueous solution interface", Colloids & Surfaces 9:297 (1984).

19. J.E. Proust, A. Baszkin & M.M. Boissonnade, "Adsorption of bovine submaxillary mucin on surface oxidized polyethylene film", J. Colloid & Interface Sci. 94:421 (1983).

20. A. Baszkin, M.M. Boissonnade, J.E. Proust, S. Tchaliovska, L. Ter-Minassian-Saraga & G. Wajs, "Silicone grafted with poly(vinyl pyrrolidone) for contact lenses. Surface properties and stability of thin tear film", J. Bioengineering 2:527 (1978).

21. A. Baszkin & D.J. Lyman, "The interaction of plasma proteins with polymers. Relationship between polymer surface energy and protein adsorption/desorption", J. Biomed. Mater. Res. 14:393 (1980).

22. D.L. Coleman, D.E. Gregonis & J.D. Andrade, "Blood materials interactions : the minimum interfacial free energy and the optimum polar/apolar ratio hypotheses", J. Biomed. Mater. Res. 16:381 (1982).

23. Y. Sakurai, T. Akaike, K. Katzaoka & T. Okano, "Interfacial phenomena in biomaterials chemistry" in "Biomedical Polymers Polymeric Materials and Pharmaceuticals for Biomedical Use", A. Nakaiima & E. Goldberg eds., Academic Press, N.Y. (1980)

POLYMERIC MICELLES AND LIPOSOMES AS POTENTIAL DRUG CARRIERS[*]

M. Emmelius, G. Hörpel, H. Ringsdorf, and B. Schmidt

Institute of Organic Chemistry, Johannes Gutenberg
University of Mainz, D-6500 Mainz / FRG

SUMMARY

In the past, a whole aray of polymeric carriers for biologically
active substances have been reported. In this field the main
interest has focused on the use of randomly solubilized polymers.
Naturally occuring amphiphilic transport proteins can be modelled
by synthetic micellar solubilized polymers. Poly(ethylene oxide)-
polypeptide block copolymers with hydrophobic and cyclophosphamide-
containing side groups are described. These micellar systems are
able to transport different hydrophobic drugs.
The development of new liposomal carriers can start from different
model considerations: Liposomes can be formed by lipids with
pharmacologically active head groups, where the pharmacon itself
constitutes the hydrophilic part of the vesicle surface. Another
method is a selective opening of partially polymerized liposomes
as stable compartments releasing entrapped material. When natural
or cleavable synthetic lipids are incorporated into polymerizable
membranes, phase-separation of the different lipid fractions may
occur. The unpolymerized components can be degraded or solubilized
and thereby can be removed from the polymerized matrix.

[*] This contribution is an extended abstract of a paper presented
at the 2nd International Conference on "Polymers in Medicine"
in Capri (Italy) on June 3-7/1985 and is partially published in
references 9) and 16).

1. NATURAL AND SYNTHETIC DRUG-CARRIERSYSTEMS

Polymers as pharmacologically active agents[1-3] are not yet
applied although their potential as selectively acting pharmaca
is quite promising[4-7]. In the past, a whole aray of polymeric
carriers for biologically active substances have been reported.
In this field the main interest has focused on the use of
randomly solubilized polymers.

Possible synthetic polymeric transport systems can be designed
taking naturally occuring carriers as a model: In vivo, a wide
variety of transport systems are engaged in a large number of
metabolic processes.

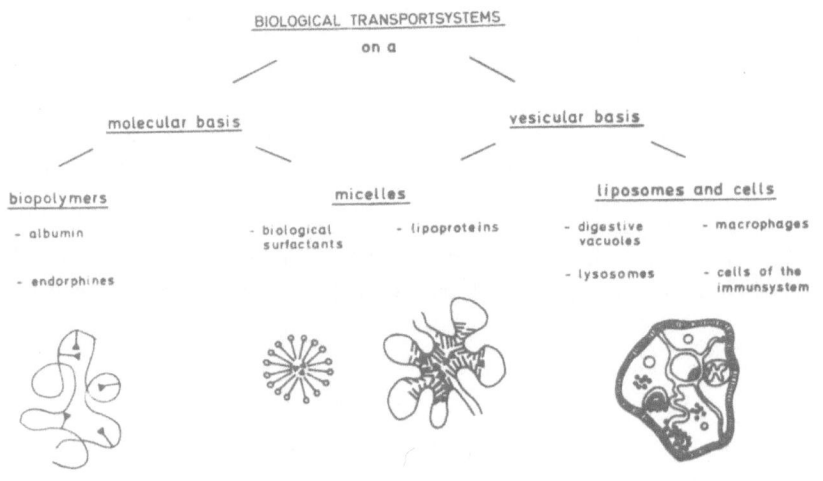

Figure 1.

Transport is achieved by such diverse carriers like randomly or
micellar oriented biopolymers, by vesicles or by cells. Biopolymers with binding capacity for different metabolic products as
well as micelles built from biological surfactants and lipoproteins can be regarded as transport systems on a molecular
basis. Whereas natural liposomes and the different cell vacuoles
and cell types are transport systems on a vesicular basis[8,9].

Improved knowledge about the function of these natural vesicles
allow us to use similar strategies for the transport of chemo-
therapeutic agents via synthetic carriers. Nowadays, chemistry
offers all the tools necessary to synthesize transport systems
on a molecular and a vesicular basis.

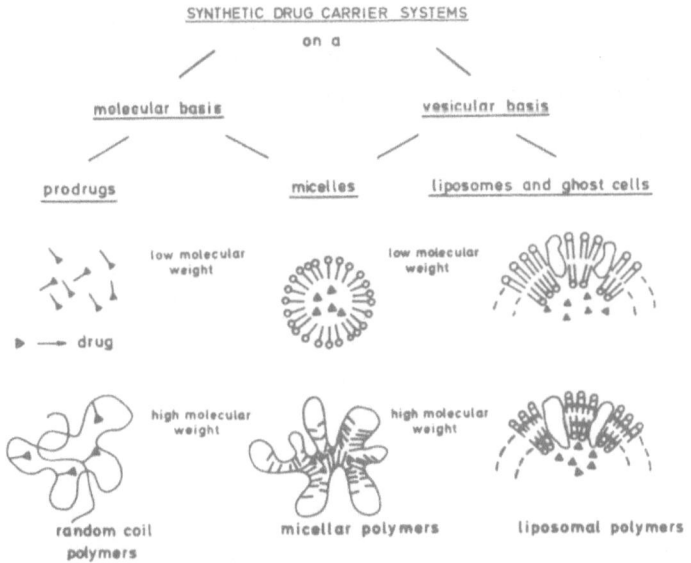

Figure 2.

Micelles from low molecular weight lipids have already been discussed very intensively[10-12]. In addition, natural transport
proteins have been mimicked by synthetic micellar solubilized
polymers, e.g. block copolymers consisting of hydrophobic and
hydrophilic parts[13-15]. Besides this, new types of liposomal
carriers have been developed and refused already[16,17].
Especially polymerized liposomes have demonstrated potential
as stable membrane and cell models[18].

2. POLYMERIC MICELLES AS CARRIERSYSTEMS

As mentioned above, micellar structures - compartments with
hydrophobic cores and hydrophilic shells - are often found in
natural transport systems[19]. For example, lipoproteins carry
hydrophobic cholesterol-esters in the blood serum. For this
reason, the synthesis of amphiphilic polymers[13-15] which spon-

taneously orientate into "polymeric micelles" in aqueous solution
is promising. Polymeric drug-carrying micelles can be prepared
by two methods: Hydrophilic, watersoluble polymers can be
partially derivatized with hydrophobic side chain units carrying
the bound drug[9,15a]. In addition, an AB-block copolymer can be
utilized. The A-block may consist of hydrophilic units while the
B-block contains the hydrophobic side chain units and the attached
drug[9,15b]. The amphiphilic nature of the drug determines the
extend to which it actually penetrates the hydrophobic interior
of the micelles.

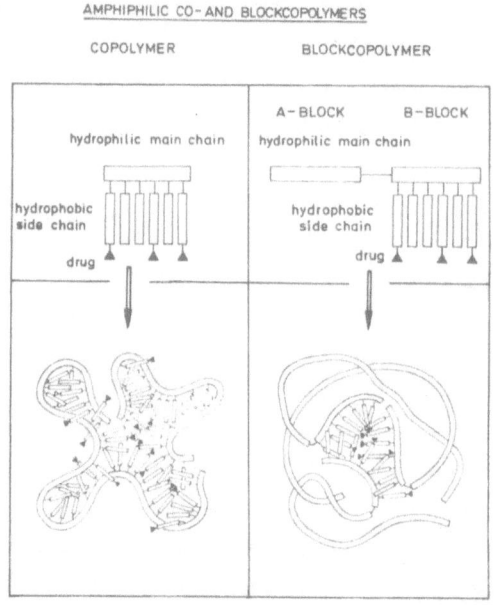

Figure 3.

Following the second approach, an AB-block copolymer has been
synthesized with poly(ethylene oxide) (PEO) as watersoluble
A-block. The B-block consists of biodegradable poly(L-lysine)
with sulfidoderivatives of cyclophosphamide (CP)[20], an alkylating antitumor agent, covalently fixed in a polymer-analogue
reaction. Normally, these cyclophosphamide derivatives are hydrolyzed rapidly to 4-hydroxy-cyclophosphamide as the active
metabolite. It was found that the rate of hydrolysis of the
micellar systems varied greatly with the increased hydrophobicity

of the carrier: the half-life for the release of 4-hydroxy-
cyclophosphamide was found to be prolonged from minutes to
several hours.

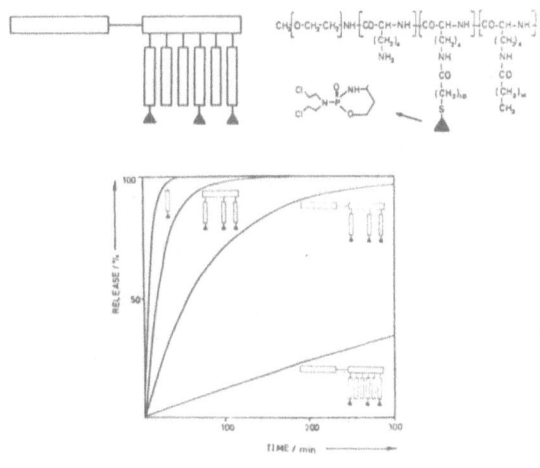

Figure 4.

In vitro studies[21] gave evidence that the cyclophosphamide
block copolymer acts as an intracellular depot for the active
metabolite of cyclophosphamide. This was demonstrated by following the dependence of the DNA-interstrand crosslinking, a method
to determine the effect of alkylating agents[22].

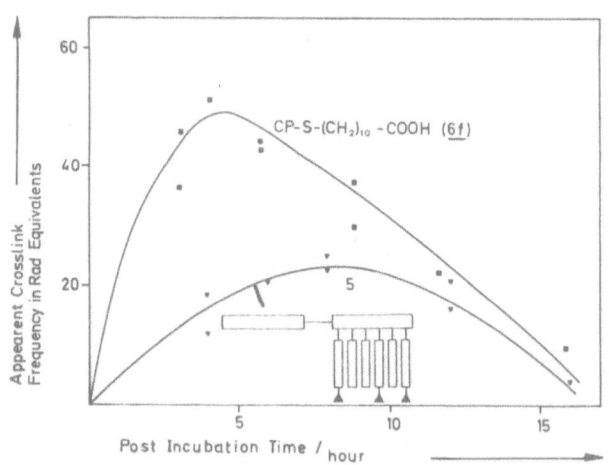

Figure 5.

For the synthesis of such liposomal drug-derivatives, different antitumor agents such as the alkylating N-Lost-derivative cyclophosphamide and the DNA-intercalating daunomycin have been used.

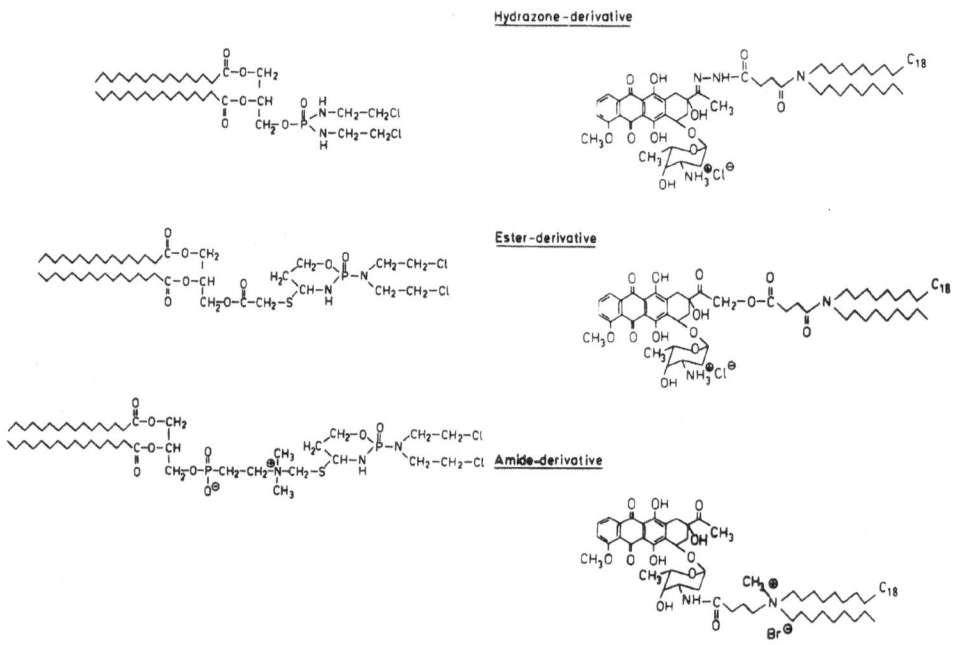

Figure 6.

The in vitro release of 4-hydroxy-cyclophosphamide has been studied using a cyclophosphamide-amphiphile in an isotropic solution (isopropanol:chremophor:water, 1:1:2) or as liposomal solution. In the isotropic solution nearly 100 % of the drug is cleaved off during one hour, whereas in the liposomal preparation the hydrolytic release shows a short increase during the first minutes and a nearly constant release for many hours.

318

L 1210 cells were treated with the low molecular weight sulfido-
derivative of cyclophosphamide (CP-S-$(CH_2)_{10}$-COOH <u>6f</u>) or with
the cyclophosphamide containing block copolymer. While free or
absorbed drug is washed away after one hour incubation, only the
drug which has penetrated through the membrane is effective. The
degree of crosslinking showed a sharp maximum for the low molecular weight drug at 4-5 h post incubation time, whereas the peak
maximum is broader for the polymeric derivative and was shifted
to longer post incubation times (8 hours). These results indicate
that cellular uptake of the block copolymer probably occurs prior
to sustained release of the active drug[9].

3. LIPOSOMAL CARRIERSYSTEMS

3.1. LIPIDS WITH PHARMACOLOGICALLY ACTIVE HEAD GROUPS

Synthetic liposomes have already been intensively investigated
as carriers[23-25]. These liposomes have been studied carrying
watersoluble drugs in their aqueous interior or lipidsoluble
substances in their bilayers.
Liposomes can also be formed by lipids having a pharmacologically
active head group[26,27] possibly fixed via a cleavable spacer.
The lipophilic modification of a drug leads to an amphiphilic
drug with varied properties and to liposomes with the drug as the
polar head group placed on the vesicle surface.

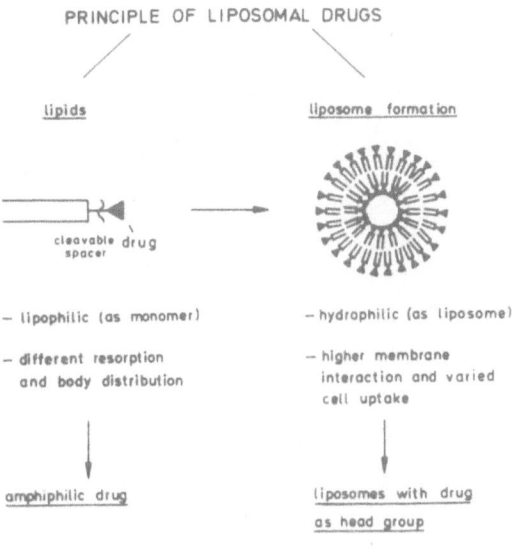

Figure 7.

IN VITRO RELEASE OF 4-HYDROXYCYCLOPHOSPHAMIDE

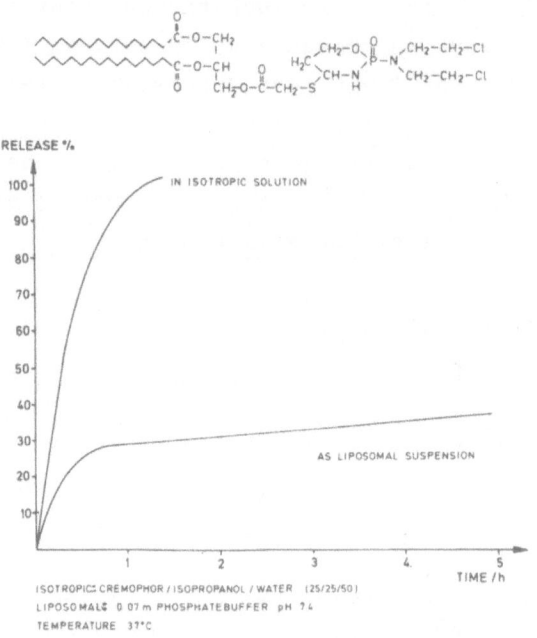

ISOTROPIC: CREMOPHOR / ISOPROPANOL / WATER (25/25/50)
LIPOSOMAL: 0.07 m PHOSPHATEBUFFER pH 7.4
TEMPERATURE 37°C

FLUORESCENCE PROPERTIES OF LIPOSOMAL-ANTHRACYCLINES

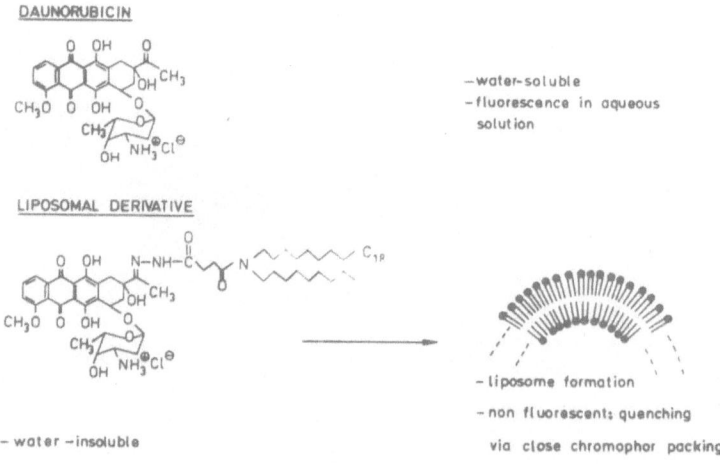

Figure 9.

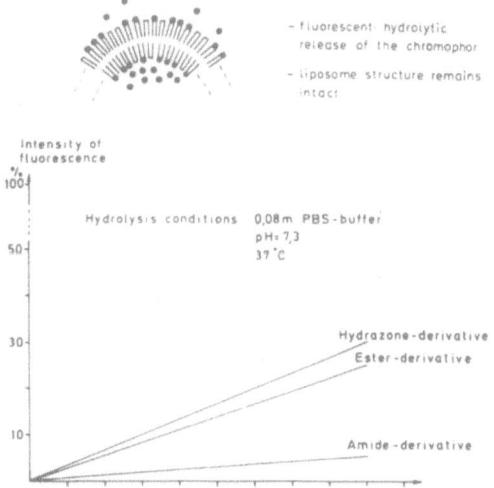

RELEASE OF ANTHRACYCLINE HEADGROUP FROM
LIPOSOMES

Figure 10.

Several lipophilic prodrugs of daunomycin with sometimes reduced
cardiotoxicity have been reported in the literature[28]. Different
linkages between the drug and the hydrophobic moiety have also
been utilized for the synthesis of liposome-forming daunomycin-
derivatives. Liposomes can be prepared from these lipophilic
daunomycin-derivatives. In contrast to a daunomycin solution,
the liposomal solution shows no fluorescence. The fluorescence
is quenched by the close packing of the chromophores on the
vesicle surface.

On prolonged incubation the liposome solution starts to fluoresce
caused by the hydrolytic cleavage of the daunomycin head group
from the vesicle surface. The intensity of the fluorescence allows
to follow the release of daunomycin with time. This is shown in
the following figure for the three daunomycin lipids with spacers
of different cleavability.

From the daunomycin-derivative with the hydrazone linkage first
results of experimental antitumor activity have been obtained.
This compound displays cytostatic properties against i.p. implanted P 388 leukemia in mice, directly comparable to daunomycin
if i.p. injected. On the other hand the same compound is inactive
when injected i.v.

3.2. SELECTIVE OPENING OF POLYMERIC LIPOSOMES

In contrast to the concept of liposomal drug derivatives discussed above phospholipid liposomes entrapping watersoluble drugs in their inner compartment have found extensive use as experimental drug carriers[23-25]. One of the shortcomings of conventional liposomes is their poor stability. Stabilization of artifical liposomes has been achieved by polymerization of saturated lipids within the liposomal membrane[16,17,29]. Polymerizable lipids are variable in their polymerizable unit and in their head group as shown in the following table.

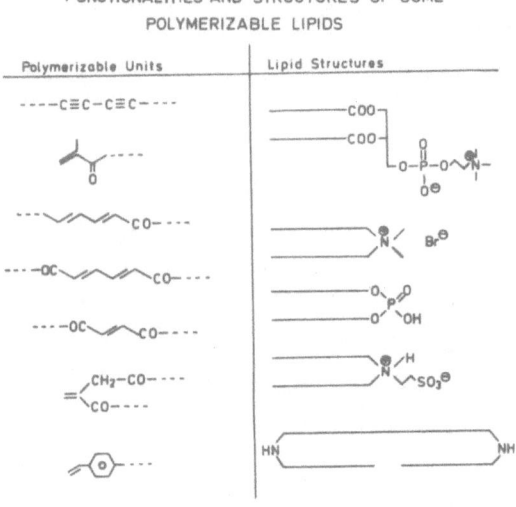

Figure 11.

Liposomes can be formed from most polymerizable lipids by ultra-sonication of the aqueous lipid dispersion above the phase transition temperatures of the lipids. The monomeric liposomes are transformed into polymeric liposomes by UV-irradiation.

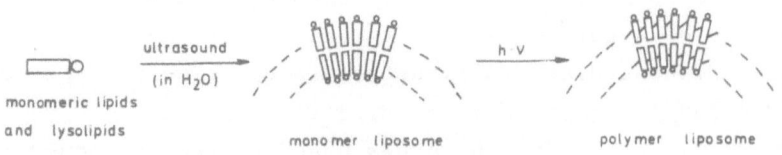

Figure 12.

The presevation of the spherical shape of liposomes after polymerization can be visualized by scanning electron microscopy.
The stability of polymeric liposomes has been found to be much
higher than that of monomeric liposomes. Their increased stability
towards detergents can be demonstrated using 6-carboxyfluorescein
(6-CF) as a marker for the membrane permeability. This technique
is explained schematically below.

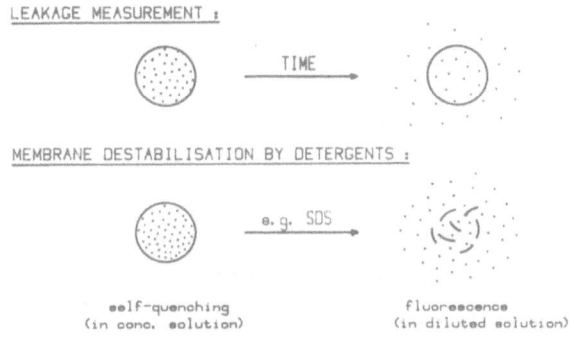

Figure 13.

Monomeric liposomes of dienoyl-lecithin are more sensitive towards
sodium dodecylsulphate (SDS)-treatment than those from a saturated
analogue, DPPC.

$$CH_3-(CH_2)_{12}-CH=CH-CH=CH-COO-CH_2$$
$$CH_3-(CH_2)_{12}-CH=CH-CH=CH-COO-CH$$
$$CH_2-O-\overset{O}{\underset{O}{\overset{\|}{P}}}-O-CH_2-CH_2-\overset{CH_3}{\underset{CH_3}{\overset{\oplus}{N}}}-CH_3$$

In contrast to this, the polymerized vesicles show only a very
small increase in 6-CF permeability up to high detergent concentrations[29k].

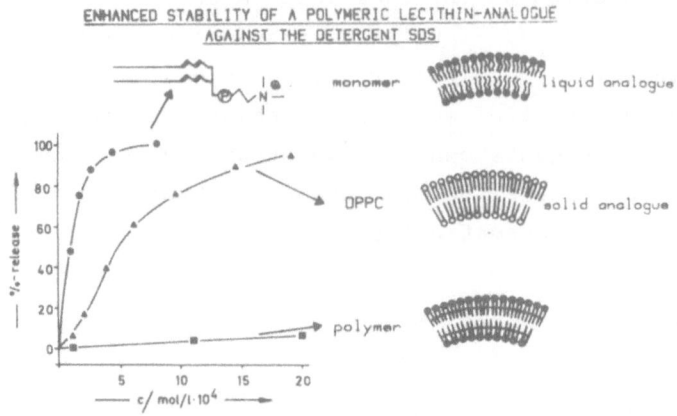

Figure 14.

For a potential use as drug carriers it is not enough to have stable liposomes only. The combination of stable polymeric and fluid domains of natural lipids in mixed membrane systems makes these liposomes more suitable for the incorporation of proteins or as carriersystems[30].

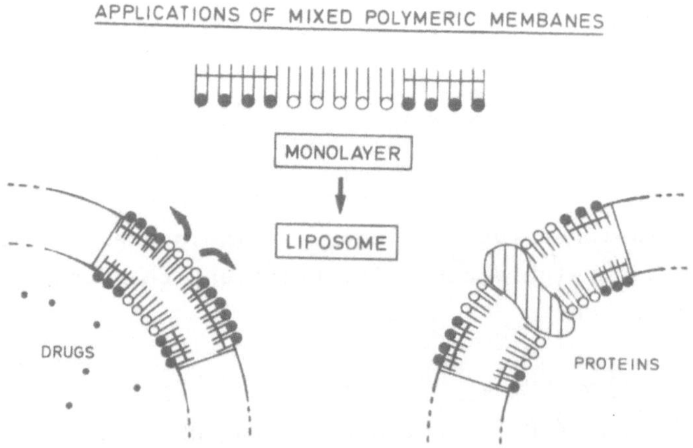

APPLICATIONS OF MIXED POLYMERIC MEMBANES

MONOLAYER

↓

LIPOSOME

DRUGS

PROTEINS

Figure 15.

In this context, it is essential to have phase-separated membranes Such systems are readily obtained from mixtures of immiscible fluorocarbon and hydrocarbon lipids having the same head group[31].

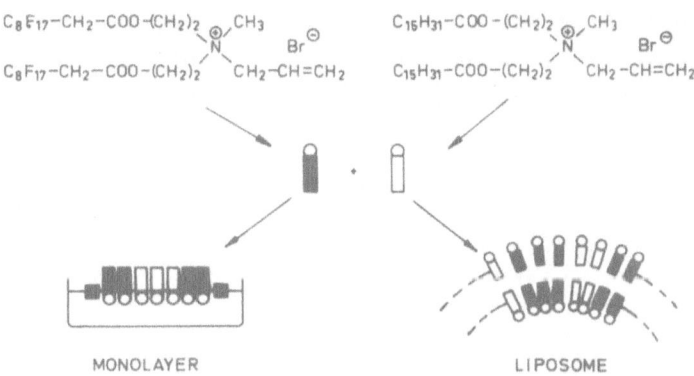

Figure 16.

Freeze-fracture electron microscopy reveals the phase-separation in liposomes. Dimyristoyl-lecithin-liposomes containing 5 mol% of fluorocarbon lipids as phase-separated units can be identified by electron microscopy. They show domains with the typical "ripple" structure of the lecithin and the smooth surface of the fluorocarbon lipid[9,31].

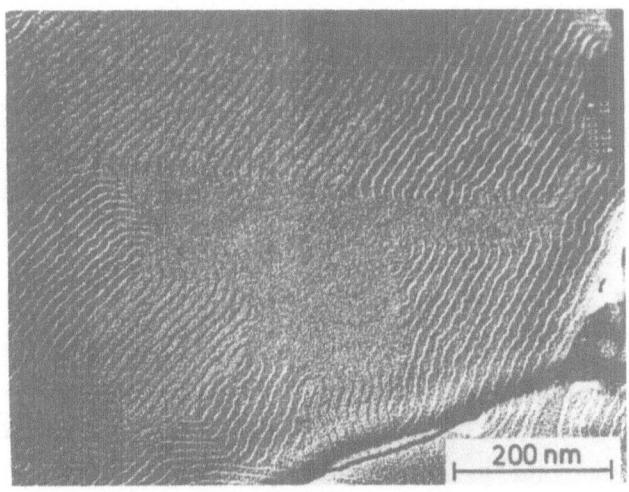

Figure 17.

Based on the investigations of phase-separated and partially polymerized mixed liposomes, methods to "uncork" polymeric liposomes have been developed[9,16]. Selective opening of the liposomes can be achieved by cleavage of the monomeric lipids to watersoluble lysolipids, as illustrated schematically in the following figure.

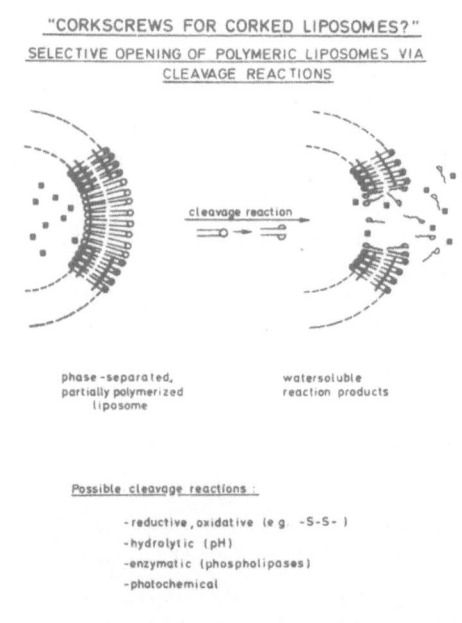

"CORKSCREWS FOR CORKED LIPOSOMES?"

SELECTIVE OPENING OF POLYMERIC LIPOSOMES VIA
CLEAVAGE REACTIONS

cleavage reaction

phase-separated,
partially polymerized
liposome

watersoluble
reaction products

Possible cleavage reactions :

- reductive, oxidative (e.g. -S-S-)
- hydrolytic (pH)
- enzymatic (phospholipases)
- photochemical

Figure 18.

Lipids containing glycolic or disulfide units in the head group have been synthesized. Treatment of these lipids with oxidizing or reducing agents, respectively, leads to the cleavage of the glycolic or disulfide moieties and the formation of watersoluble lysolipids[32].

326

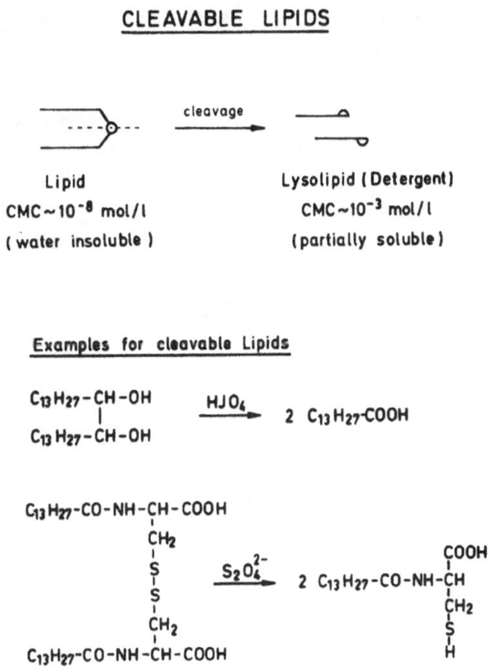

CLEAVABLE LIPIDS

Lipid
CMC~10^{-8} mol/l
(water insoluble)

Lysolipid (Detergent)
CMC~10^{-3} mol/l
(partially soluble)

Examples for cleavable Lipids

$$C_{13}H_{27}-CH-OH$$
$$| \quad \xrightarrow{HJO_4} \quad 2 \; C_{13}H_{27}-COOH$$
$$C_{13}H_{27}-CH-OH$$

$$\begin{array}{l}
C_{13}H_{27}-CO-NH-CH-COOH \\
\qquad\qquad\qquad | \\
\qquad\qquad\quad CH_2 \\
\qquad\qquad\qquad | \\
\qquad\qquad\qquad S \\
\qquad\qquad\qquad S \\
\qquad\qquad\qquad | \\
\qquad\qquad\quad CH_2 \\
\qquad\qquad\qquad | \\
C_{13}H_{27}-CO-NH-CH-COOH
\end{array}
\quad \xrightarrow{S_2O_4^{2-}} \quad
\begin{array}{l}
\qquad\qquad\qquad COOH \\
\qquad\qquad\qquad | \\
2 \; C_{13}H_{27}-CO-NH-CH \\
\qquad\qquad\qquad | \\
\qquad\qquad\qquad CH_2 \\
\qquad\qquad\qquad | \\
\qquad\qquad\qquad S \\
\qquad\qquad\qquad | \\
\qquad\qquad\qquad H
\end{array}$$

Figure 19.

The "uncorking" of partially polymerized phase-separated liposomes
can be proven by the release of entrapped dyes as well as by
scanning electron microscopy. Partially polymerized liposomes
have been synthesized from a butadienoyl lipid with taurine head
group and a cystine derivative containing fluorocarbon chains.
The release of eosin from the polymerized liposome was minimal,
but increased rapidly after addition of the disulfide cleavage
agent, dithiothreitol (DTT). In the monomeric liposomes, the
"holes" can be cured by the rapidly diffusing lipids trapping
the rest of the eosin.

RELEASE OF EOSIN FROM MIXED POLYMERIC
LIPOSOMES DURING REDUCTIVE CLEAVAGE

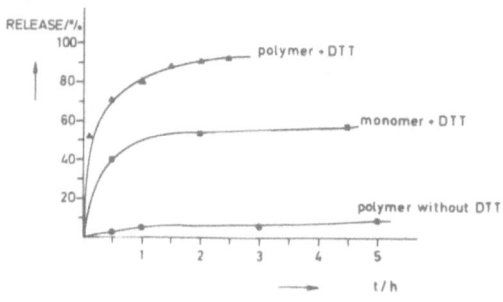

MONOMER:

$CH_3-(CH_2)_9-CH=CH-CH=CH-COO$ ⏜●⏜ H
$CH_3-(CH_2)_9-CH=CH-CH=CH-COO$ ⏝ N ⏝ $SO_3^\ominus$

CLEAVABLE LIPID:

$C_8F_{17}-CH_2-CO-NH-CH-COCH$
$\quad\quad\quad\quad\quad\quad\quad\quad\quad\quad S$
$\quad\quad\quad\quad\quad\quad\quad\quad\quad S$
$C_8F_{17}-CH_2-CO-NH-CH-COOH$

CLEAVAGE REAGENT: $HS-CH_2-CHOH-CHOH-CH_2-SH$ (DTT)

Figure 20.

This "uncorking" process has a drastic effect on the appearance
of the liposomes as demonstrated by scanning electron microscopy
before and after pulling the cork[33].

Liposomes before treatment with DTT:

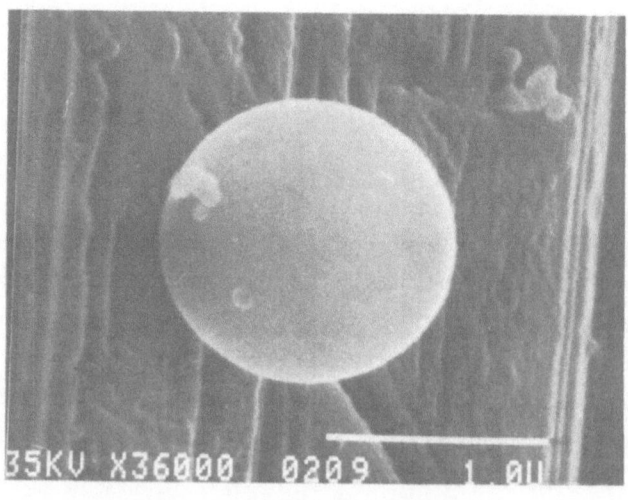

Figure 21.

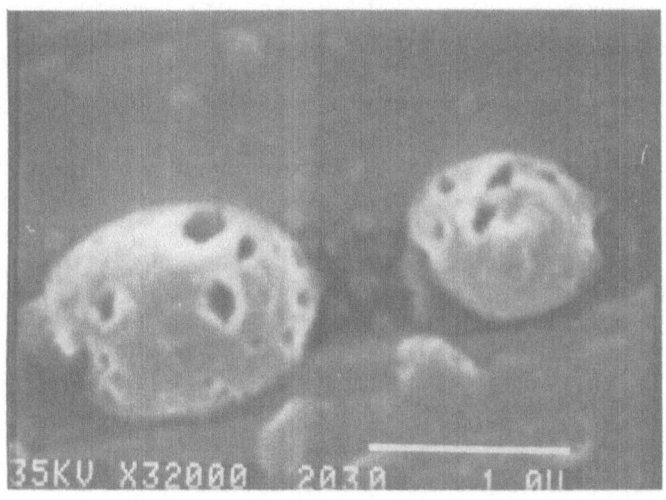

Figure 22.

Still far away from any application aspects the "uncorking" of partially polymerized liposomes opens interesting possibilities for the manipulation of biomembrane models.

REFERENCES

1. K. Dušek, ed. "Advances in Polymer Science", Springer Verlag Berlin, Heidelberg New York 57 (1984).

2. a) Ch.G. Gebelein, Ch.E. Carraher, Jr., eds. "Bioactive Polymeric Systems - An Overview", Plenum Press New York and London 1985; b) J. Kahovec, B. Sedlácek, eds. "Polymers in Medicine and Biology", Makromol.Chem., Suppl. 9, 1 (1985).

3. E. Chiellini, P. Guisti, eds. "Polymers in Medicine - Bio-medical and Pharmacological Applications", Polymer Science and Technology, Vol. 28, Plenum Press New York and London 1983

4. G. Gregoriadis, ed. "Drug Carriers in Biology and Medicine", Academic Press, London 1979.

5. E. Goldberg, L. Donaruma, O. Vogl, eds. "Targeted Drugs", Wiley & Sons, New York 1983.

6. A. Trouet, Eur.J.Cancer 14, 105 (1978).

7. D.S. Zaharko, M. Przybylski, V.T. Oliverio, Methods in Cancer Res. 16, 347 (1979).

8. L. Gros, H. Ringsdorf, H. Schupp, Angew.Chem. 93, 311 (1981); Angew.Chem.Int.Ed.Engl. 20, 305 (1981).

9. H. Bader, H. Ringsdorf, B. Schmidt, Angew.Makromol.Chem. <u>123/124</u>, 457 (1984).

10. J.H. Fendler, E.J. Fendler, eds. "Catalysis in Micellar and Macromolecular Systems", Academic Press, New York 1975.

11. K.L. Mittal, ed. "Micellization, Solubilization and Mikroemulsions" Vols. 1 and 2, Plenum Press, New York 1977.

12. S. Yalkowsky, ed. "Techniques of Solubilization of Drugs", Marcel Dekker, New York 1981.

13. D. Attwood, A.T. Florence, eds. "Surfactant Systems - Their Chemistry, Pharmacy and Biology", Chapman and Hall, London and New York 1983.

14. P. Speiser, Österreichische Apotheker-Zeitung <u>35</u>, 805 (1981).

15. a) W. Klesse, Dissertation, University Mainz 1982;
    b) G. Hörpel, Dissertation, University Mainz 1983.

16. H. Bader, K. Dorn, B. Hupfer, H. Ringsdorf, Advances in Polymer Sciences <u>64</u>, 1 (1985).

17. J.H. Fendler, ed. "Membrane Mimetic Chemistry", Wiley & Sons, New York 1982.

18. A. Akimoto, K. Dorn, L. Gros, H. Ringsdorf, H. Schupp, Angew. Chem. <u>93</u>, 108 (1981); Angew.Chem.Int.Ed.Engl. <u>20</u>, 90 (1981).

19. J.C. Osborne, H.B. Brewer, Adv.in Prot.Chem. <u>31</u>, 253 (1977).

20. a) G. Peter, T. Wagner, H.J. Hohorst, Cancer Treat.Rep. <u>60</u>, 429 (1976); b) T. Hirano, W. Klesse, H. Ringsdorf, Makromol. Chem. <u>180</u>, 1125 (1979).

21. L.C. Erickson, G. Hörpel, K.W. Kohn, H. Ringsdorf, D.S. Zaharko, in preparation.

22. K.W. Kohn, R.A.G. Ewig, L.C. Erickson, L.A. Zwelling, in: "DNA-Repair" (E.C. Friedberg, P.C. Hanawalt, eds.), Marcel Dekker, New York and Basel, Vol. 1, Part B, Chap. 29, p. 379 (1981).

23. D. Papahadjopoulos, ed. "Liposomes and their Uses in Biology and Medicine", Ann.N.Y.Acad.Sci. <u>308</u>, 1 (1978).

24. G. Gregoriadis, A.C. Allison, eds. "Liposomes in Biological Systems", Wiley & Sons, New York 1980.

25. C. Nicolau, G. Poste, eds. "Liposomes in vivo", Biol.Cell, Vol. 47 (1983).

26. a) B. Rosemeyer, M. Ahlers, B. Schmidt, F. Seela, Angew.Chem. <u>97</u>, 500 (1985); b) B. Schmidt, Dissertation, University Mainz 1985.

27. a) M. MacCoss, J.J. Edwards, T.M. Seed, S.P. Spragg, Biochim. Biophys.Acta <u>719</u>, 544 (1982); b) A. Garzon-Aburbeh, J.H. Poupaert, M. Claesen, P. Dumont, G. Atassi, J.Med.Chem. <u>26</u>, 1200 (1983).

28. R.S. Jaenke, D. Depres-De Campeneere, A. Trouet, Cancer Res. <u>40</u>, 3530 (1980).

29. a) H.H. Hub, B. Hupfer, H. Koch, H. Ringsdorf, Angew.Chem.
    92, 962 (1980), Angew.Chem.Int.Ed.Engl. 19, 938 (1980);
    b) D.S. Johnston, S. Sanghera, D. Chapman, Biochim.Biophys.
    Acta 602, 213 (1980); c) M. Pons, C. Villaverde, D. Chapman,
    Biochim.Biophys.Acta 730, 306 (1983); d) S.L. Regen, B. Czech,
    J.Am.Chem.Soc. 102, 6638 (1980); e) A. Kusumi, M. Singh,
    D.A. Tirell, G. Oehme, A. Singh, N.K.P. Samuel, J.S. Hyde,
    S.L. Regen, J.Am.Chem.Soc. 105, 2975 (1983); f) P. Tundo,
    D.G. Kippenberger, P.H. Klahn, N.E. Prieto, T.C. Jao,
    J.H. Fendler, J.Am.Chem.Soc. 104, 456 (1982); g) D. Kippen-
    berger, K. Rosenquist, L. Odberg, J.H. Fendler, J.Am.Chem.Soc.
    105, 1129 (1983); h) K. Dorn, R.T. Klingbiel, D.P. Specht,
    P.N. Tyminski, H. Ringsdorf, D.F. O'Brien, J.Am.Chem.Soc.
    106, 1627 (1984); i) D.F. O'Brien, T.H. Whitesides, R.T. Kling-
    biel, J.Polym.Sci.Polym.Lett.Ed. 19, 95 (1981); j) B. Hupfer,
    H. Ringsdorf, Chem.Phys.Lipids 33, 263 (1983); k) B. Hupfer,
    H. Ringsdorf, H. Schupp, Chem.Phys.Lipids 33, 355 (1983).

30. a) R. Büschl, B. Hupfer, H. Ringsdorf, Makromol.Chem., Rapid
    Commun. 3, 589 (1982); b) H. Gaub, H. Sackmann, R. Büschl,
    H. Ringsdorf, Biophys.J. 45, 725 (1984).

31. R. Elbert, Th. Folda, H. Ringsdorf, J.Am.Chem.Soc. 106, 7687
    (1984).

32. Th. Folda, Dissertation, University Mainz 1983.

33. Th. Folda, J. Lando, H. Ringsdorf, in preparation.

IMPROVED ACTIVATION OF MACROPHAGES BY POLYANIONIC POLYMERS

ENCAPSULATED IN MANNAN DERIVATIVE-COATED LIPOSOMES

Raphael M. Ottenbrite[*], and Junzo Sunamoto[**]

*Department of Chemistry and Massey Cancer Center
Virginia Commonwealth University, 1001 W. Main Street
Richmond, Virginia 23284; **Department of
Industrial Chemistry, Faculty of Engineering, Nagasaki
University, 1-14, Bunkyo-machi, Nagasaki 852, Japan

Introduction

Both of the authors of this article attended the first international symposium on "Polymers in Medicine", which was held in Porto Cervo, Sardinia, Italy, in 1983. Ottenbrite presented a paper on the synthesis and biological characterization of several polyanionic polymers such as the copolymers of itaconic acid and styrene, maleic anhydride and styrene, and maleic anhydride and cyclohexyl-1,3-dioxepin. He reported that some of these copolymers had the unique feature of exhibiting immunomodulator activity(1). The other author, Sunamoto, presented a paper at the same meeting concerning a new technique for stabilizing liposomes to be used effectively as drug delivery systems(2). He had developed a unique method of coating liposomes with derivatives of naturally occuring polysaccharides. This process not only made the liposomes more stable against chemical and/or enzymatic stimuli, but also targetable to specific cells, such as alveolar macrophages and specific lung tissue.

For immunotherapy, it is most important to effect activation, and not just stimulation, of macrophages; this means that the combination of a strong immunopotentiator and a suitable carrier that is targetable to macrophages could be very effective. After some discussion, we, Ottenbrite and Sunamoto, planned a collaboration to evaluate the encapsulation of synthetic polyanionic polymer immunomodulators within a polysaccharride-coated liposome. We expected an improvement in macrophage activation, thereby leading to an increase in the cytotoxic efficiency of these polymer drugs. If successful, this technique could provide a viable method in the immunotherapy for cancer and/or infectious diseases.

In this publication, "Improved Activation of Macrophages by Polyanionic Polymers Encapsulated in Mannan Derivative-Coated Liposomes", we would like to report the progress achieved through our international collaboration. We carefully worked out the following goals, with each participant assuming specific responsibilities for:

(1) the molecular design, synthesis of monomers, and polymerization,
(2) the fractionation and characterization of the polymers obtained,
(3) the biological testing of the polymers for physiological activity and toxicity,
(4) the determination of the compatibility of these polymers with liposomes,
(5) the molecular design and synthesis of polysaccharide derivatives,
(6) the coating of the surface of liposomes with the chosen polysaccharide derivatives,
(7) the characterization of polysaccharide-coated liposomes in vitro and in vivo,
(8) the encapsulation of polymeric drugs in polysaccharide-coated liposomes,
(9) the evaluation of the effects of the polymeric drugs encapsulated in polysaccharide-coated liposomes on such biological activities as superoxide production and tumor cytotoxicity,
(10) the proliferation of bacteria in phagocytes in the presence of polymeric drugs and antibiotics encapsulated in polysaccharide-coated liposomes,
(11) the evaluation of liposome encapsulated polymeric drugs in animals.

At this time, we have accomplished all of these objectives, with the exception of the last two, which are now in progress.

Biological Activity of Anionic Polymers

Both naturally occuring and synthetic anionic and cationic polymers have been found to exhibit in vivo inhibitory effects on the growth of viruses and neoplasia(3,4). The potential use of synthetic anionic polymers in biological systems first received attention in the early 1960's with the synthesis of the cyclic copolymer of divinyl ether and maleic anhydride.

This polymer is referred to as pyran due to the formation of six-membered rings along the polymer backbone during polymerization. Recent ^{13}C NMR studies have indicated, however, that about 50% of the rings formed are actually five-membered(5). From an NIH screening process, pyran was found to exhibit a variety of biological activities(6), and was designated as NSC 46015 by the National Cancer Institute. Pyran is able to induce interferon(7), activate macrophage(8), act against a number of lethal viruses, including Friend and Rauscher leukemia(9), exhibit antifungal(10) and antibacterial(11) activity, stimulate immune responses, and affect anticoagulant properties(12). Pyran also inhibits adjuvant activity(13) and has been reported to enhance plutonium removal from the liver(14).

Many of the biological activities exhibited by polyanionic polymers may be related to macrophage participation. For example, macrophages can be activated to tumor cytotoxicity in vivo by a number of agents: these include natural and synthetic polynucleotides, C. parvum and BCG bacterium, muranyl dipeptide, and synthetic polycarboxylic acid polymers(3). Although the process for antitumor activity has not been elucidated, several mechanisms, such as (a) macrophage activation, (b) stimulation of natural killer cell population, and (c) activation of tumor immunity process have been proposed.

Due to the toxicity of pyran, its clinical use as an anti-neoplastic agent was initially curtailed. Subsequent work, however, has focused on how to reduce these side effects. One of the first parameters to be investigated was that of molecular weight. Kaplan documented that the toxicity of pyran increased with the molecular weight of the polymer administered(15). Consequently, several narrow polydispersed fractions of pyran of known molecular weight were prepared and evaluated for biological activity. The evaluation indicated that the level of serum glutamic pyruvate transaminase, which is a measure of liver damage, increases with an increase in the molecular weight of pyran tested. However, the level of activity against Lewis lung and Ehrlich ascites tumor remains relatively the same(16).

We separated a sample of pyran copolymer preparation (XA-124-177) into two lower molecular fractions, (1-10K) and (10-30K). Pharmacological evaluations showed significant reduction in splenomegaly, hepatomegaly, endotoxin sensitization, and inhibition of mixed microsomal enzyme reactions, while significantly increasing the LD_{50}(17). The antitumor activity against Lewis lung carcinoma of these two lower molecular weight fractions was comparable to the parent material while the antiviral activity against encephalomyocarditis was absent in the 1-10K fraction and reduced to 30% of that of the parent material in the 10-30K fraction. The correlation between toxicity and molecular weight of pyran prompted a reinvestigation of pyran as a potential tumoricidal drug. A narrow band low molecular weight form of pyran is presently undergoing phase II clinical trials.

The Synthesis and Evaluation of Some New Polycarboxylic Acid Polymers

The potential use of synthetic anionic polymers as viable tumoricidal agents has stimulated research in the area of immunostimulators. We have synthesized several polyanionic polymers with similar functional groups as pyran but with specific molecular weights, hydrophobicity, surface charges, and chain rigidity. These were evaluated in order to determine

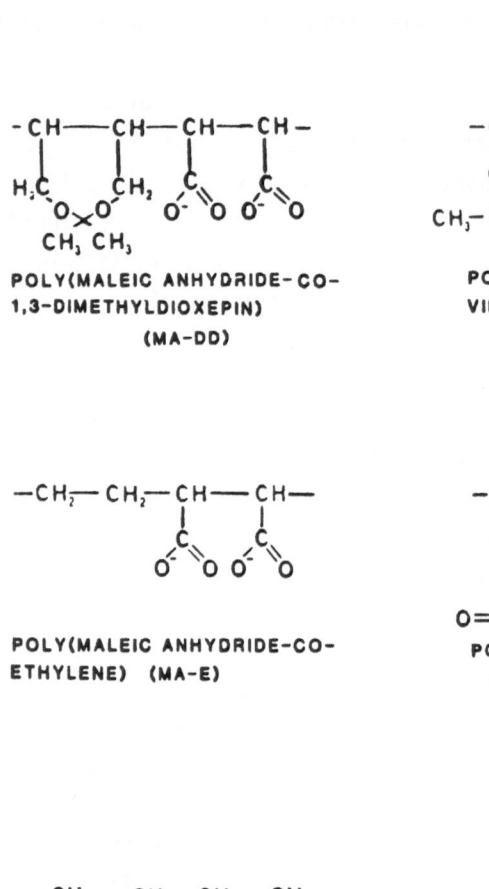

POLY(MALEIC ANHYDRIDE-CO-
1,3-DIMETHYLDIOXEPIN)
(MA-DD)

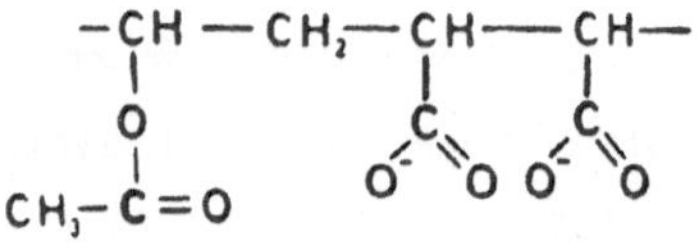

POLY(MALEIC ANHYDRIDE-CO-
VINYL ACETATE)
(MA-VA)

POLY(MALEIC ANHYDRIDE-CO-
ETHYLENE) (MA-E)

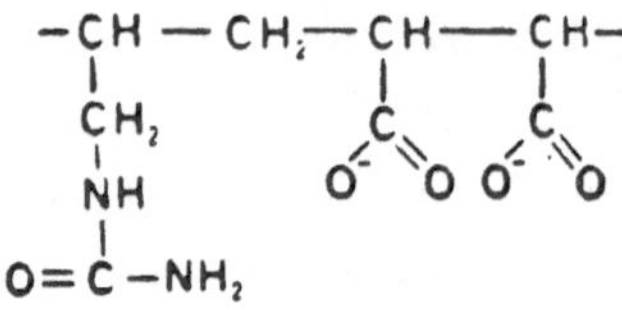

POLY(MALEIC ANHYDRIDE-CO-
ALLYL UREA) (MA-UA)

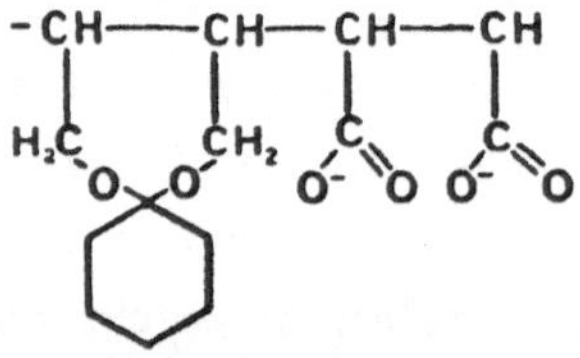

POLY(MALEIC ANHYDRIDE- CO-
CYCLOHEXYL-1,3-DIOXEPIN)
(MA-CDA)

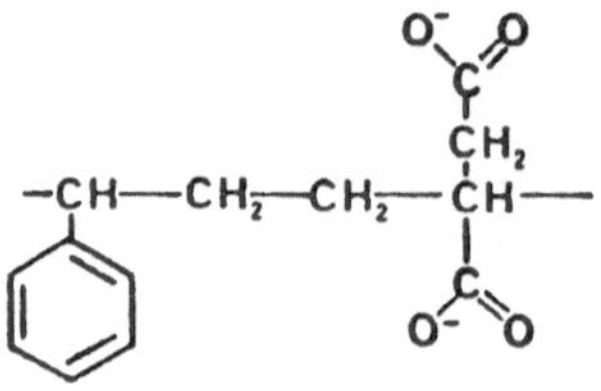

POLY(ITACONIC ACID- CO-STYRENE)
(IA-ST)

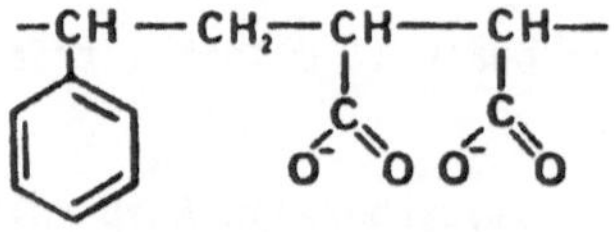

POLY(MALEIC ANHYDRIDE-CO-
STYRENE) (MA-ST)

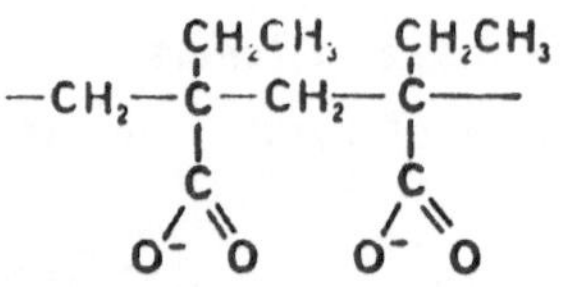

POLY(ETHACRYLIC ACID) (EAA)

Figure 1. Structures of Synthetic Polycarboxylic Acid Polymers.

336

whether structure-reactivity relationships exist with respect to macrophage activation and anti-tumor activity. This study was of particular interest since earlier research with poly(acrylic acid) and poly(methacrylic acid), which also have high carboxylic acid functionality on the polymer backbone, indicated behavior similar to other biological response modifiers(18). Our main goal was to obtain agents that would selectively modulate macrophage function and exhibit high therapeutic efficacy with respect to antitumor activity.

Maleic anhydride was used as the comonomer for many of the carboxylic acid polyanionic polymers prepared in this investigation. Although maleic anhydride does not homopolymerize, it does readily copolymerize with other monomers and affords a high density of carboxylic acid groups along the polymer chain after hydrolysis. The polarity of maleic anhydride makes it an excellent electron acceptor molecule and so readily forms complexes with electron donating π-systems. These maleic anhydride complexes are usually highly susceptible to free radical copolymerization. Studies have shown that the copolymers formed are invariably 1:1 in compostion and structural analyses indicate that the monomers are added to the polymeric chain in an alternating mode. Thus, not only does the resultant polymer have a high carboxylic acid content, but it has a regular structure as well. The molecular weight of the polymers was controlled by using appropriate monomer concentrations, solvents, initiators and chain transfer agents.

Some of the polymers prepared and evaluated are listed in Figure 1. The structures were varied in hydrophilic and hydrophobic character, as well as in the number of carboxyl groups per repeating unit. The molecular weight of the polymers was kept low and the resultant polymers, after hydrolysis in 0.2 N NaOH solution, were fractionated using appropriate Amicon Diaflo[R] ultrafiltration membranes (Amicon Corp. Lexington, Mass). Three fractions were usually collected: 1,000-10,000, 10,000-30,000, and greater than 30,000(19). The purified polymers were analyzed by ^{1}H and ^{13}C nuclear magnetic resonance spectroscopy, infrared spectroscopy, and carbon-hydrogen analysis(20).

In Vitro Induction of Cytotoxic Macrophages with Polyanionic Polymers

The biological activity of specific interest to us was the ability of these synthetic carboxylic acid polymers to activate macrophage cells to tumor cytotoxicity. In the initial experiments, the test polymers were administered intraperitoneally at a dosage of 50 mg/kg body weight. The protocols and drug regimens used for initial screening were based on previous studies by Kaplan(19) for the elicitation of tumoricidal macrophages by the conventional activating agents pyran and C. parvum. Peritoneal exudate macrophages were routinely harvested seven days after the administration of the test agent and tested for their cytotoxic capacity against Lewis lung carcinoma in both morphologic and ^{3}[H]-thymidine release assays. Pyran (50 mg/kg) and C. parvum (17.25 mg/kg) elicited macrophages were included in each experiment as positive controls. Base controls consisted of thioglycollate (1 ml of 10% solution, 3 days post-injection) and normal or saline elicited macrophages (24-48 hours post-injection)(21). The ability of the test polymers to induce cytotoxic macrophages determined by the [^{3}H]-thymidine assay is shown in Figure 2.

The data represents the mean of triplicate samples in three experiments minus the percent specific cytotoxicity obtained with thioglycollate peritoneal exudate cells. Percent specific cytotoxicity was calculated from the following formula:

$$\frac{\text{test cpm - spontaneous release}}{\text{maximal release - spontaneous release}} \times 100 = \% \text{ cytotoxicity}$$

The cytotoxic data was obtained with effector-target cell ratios of 20:1. Similar cytotoxicity data was obtained by morphological assays. The cytotoxicities observed for pyran, C. parvum, MA-CDA, IA-ST, and MA-VA elicited macrophage populations were significantly greater than either normal or thioglycollate elicited macrophages. EAA initiated moderate cytotoxic activity while MA-ST, MA-AU, and MA-E did not initiate any

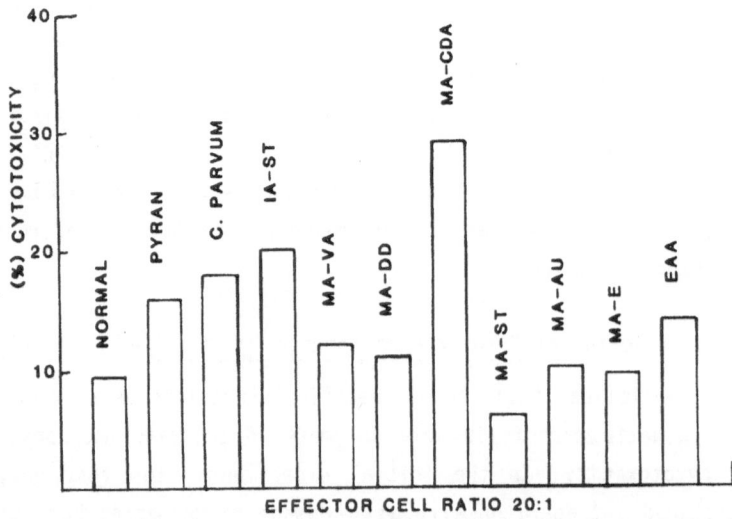

Figure 2. Polycarboxylic Acid Polymer Activation of Macrophages and Cytotoxicity of Lewis Lung Carcinoma.

observable cytotoxic activity (see Table 1 for structures of these polymers).

Those polymers that exhibited the best cytotoxic induction capacity, MA-CDA and IA-ST, possess a lipophilic group and enhanced surface charge properties compared to the other polymers. Although MA-ST, a very low molecular weight sample (MW 1600), has a lipophilic benzene ring, it did not induce cytotoxic macrophages under these experimental conditions. In a previous study, using a higher molecular weight sample, MA-ST exhibited anti-tumor activity against Ehrlich ascites tumor in vivo(20). Macrophages induced by test polymers as well as controls demonstrated a dose response in their cytotoxic activity against Lewis lung. The greatest differences between the groups were observed at an E:T of 20:1 with no cytotoxicity seen at the ratio of 1:1. It was concluded from these studies that MA-CDA and IA-ST have the greatest potential for cytotoxic induction(22). In addition, these test polymers, as well as pyran and C. parvum, elicited peritoneal exudate cells, are not cytotoxic to normal cells. Macrophages elicited with these agents demonstrated no activity against newborn mouse fibroblasts(21).

Kinetics of Induction of Tumoricidal Macrophages

Based on our previous observations of the activity elicited by the test polyanionic polymers, MA-CDA and IA-ST were chosen for further evaluation. Pyran was chosen as the control and MA-ST was employed as the negative control for the polymer group.

The kinetics of induction of tumoricidal macrophages by selected test polymers are shown in Figure 3. Several different patterns of induction were observed. Macrophages elicited with MA-CDA and pyran demonstrated similar kinetics, with maximal cytotoxicity on day 7. This corresponds to the time course of cytotoxic induction previously observed with pyran and C. parvum(15). The cytotoxicity induced with IA-ST peaked at day 5. Interestingly, MA-ST demonstrated significant cytotoxicity on day 3 which may explain its previously observed protective activity against Ehrlich ascites tumor in vivo(22). The cytotoxic activity present in the resident population was consistent with the observed levels of natural killer activity in peritoneal exudate cells(23) and with levels of spontaneous cytotoxicity to certain tumor cell lines(24).

In Vivo Studies of Tumor Growth Inhibition

After observing that MA-CDA and IA-ST exhibited the ability to activate peritoneal macrophases to tumoricidal capacity in vitro, further

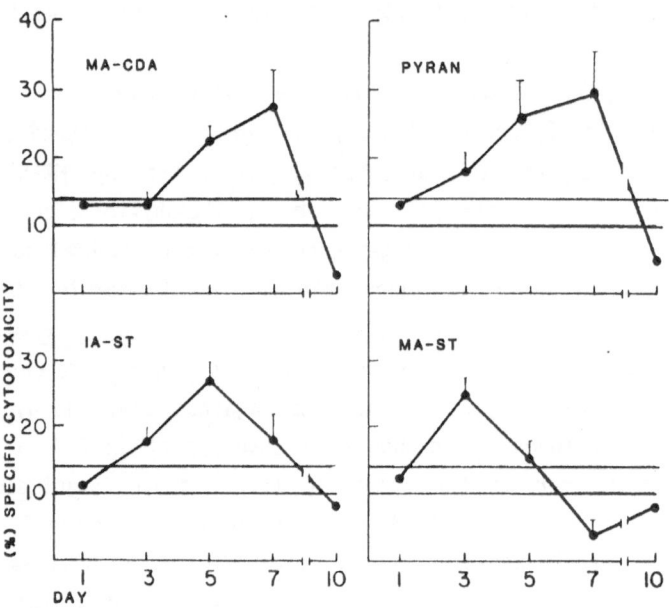

Figure 3. Time of Maximum Macrophage Activation against Lewis Lung Carcinoma

experiments were performed to determine if MA-CDA and IA-ST had antitumor capacity in vivo. Mice were subcutaneously injected with 2×10^4 freshly trypsinized Lewis lung tumor cells on day 0. MA-CDA, IA-ST, or saline was administered at dosages of 25, 50, and 100 mg/kg on days 1, 2, and 3 after tumor inoculation. Mouse survival was followed for 90 days and the results are listed in Table 1.

The mean survival time (MST) of mice that received tumor alone was 40 days. All of the mice in this group succumbed to the tumor. The MST of mice that received MA-CDA and IA-ST at all dosages was significantly increased. Not only did MA-CDA increase the MST of mice to 64 and 69 days at dosages of 50 mg/kg and 100 mg/kg, respectively, but two and three of the original five mice receiving these respective dosages were still alive on day 90, the termination time of the experiment. IA-ST, on the other hand, proved to be very toxic: for example, at a dosage of 100 mg/kg, 3 mice out of 5 died within 3 days after the injection. However, one of the remaining animals lived for 80 days and the other lived until the end of the experiment (90 days) with no evidence of tumor. These data are particularly interesting since this is the first time that a synthetic polymer was able to produce animals that survived after being challenged with Lewis lung carcinoma.

TABLE 1.   In Vivo Evaluation of the Effect of Polymers
on the Survival Time of Tumor Bearing Mice

Treatment	25 mg/kg		50 mg/kg		100 mg/kg	
	MST[a]	S/T[b]	MST	S/T	MST	S/T
None	40	0				
C. parvum	52	1				
Pyran	--	-	57	1		
MA-CDA	52	0	64	2	69	3
IA-ST[c]	52	1	48	1	85	1

a) Mean Survival Time, days
b) Mice from a group of five surviving 90 days
c) This polymer was very toxic, and most animals died
   within 3 days of injection. The MST values are based
   on 3 mice surviving the initial 25 and 75 mg/kg dose
   and 2 mice surviving the 100 mg/kg dose.

Polysaccharide-Coated Liposomes as Drug Carriers

Liposomes are considered potentially convenient carriers for encapsulated, water-soluble and high molecular weight drugs. They would be
especially advantageous if they could be made more stable and targetable.
Sunamoto and his co-workers have developed an improved technique for
producing stable liposomes for animal systems(25-28). The technique
involves coating the outer surface of egg lecithin liposomes with polysaccharide derivatives such as amylopectin or mannan. This is a non-
covalent molecular assembly of an artificial cell wall on the artificial
cell. In addition, these polysaccharide-coated liposomes have unique
targetability to specific tissues and cells(26-29). For example, polysaccharide-coated liposomes, intravenously injected into guinea pigs
showed a significantly high accumulation of liposomes in the lung when
compared with conventional non-polysaccharide-coated liposomes(28). The
efficiency of liposomal uptake by the alveolar macrophages was affected by
the structure of the polysaccharide employed(25,26).

Liposomes are usually rapidly cleared by macrophages in the reticulo-
endothelial system, which in most cases causes difficulty in the utilization of liposomes as drug carriers. However, when considering the delivery
of drugs that act as immunomodulators, interferon inducers, or antibiotics, this is an advantage. In these situations, the macrophages could
serve as processing cells, appropriately presenting certain antigens to
the immune system.

Interaction of Polymer Drugs with Liposomal Membranes

In order to encapsulate a drug in liposomes, it is necessary to first determine the interaction between the drug candidate and the liposomal membrane. Considering that a liposome is basically a cell model, one can also obtain information on the systemically administered interaction of drugs with the membranes of intact cells. It is known that most drugs, especially water-soluble drugs, are easily released from the interior of liposomes. This release is due to the facile destruction of the liposomes. Therefore, to obtain stable liposomes for drug delivery, the drug candidate for encapsulation is required not to perturb the liposomal membranes during and after encapsulation.

We investigated the lysis of liposomes by the polyanionic polymers that are listed in Figure 1. Recent correlation between the polymer-induced lysis of liposomes and the pharmacological activity of these polymeric drugs has been reported earlier(30,31). Small unilamellar vesicles (SUV) of egg lecithin containing the water soluble fluorescent probe, carboxyfluorescein (CF, Eastman Kodak, Rochester, New York), were prepared and isolated on a Sepharose 4B column. Appropriate quantities of the polyanion polymer drugs (1-2 mg/ml) were added to the liposome suspension in which the lipid concentration was adjusted to 1.3×10^{-4} M. The polymer induced CF-release was followed as a function of time by monitoring an increase in the fluorescence intensity at 520 nm on excitation at 470 nm.

A typical result in the kinetics of the polymer-induced CF-release from liposomes is shown in Figure 4. Although the data are not shown in this figure, other polymers, such as MA-DD, MA-E, MA-VA, MA-CDA, MA-AU, EAA, and pyran, (see Figure 1 for structures), did not cause any significant damage to the liposomal membranes under these experimental conditions. Only IA-ST caused damage to the liposomes with significant polymer-induced CF-release (Figure 4). In addition, both the relative rate of lysis and the extent of percent CF-release were found to be directly related to molecular weight of the IA-ST (Figure 4). This molecular weight dependency of the IA-ST-induced CF-release was inconsistent with the observed cytotoxicity.

A strong interaction of IA-ST with liposomal membranes was confirmed by fluorescent depolarization studies. These studies showed that IA-ST was adsorbed onto liposomal membranes, making the membrane less fluid in both the hydrophobic domain and the hydrophilic region close to the surface of bilayers. All the other polyanionic polymers studied exhibited no significant changes in the membrane fluidity. These results are con-

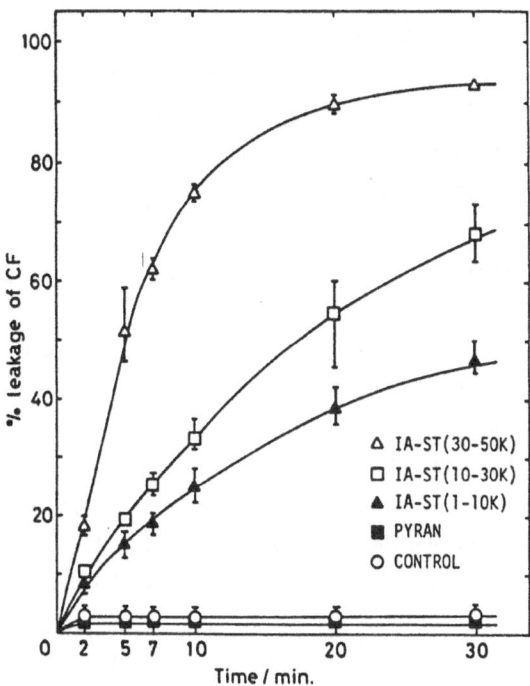

Figure 4. Polyanionic polymer-induced damage of liposomes as monitored by CF-release from liposomes. Polymers (2mg/60μl) were added to 0.94 ml of liposome suspension (1.3 x 10^{-4}M) at 37.0°C and pH 8.6.

sistent with the polymer-induced CF-release experiments. It is interesting that IA-ST showed this strong perturbation to the liposomal membrane, since it was not only cytotoxic to tumor cells, but extremely toxic in general (LD$_{50}$ 100 mg/kg). These data suggest that the strong toxicity upon systemic administration of IA-ST may be the result of gross membrane interaction.

From the results of these studies, we classified the polyanionic polymers into three general categories:

1) Polymers that show strong tumor cytotoxicity and perturb liposomal membranes, such as IA-ST (we could not encapsulate this polymer in liposomes because of the destruction of the liposomes during the encapsulation process).
2) Polymers that show tumor cytotoxicity but do not perturb liposomal membranes, such as MA-CDA and pyran (these polymers are considered to be the best candidates for encapsulation in liposomes).
3) Polymers that show neither cytotoxicity nor perturbation of liposomal membranes, such as MA-AU and EAA.

The polymers in the third classification were also considered suitable candidates, even though these polymeric drugs did not show any significant cytotoxicity when administered alone. This could be expected if the macrophage cells were not able to effectively internalize them in the free state. Therefore, it was considered that if these polymers were encapsulated and successfully transferred into phagocytic cells by liposomes they may exhibit some cytotoxicity.

Utilization of Polysaccharide-Coated Liposomes as a Targetable Carrier for Polyanionic Polymers

Sunamoto and his co-workers have shown that coating the outer surface of liposomes with polysaccharide derivatives (Figure 5) produces vesicles that are much more stable, not only under laboratory conditions, but, also, in vivo environments(25). Furthermore, these coated liposomes are targetable toward specific cells and tissues, depending on the chemical structure of polysaccharides employed. The stabilizing efficiency of coated liposomes was ascertained by several techniques:

1) isolation of polysaccharide-coated liposomes by gel-filtration,
2) increased barrier function against the water soluble probe, carboxy-fluorescein, encapsulated in the interior of liposomes. (increased stability was found even in the presence of plasma and serum), and
3) increased resistance to phospholipase and lipoxygenase, by which liposomes can be protected from enzymatic lysis in vivo.

The targetability of the polysaccharide-coated liposomes was ascertained by two methods:

1) qualitative and quantitative phagocytosis of liposomes with alveolar macrophages, monocytes, and neutrophiles, and
2) tissue distribution of liposomes when intravenously administrered to guinea pigs.

Effective Internalization of Polysaccharide-Coated Liposomes with Phagocytes

The internalization of liposomes by phagocytes can be confirmed by several methods. First, the phagocytosis of the liposomes by alveolar macrophages of guinea pigs was directly determined by electron microscopy. The internalization of liposomes by the macrophages was observed after incubation for 30 minutes. The subsequent digestion of these liposomes in the cell lysosomes was observed after further incubation. The phagocytosis was also determined by fluorescence microscopy for which two different fluorescent probes were employed: a water-soluble carboxyfluorescein, and a water-insoluble terbium tris(acetyl)acetonate dihydrate. From these investigations, we concluded:

(1) The mechanism of internalization of liposomes by phagocyte cells is probably by phagocytosis, not by fusion.
(2) When a water-soluble drug is encapsulated in a liposomal drug de-

344

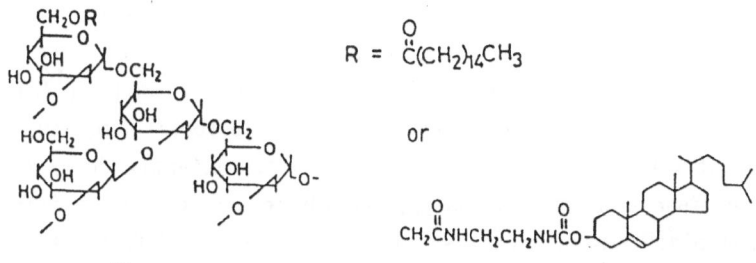

Figure 5. Polysaccharide derivatives used for coating of liposome surface.

livery system, the water-soluble drug is effectively delocalized in the cell cytosol by, first, cell phagocytosis and, then, lysis of the liposome carrier.

(3) The polysaccharide-coated liposomes are more effectively internalized by phagocytic cells than the conventional liposomes without a poly-saccharide coat. (This mechanism of internalization was further ascertained by the lack of observed internalization of the liposome at 4.0°C and in the presence of cytochalasin B at 37.0°C. and 4.0°C.

Quantitative investigations of the efficiency of phagocytosis of the liposomes were carried out through both radioisotope techniques, using $[^{14}C]$-DPPC labeled liposomes, and by flow cytofluorometry, using CF-loaded liposomes. In both cases, we have found that mannan derivatives, such as CHM-200-2.4 (cholesterylcarbonylaminoethyl mannan with a molecular weight of 200,000 and 2.4 cholesteryl moieties per hundred mannose units), showed the most effective internalization of all the polysaccharide derivatives evaluated (Figure 5). The results are shown in Figure 6 for the radioiso-tope investigation and in Figure 7 for the flow cytofluorometry investigation.

Tissue Distribution of Mannan Derivative-Coated Liposomes

From a number of studies, it has been ascertained that the majority of the liposomes injected intravenously are retained in the liver and spleen, regardless of size, structural characteristics (unilamellar or multilamellar), or lipid composition. Consequently, the metabolic fate and tissue distribution of liposomes coated with the mannan derivative CHM-200-1.7, was investigated by monitoring the radioactivity of $[^{14}C]$-CoQ$_{10}$ and $[^3H]$-inulin in the blood and specific organ tissues. The $[^3H]$-inulin was encapsulated in the interior water phase, while $[^{14}C]$-CoQ$_{10}$ was entrapped in the lipid bilayer of the liposomal membrane. The blood radioactivity, determined from samples intermittantly drawn after intra-venous injection of the polysaccharide-coated liposomes into the male guinea pigs' femoral vein, was rapidly cleared. In fact, the radioactive probes were completely depleted from the blood within 2 hours after injection.

In addition, we investigated the urinary excretion of the doubly labeled liposomes with $[^3H]$-inulin and $[^{14}C]$-CoQ$_{10}$(28). When free $[^3H]$-inulin aqueous solution was injected intravenously, the radioactivity was quickly cleared from the blood stream and about 90% of the dose appeared in the urine within 3 hours. On the other hand, the urinary excretion rate of $[^3H]$-inulin encapsulated in the polysaccharide-coated liposomes was much slower and only 21% of the dose given appeared in the urine within 24 hours after injection. The tissue distribution of the doubly labeled

346

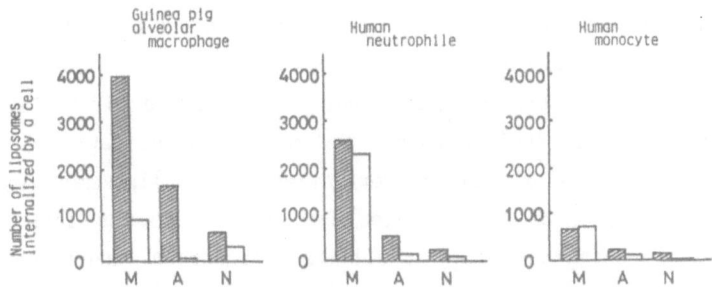

Figure 6. Internalization efficiency of liposomes into three kinds of phagocytic cells monitored by RI method using [^{14}C]-DPPC labeled liposomes. Liposomes [☐ , 3.9 x 10^{11} LUVs/ml and ▨ , 3.9 x 10^{10} LUVs/ml) were incubated with phagocytes (1.5 x 10^6 cells/ml) at 37.0°C for 60 min. M, mannan-coated LUV; A, amylopectin-coated LUV; and N, non-coated LUV.

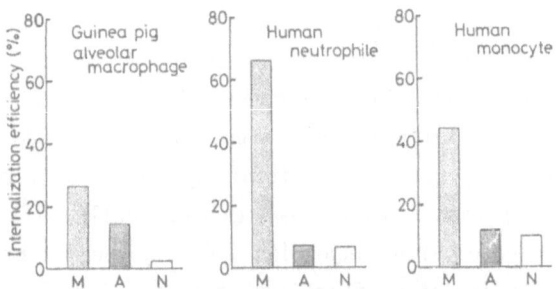

Figure 7. Internalization efficiency of liposomes into three kinds of phagocytic cells monitored by flow cytofluorometry using CF-loaded liposomes. Liposomes (5.2 x 10^{10} LUVs/ml) were incubated with phagocytes (2.0 x 10^6 cells/ml) at 37.0°C for 60 min. M, mannan-coated LUV; A, amylopectin-coated LUV; and N, non-coated LUV.

liposomes at 0.5 and 24 hours after administration is listed in Table 2.

We concluded, from this data, that the polysaccharide-coated liposomes are relatively stable, not only in blood circulation, but also in organ tissues. Another noteworthy result is the fact that there is a high accumulation of mannan derivative-coated liposomes in the lungs compared to that of the uncoated liposomes. Based on this information, studies are being conducted to utilize this unique lung-targeting characteristic.

Evaluation of Biological Activities of the Polymeric Drugs Encapsulated in Mannan Derivative-Coated Liposomes

From the results obtained in the above investigations, we decided to encapsulate the polyanionic copolymer, poly(maleic anhydride-co-cyclohexyl-1,3-dioxepin) (MA-CDA) within CHM-200-1.7 coated egg lecithin LUV. MA-CDA was chosen because it has a specifically high cytotoxicity to tumor cells and is stable in the liposome; CHM-200-1.7-coated liposomes were chosen because they are very effectively internalized by phagocytes such as alveolar macrophages, monocytes, and neutrophiles.

The preparation of CHM-200-1.7 has been previously described(25). The substitution degree of the cholesteryl moiety per 100 mannose units was determined by ^{1}H-NMR. The CHM-200-1.7-coated liposomes of egg lecithin were prepared by essentially the same procedure as that adopted for conventional liposomes without any polysaccharide coating. A 1.0 ml saline containing 7.5 mg of CHM-200-1.7 was added to 4.0 ml of liposome suspension which was prepared beforehand from 37.5 mg of purified egg lecithin and contained 25-50 mg of MA-CDA.

First, the activation of macrophages was evaluated by monitoring superoxide production elicited by the activated peritoneal macrophages of mice subjected to the intraperitoneal injection of MA-CDA-loaded in liposomes. A 100 µl solution containing 500 mg of MA-CDA, free or encapsulated in liposomes, was intraperitonially injected into three male mice (C57BL/6, 5-7 g each). Peritoneal exudate cells (PEC) were collected beginning 30 minutes after injection and at given intervals for 24 hours. Macrophages were isolated after incubation for 1 hour at 37.0°C under 5% CO_2 on a plate coated with heat inactivated FCS. After washing the cells sequentially with phosphate buffered saline buffer, 0.05% EDTA, and 10% heat inactivated FCS. The cells were re-suspended in KRP (2 X 10^5 cells/ml). An aliquot of the cell suspension (0.98 ml) was incubated at 37°C for 2 minutes with 10 µl of 2 mM glucose (25 mg/ml) and 10 µl of 100 µM cytochrome C. Immediately after this, aqueous phorbol 12-myristate-

Table 2. Tissue distribution of liposomes doubly labeled with [^{3}H]-inulin and [^{14}C]-CoQ$_{10}$ at 30 min and 24 hr after intravenous administration into guinea pigs

Tissue	Time after injection (hr)	Control LUV		CHM-200-1.7-coated LUV	
		[^{3}H]-Inulin	[^{14}C]-CoQ	[^{3}H]-inulin	[^{14}C]-CoQ
Brain	0.5	0.08	0.09	0.01	0.13
	24	0.02	0.02	0.02	0.03
Heart	0.5	0.37	0.43	0.20	0.25
	24	0.09	0.38	0.03	0.20
Lung	0.5	3.10	30.90	34.90	67.10
	24	0.20	5.60	0.87	12.50
Spleen	0.5	7.00	10.80	1.40	2.80
	24	20.80	19.50	3.20	7.90
Liver	0.5	4.50	22.20	10.30	24.30
	24	16.80	53.90	21.40	54.30
Kidney	0.5	2.30	0.60	6.80	0.40
	24	0.72	0.31	1.40	0.23
Adrenal Gland	0.5	0.02	0.03	0.009	0.00
	24	0.008	0.32	0.005	0.17

% of the dose in each tissue

Values are means ±S.E. of 3 guinea pigs
* Significantly different from control LUV, p<0.05, Student's t-test
** p<0.01, Student's t-test

13-acetate (PMA) solution (5 µg/10 µl) was added, and the increase in the absorbance at 550 nm was followed. Table 3 is a summary of the data presented as the percent increase of superoxide production 2 hours after injection compared with the control group. The injection of MA-CDA alone did not change superoxide production in this time period. However, the cholesteryl derivatized mannan (CHM-200-1.7) and the CHM-200-1.7-coated liposomes elicited some superoxide enhancement. The maximum increase in superoxide production, 300% greater than control or free MA-CDA, was observed when MA-CDA was encapsulated in the CHM-200-1.7-coated LUV.

Based on these successful superoxide experiments, the tumorcidal properties of murine alveolar macrophages activated by MA-CDA-loaded liposomes coated by CHM-200-1.7 were examined (33). A 100-200 µl solution containing free MA-CDA or MA-CDA encapsulated by liposomes was intra-venously injected into 3 mice (C57BL/6, male, body weight 5-7 g). After activation _in vivo_ for 1, 3, 5, and 7 days, alveolar macrophages were collected with 50 ml of a phosphate buffered saline by bronchio alveolar lavage (BAL)(32). The alveolar macrophages collected were attached to a Falcon Microtest III plate (2 X 10^5 cells/well). Within 60 minutes after the macrophage monolayers were prepared, 100 µl of target tumor cell suspensions (Lewis Lung carcinoma cell line (3LL), 1 X 10^4 cells/well) were added. Plates were incubated for 48 hours at 37.0°C under 5% CO_2, 0.1 µCi [^{3}H]-thymidine/50 µl (New England Nuclear, Boston, Mass.) was added to the cells 16 hours before the completion of the culture. On

TABLE 3. Superoxide Production of Mouse Peritoneal Macrophages Activated by MA-CDA for 2hr _In Vivo_

	0^{2-} production nmol/min/2 X 10^5 cells	% increase in 0^{2-} production
MA-CDA (500µg/100µl) in CHM-LUV[a]	6.99	315
MA-CDA (500µg/100µl) in LUV[b]	5.62	250
CHM-LUV[a]	5.76	254
CHM (0.15mg/100µl)	3.15	142
MA-CDA (500µg/100µl)	2.05	92
control (PBS)	2.22	100

a) egg lecithin LUV coated by CHM-200-1.7
b) conventional liposomes without polysaccharide coat

termination of the culture, the plates were washed to remove the unincorporated [³H]-thymidine and the thymidine uptake by 3LL was evaluated on an Aloka III counter. Percentage cytotostasis was determined by the following expression:

$$100 \times \frac{\text{CPM(normal macrophages + 3LL) - CPM(activated macrophages + 3LL)}}{\text{CPM(3LL alone)}}$$

The kinetic data for the cytostatic effects of murine alveolar macrophages, activated by two different concentrations of MA-CDA at the constant E:T (effector cells to target cells ratio) of 20:1, is listed in Figures 8A and 8B. When 500 μg of MA-CDA per mouse were administered in tests for macrophage activation _in vivo_, the cytotostasis elicited by the polymer drug encapsulated in the mannan-coated liposomes showed maximum activity 5 days after i.v. injection. Most importantly, the cytostatic effect observed was about 4 times greater than cytotostasis elicited by the free drug alone (Figure 8A).

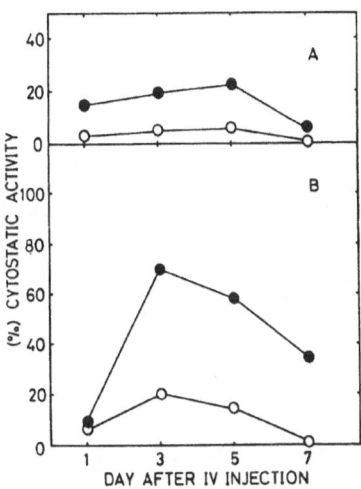

Figure 8. Kinetics of the cytostatic effect of mouse alveolar macrophage on 3LL when macrophages were activated _in vivo_ by i.v. injection of MA-CDA (A, 500μg/mouse and B, 1000μg/mouse). Date represents cytostatic activity at E:T=20.1. ●, MA-CDA encapsulated in CHM-200-2.4-coated LUV and ○, free MA-CDA.

The cytotostasis, based on the [3H]-thymidine uptake method, increased from 6% for the free MA-CDA to 24% for the liposome encapsulated MA-CDA. Furthermore, when the dosage of MA-CDA was increased from 500 to 1000 mg per mouse, the relative cytostasis of the drug encapsulated in the mannan-coated liposomes, increased about 3.5 fold. More importantly, cytostatic activity significantly increased from 20% cytostasis for the free MA-CDA to 70% with the liposome encapsulated MA-CDA at the higher dose and the maximum activity was observed 3 days after injection (Figure 8B). Thus, higher doses are considerably more effective in the in vivo activation of macrophages.

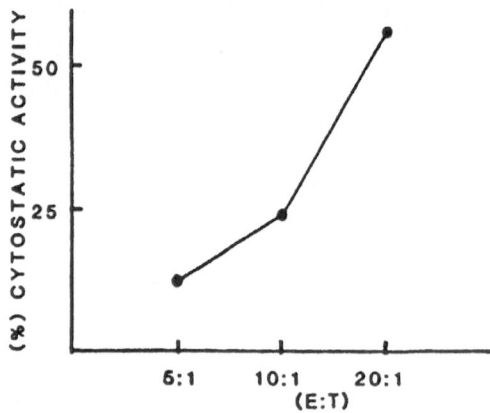

Figure 9. Cytostatic activity of alveolar macrophage on 3LL at different E:T ratios. Macrophages were harvested 5 days after i.v. injection of MA-CDA (1000µg/mouse) encapsulated in CHM-200-2.4-coated LUV.

Another interesting observation was that the increase in the E:T-ratio brought about by the encapsulated MA-CDA in the mannan-coated liposomes exhibited a larger enhancement of cytostatic activity than was expected for this system (Figure 9).

Conclusions

Several new polycarboxylic acid copolymers capable of eliciting macrophages to become cytotoxic to Lewis lung carcinoma have been developed. The polymers with hydrophobic character associated with the polymer chain were found to be the most active. The two hydrophobic groups that appear to play the most important role are the phenyl and the cyclohexyl groups. The two polymers that exhibit the best activity are poly(maleic anhydride-co-cyclohexyl-1,3-dioxepin) (MA-CDA) and poly(itaconic acid-co-styrene) (IA-ST). Both of these polymers, in the _in vitro_ morphological and [^{3}H]-thymidine evaluations against Lewis lung, exhibited greater macrophage activition than either pyran or C. _parvum_. More significantly, in _vivo_ evaluations of both polymers produced increased animal life spans, as well as survivors to Lewis lung carcinoma challenges. No other synthetic polycarboxylic acid polymer has been reported to effect the survival of animals challenged with Lewis lung carcinoma.

Liposomes were found to be suitable carriers for these poly-carboxylic acid copolymers which were found to be readily internalized by macrophages. Moreover, the technique of coating the surface of the liposomes with polysaccharides, such as mannan and amylopectin further increased both their stability in many environments and their rate of uptake by phagocytic cells such as the macrophages. It was found that the structure of the polyanionic polymer plays a significant role since MA-CDA could be encapsulated in the liposome; but IA-ST could not. IA-ST rapidly perturbed the liposome, which prevented effective encapsulation; this perturbation was both concentration and molecular weight dependent. Further studies indicated that the IA-ST binds to the liposomal membrane: this effect could also be the cause of the general toxic effect observed with this agent ($LD_{50} > 100$ mg/kg). Other polycarboxylic acid copolymers, such as pyran, poly(maleic anhydride-co-styrene), and poly(maleic anhydride-co-allyl urea) were also effectively encapsulated and gave stable liposomes.

The biological evaluation results of MA-CDA encapsulated in cholesterol derivatized mannan coated liposomes (CHM-LUV) were significant. Superoxide production of mouse peritoneal macrophages activated by MA-CDA encapsulated in CHM-LUV was increased by more than 300%, compared to both control and free MA-CDA 2 hours after injection. After 24 hours, the free MA-CDA also showed enhanced superoxide production (150%), and the CHM-LUV encapsulated MA-CDA was still significantly higher (180%).

Following the success of the superoxide experiments, the _in vitro_ tumoricidal activity of murine alveolar macrophages activated by MA-CDA-loaded CHM-LUV were examined. The kinetic data, based on the [^{3}H]-thymidine uptake method for encapsulated MA-CDA, showed a significant increase in cytotostatic activity to Lewis lung carcinoma. The cytostatic activity increased from 6%, for the free MA-CDA, to 24%, for the CHM-LUV encapsulated MA-CDA at a 500 µg/animal dose. When the dose was increased to 1000 µg/animal, the Lewis lung cytostasis was 20% for the free MA-CDA, and 70% for the CHM-LUV encapsulated MA-CDA. Therefore, increased doses are much more effective in the _in vivo_ activation of macrophage.

In conclusion, our collaboration, which began at the symposium on "Polymers in Medicine", held in Porto Chervo, in 1983, has been highly successful, and we are delighted to have had the opportunity to report the results of our work at the second symposium on "Polymers in Medicine", which was held in Capri in 1985.

Acknowledgements

The authors wish to thank all their colleagues who participated in these studies, in particular, A. Kaplan, T. Kuus, T. Sato, and M. Oka. We also wish to thank the organizers of the symposium and this book, particularly E. Chiellini, and Raphael M. Ottenbrite's daughter, Shelley Ottenbrite, for typing the manuscript.

References

1. R.M. Ottenbrite, K. Kuus, and A.M. Kaplan, Polymers in Medicine, eds., E. Chiellini, and P. Giusti, Plenum Publishing Co., New York, p.3 (1982).

2. J. Sunamoto, K. Iwamoto, M. Takada, T. Yuzuriha, and K. Katayama, Polymers in Medicine, eds., E. Chiellini, P. Giusti, Plenum Publishing Co., p.157 (1984).

3. R.M. Ottenbrite, and G.B. Butler, Anticancer and Interferon Agents, Marcel Decker, Inc.,New York, (1984).

4. L.G. Donaruma, R.M. Ottenbrite, and O. Vogal, eds., Anionic Polymeric Drugs, John Wiley and Sons, New York, (1980).

5. W.J. Freeman, and D.S. Breslow, in Biological Activities of Polymers, eds., C.E. Carraher, and C.G. Gebelein, ACS Symp. Ser. 186, American Chemical Society, Washington, D.C., p.243 (1982).

6. W. Regelson, A. Munson, and W. Wooles, Int. Symp. Stand. Interferon and Interferon Inducers, London, 1969; Symp. Series Immunobiol. Stand., 14, Karger, Basel, p.227 (1970).

7. A.M. Kaplan, R.M. Ottenbrite, W. Regelson, R. Carchman, P. Morahan, and A. Munson, in Handbook of Cancer and Immunology, 5, ed. H. Walters, Garland, New York, p.135 (1978).

8. R.M. Schultz, J.D. Papamatheakis, and M.A. Chirigos, in Immune Modulation and Control of Neoplasia by Adjuvant Therapy, ed. M.A. Chirigos, Raven Press, New York, p.459 (1978).

9. A.E. Munson, W. Regelson, W. Lawrence, and W.R. Wooles, J. Reticuloendothel. Soc., 7, p.375 (1970).

10. F.F. Pindak, Infect. Immun.; 1, 271 (1970).

11. P.S. Morahan, W. Regelson, and A.E. Munson, Antimicrob. Agents Chemother., 3, 16 (1972).

12. P.S. Roberts, W. Regelson, and B. Kingsbury, J. Lab. and Chemical Med., 2, 822 (1973).

13. M.A. Chirigos, ed., Control of Neoplasia by Modulation of the Immune System, 3, Raven Press, New York, (1977).

14. M.W. Rosenthal, M.W. Argonne National Laboratory, Argonne, Illinois, personal communication.

15. A.M. Kaplan, in Anionic Polymeric Drugs, eds., L.G. Donaruma, R.M. Ottenbrite, and O. Vogl, John Wiley and Sons, New York, p.227 (1980).

16. P.S. Morahan, D.W. Barnes, and A.E. Munson, in Anionic Polymeric Drugs, eds., L.G. Donaruma, R.M. Ottenbrite, and O. Vogl, John Wiley and Sons, New York, (1980).

17. R.M. Ottenbrite, E.M. Goodell, and A.E. Munson, Polymer, 18, 461 (1977).

18. K. Takemoto, R.M. Ottenbrite, and I.Inaki, eds., Functionalized Polymers, Marcel Dekker, New York, (in press).

20. R.M. Ottenbrite, and A.M. Kaplan, Polymer Preprints, 23, 205 (1982).

21. K. Kuus, PhD., Dissertation, "Macrophage Activation by Unique Polyanionic Polymers", Virginia Commonwealth University, 1983.

22. K. Kuus, R.M. Ottenbrite, A.M. Kaplan, J. Bio. Resp. Mod., 4, 46 (1985).

23. S.A. Cohen, D. Salazar, and J.P. Nolan, J. Immunol., 129, 495 (1982).

24. K.C. Lee, M. Wong, and D. McIntyre, J. Immunol., 126, 2474 (1981).

25. J. Sunamoto, K. Iwamoto, M. Takada, T. Yuzuriha, and K. Katayama, Polymers in Medicine, eds., E. Chiellini and P. Giusti, Plenum Publishing Co., p.157 (1984).

26. J. Sunamoto, K. Iwamoto, M. Takada, T. Yuzuriha, and K.Katayama, Recent Advances in Drug Delivery Systems, eds., J.M. Anderson, and S.W. Kim, Plenum Publishing Co., New York, p.153 (1984).

27. J. Sunamoto, M. Goto, T. Iida, K. Hara, A. Saito, and A. Tomonaga, Receptor-Mediated Targeting of Drugs, eds., G. Gregoradis, G. Poste, J. Senior, and A. Trouet, Plenum Publishing Co., New York, p.359 (1985).

28. M. Takada, T. Yuzuriha, K. Katayama, K. Iwamoto, and J. Sunamoto, Biochem. Biophys. Acta., 802, 237 (1984).

29. I.J. Fidler, S. Sone, W.E. Fogler, and Z.L. Barnes, Proc. Nat. Acad. Sci., USA, 78, 1680 (1981).

30. J. Sunamoto, T. Sato, and R.M. Ottenbrite, in Current Topics in Polymer Science, eds., R.M. Ottenbrite, S. Inoue, and L.A. Utracki, McMillan Pub. Co., (in press).

31. R.M. Ottenbrite, J. Sunamoto, T. Sato, and M. Oka, Polymer Preprints, 26, 212 (1985).

32. P.G. Holt, J. Immuno. Methods, 27, 189 (1979).

33. J.J. Reinehart, P. Lange, B.J. Gormus, and M.E. Kaplan, Blood, 52, 211 (1978).

CONTROL OF POLYMER SURFACE EROSION BY THE USE OF EXCIPIENTS

Jorge Heller

Polymer Sciences Dept.
SRI International
Menlo Park, CA 94025

INTRODUCTION

Polymers that contain hydrolytically labile linkages in their backbone can hydrolyze by two different mechanisms. These are bulk erosion and surface erosion. In a bulk-eroding polymer, the hydrolytic process occurs throughout the bulk of the material, whereas in surface erosion the hydrolysis is confined to the outer surface of a solid device (1).

The distinction between these two processes is important because kinetics of release of a therapeutic agent homogeneously dispersed in the bioerodible polymer is significantly different for these two processes. If a polymer is undergoing a bulk-erosion process, the predominant mode of drug release is simple diffusion and drug release kinetics show typical first-order dependence. However, as erosion of the polymer becomes significant, the diffusional release can be significantly affected by the erosion process and the drug release profile will depend on the erosional contribution to drug release.

Thus, if rate of polymer hydrolysis is so slow that all the drug has been released before significant polymer erosion takes place, then a pure first-order drug release profile will be observed. However, if significant polymer erosion takes place during the diffusional release of the drug, then the diffusional release profile will be significantly altered; in an extreme case, all the drug can be abruptly released if polymer erosion is complete before all the drug has diffused out of the polymer.

Because the relative contributions of these two processes change with time, drug release is usually not predictable or amenable to precise control. Nevertheless, drug delivery systems based on bulk-eroding aliphatic polyesters such as glycolic, lactic, or their copolymers are under active development and useful delivery systems are evolving (2).

If a polymer is undergoing a surface-erosion process and if the drug is well immobilized in the polymer so that diffusional release is not important, then drug release is determined completely by erosion of the matrix. That is, as surface layers of the polymer solubilize by the hydrolytic process, drug contained in that layer is released; however, because layers immediately adjacent to the eroding layer remain essentially intact, drug release is confined to the outer layer only.

Other than predictability, drug release by polymer surface erosion has a number of important consequences. These are: (a) rate of drug release is directly proportional to the concentration of the drug in the matrix, (b) rate of drug release is directly proportional to the total surface area and (c) lifetime of the device is directly proportional to the thickness of the device. However, unless a thin slab or other specialized geometries are used, rate of drug release will decline as the total surface area of the device diminishes as a consequence of the erosion process.

One means of achieving surface erosion is to use polymers that have pH-labile bonds in the backbone and then use excipients physically incorporated into the polymer to lower the pH in the surface layers of the device relative to the device interior; this induces a hydrolysis process in the surface layers.

Three types of polymer can produce surface-eroding matrices by means of physically incorporated excipients. These are polyacetals (3), polyketals (4), and poly(ortho esters) (5), which are schematically represented below.

$$\begin{array}{ccc}
\overset{\displaystyle R}{\underset{\displaystyle H}{\left[R\!-\!O\!-\!\overset{|}{\underset{|}{C}}\!-\!O\right]_n}} &
\overset{\displaystyle R}{\underset{\displaystyle R}{\left[R\!-\!O\!-\!\overset{|}{\underset{|}{C}}\!-\!O\right]_n}} &
\overset{\displaystyle OR}{\underset{\displaystyle R}{\left[R\!-\!O\!-\!\overset{|}{\underset{|}{C}}\!-\!O\right]_n}} \\[2em]
\text{polyacetal} & \text{polyketal} & \text{poly(ortho ester)}
\end{array}$$

These polymers are stable in base, erode at varying rates depending on structure at the physiological pH of 7.4, and undergo a rapidly accelerating rate of hydrolysis as the pH is lowered. As with all polymers, rate of hydrolysis depends not only on the intrinsic reactivity of the hydrolytically labile bonds, but also on the hydrophobicity of the matrix.

In using excipients, two fundamentally different methods can be used. In one, an acidic excipient is used to accelerate rate of hydrolysis in the surface layers. In the other, a basic excipient is used to stabilize the interior of the matrix, which allows the uncatalyzed polymer erosion to take place only in the surface layers. The use of acidic excipients is useful in producing devices having lifetimes varying from hours to about one month, whereas use of basic excipients has the potential of producing devices having useful lifetimes of years.

In this chapter we will discuss work with poly(ortho esters) using acidic and basic excipients. Synthesis of linear and crosslinked polymers as well as device preparation have been described previously (6,7).

USE OF ACIDIC EXCIPIENTS

When acidic excipients are used, the following process will take place when a highly hydrophobic matrix containing incorporated excipient is placed in an aqueous environment: water will slowly intrude into the polymer and activate the excipient either by a simple dissolution process of an incorporated acidic salt such as calcium lactate or by a hydrolytic activation of an incorporated latent catalyst such as acid anhydride. In either case a lowered pH will occur in the reaction zone and catalyzed hydrolysis of the matrix will occur within that zone.

However, as shown in Figure 1, the rate at which water intrudes into the polymer is an extremely important consideration, and details of the erosion process are determined by the relative movement of two fronts: V_1, the rate of water intrusion and V_2, the rate of polymer erosion.

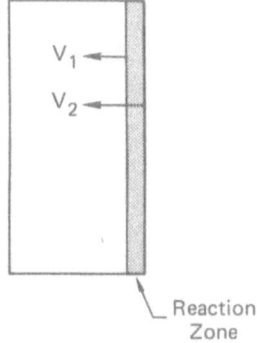

Fig. 1   Schematic representation of water intrusion and erosion for one side of a bioerodible device.

If $V_1 > V_2$, the thickness of the reaction zone will gradually increase with time and eventually the matrix will become completely permeated by water, thus leading to an eventual bulk erosion process. If $V_2 > V_1$, a surface erosion process will take place, but the rate of polymer erosion will be completely determined by the rate at which water intrudes into the polymer.

Long device lifetime necessitates very slow polymer erosion. Clearly, even with the most hydrophobic polymer, it is unlikely that devices with incorporated acidic excipients having a very long useful lifetime can be constructed because water will permeate a small implant-sized device in a matter of a few weeks. Thus a bulk erosion process and ultimate device disintegration are inevitable.

However, useful devices having lifetimes from hours to about one month can be prepared by using excipients such as acid anhydrides; in the presence of water, these anhydrides hydrolyze to the corresponding diacid (8). As expected, rate of polymer erosion and hence drug release depends on the pKa of the diacid and its concentration in the matrix. This dependence is shown in Figure 2 for 2,3-pyridine dicarboxylic anhydride and phthalic anhydride.

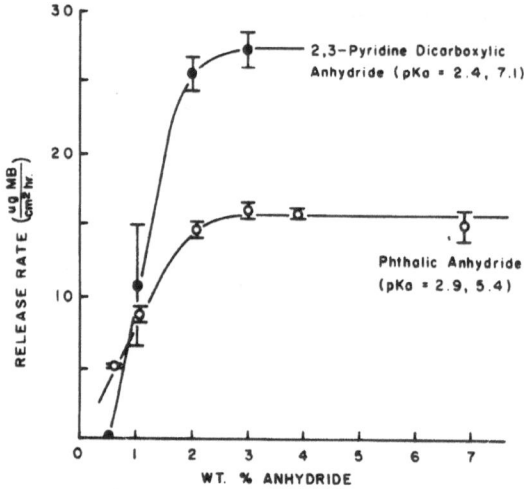

Fig. 2   The effect of anhydride content on the methylene blue (MB) release rate from 65:35 HD/t-CDM disks containing 0.2% MB (37°C, pH 7.4).
(Reprinted from R. V. Sparer, C. Shih, C. D. Ringeisen, and K. J. Himmelstein, Controlled release from erodible poly(ortho ester drug delivery systems, J. Controlled Release 1 (1984) 23-32. With permission).

The rate of release of the marker, methylene blue, physically incorporated into the polymer, clearly depends on both the pKa of the diacid hydrolysis product and the amount of incorporated anhydride. However, above about 2 wt% anhydride concentration, erosion rate reaches a limiting value and further increase in anhydride concentration does not affect rate of polymer hydrolysis. Presumably at that point, V_1, the rate of water intrusion into the matrix, becomes rate-limiting.

Convincing evidence for a surface erosion process is presented in Figure 3, which shows the concomitant release of an incorporated marker drug, methylene blue, release of the excipient hydrolysis product succinic acid, and total weight loss of the device. Thus, the release of an incorporated drug from an acid anhydride catalyzed erosion of poly(ortho esters) can be unambiguously described by a polymer surface erosion mechanism.

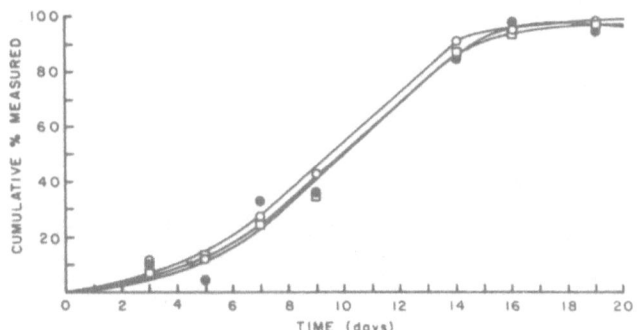

Fig. 3   The appearance of methylene blue (MB, o) and 1,4-^{14}C succinic acid (□), and the disk weight loss (●) as they were recorded from the erosion of 50:50 HD/t-CDM containing 0.1% 1,4-^{14}C succinic anhydride and 0.3% MB (37°C, pH 7.4). (Reprinted from R. V. Sparer, C. Shih, C. D. Ringeisen, and K. J. Himmelstein, Controlled release from erodible poly(ortho ester) drug delivery systems, J. Controlled Release 1 (1984) 23-32. With permission).

Release kinetics can be described in terms of an initial lag phase, a zero-order release phase, and a depleting phase. The lag phase characterizes a time period during which water begins to intrude into the polymer and activates the latent catalyst, which then initiates polymer erosion. Because thin disks (typically 8 mm diameter and 0.8 mm thick) were used, device geometry remains essentially unchanged and zero-order drug release is achieved. The depletion phase characterizes a decrease in device size and possibly some depletion of the incorporated excipient.

As mentioned previously, drug release from surface-eroding polymers has other important consequences. Those are (a) rate of drug release is directly proportional to drug loading, (b) lifetime of the device is directly proportional to device thickness, and (c) rate of drug release is directly proportional to total surface area. These expectations are fully verified in Figures 4, 5, and 6. Thus, the acid anhydride catalyzed erosion of poly(ortho esters) is a textbook case of surface erosion.

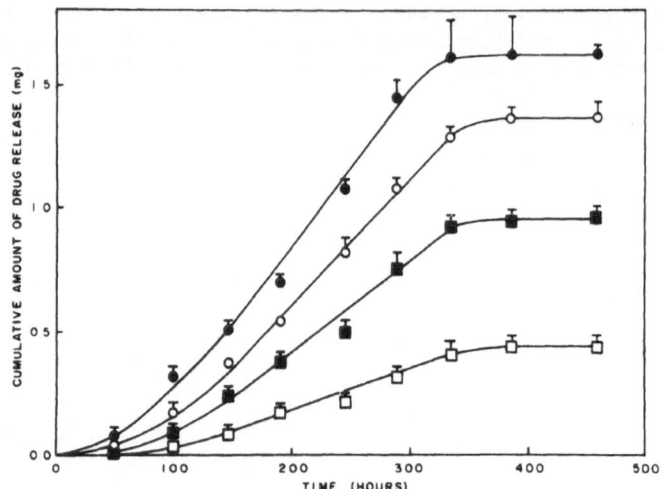

Fig. 4   The effect of drug loading on the drug release profile from
         50:50 HD/t-CDM poly(ortho ester) disks containing 0.2%
         poly(sebacic anhydride) (37°C, pH 7.4). Drug loadings: 8%
         w/w (●), 6% (o), 4% (■), 2% (□).
         (Reprinted from R. V. Sparer, C. Shih, C. D. Ringeisen, and
         K. J. Himmelstein, Controlled release from erodible poly(ortho
         ester) drug delivery systems, J. Controlled Release 1 (1984)
         23-32. With permission).

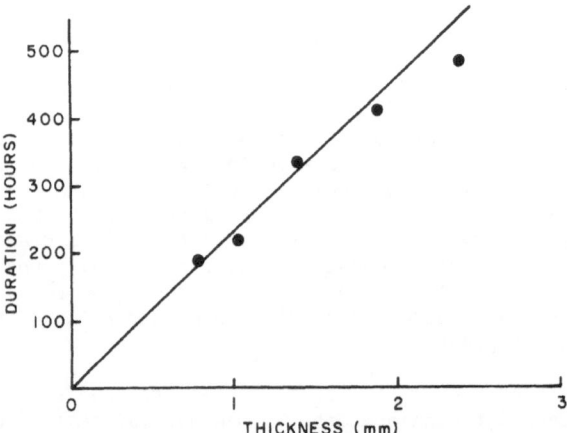

Fig. 5   The effect of poly(ortho ester) disk thickness on the duration
         of drug release from 50:50 HD/t-CDM containing 0.2%
         poly(sebacic anhydride) and 4% drug (37°C, pH 7.4).
         (Reprinted from R. V. Sparer, C. Shih, C. D. Ringeisen, and
         K. J. Himmelstein, Controlled release from erodible poly(ortho
         ester) drug delivery systems, J. Controlled Release 1 (1984)
         23-32. With permission).

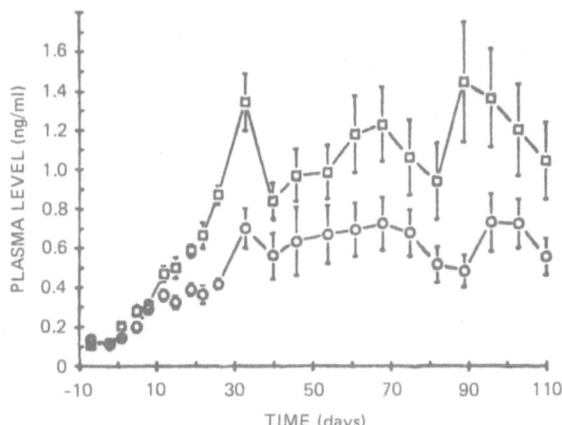

Fig. 6   The effect of poly(ortho ester) disk surface area on the drug
         release rate from 50:50 HD/t-CDM containing 0.2% poly(sebacic
         anhydride) and 4% drug (37°C, pH 7.4).
         (Reprinted from R. V. Sparer, C. Shih, C. D. Ringeisen, and
         K. J. Himmelstein, Controlled release from erodible poly(ortho
         ester) drug delivery systems, J. Controlled Release 1 (1984)
         23-32. With permission).

USE OF BASIC EXCIPIENTS

Because poly(ortho esters) are stable in base the use of basic
excipients will prevent bulk polymer erosion even though the hydrophobic
matrix is penetrated by small amounts of water. Thus, using this
methodology, surface erosion can be achieved by relying on the slow,
noncatalyzed erosion of the polymer that takes place in the surface
layers of the device where the basic excipient is neutralized. The
neutralization very likely occurs by water intrusion into the matrix and
diffusion of the slightly water-soluble basic excipient out of the
device where it is neutralized by the external buffer. As a consequence
of this process, a base-depleted layer develops at the outer surface of
the device and polymer erosion occurs in that base-depleted layer. This
process is shown schematically in Figure 7 for a device containing
$Mg(OH)_2$.

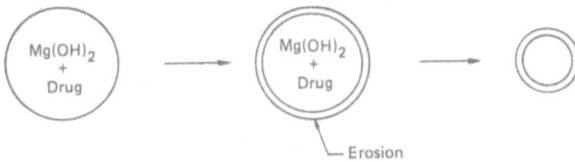

Fig. 7   Schematic representation of erosion using $Mg(OH)_2$ excipient.

Figure 8 shows cumulative weight loss and cumulative drug release
from crosslinked poly(ortho ester) rods containing 30 wt% levonorgestrel
and 7 wt% Mg(OH)$_2$ implanted subcutaneously in rabbits and explanted at
2-week intervals (7). Drug release was determined by assaying residual
drug in the device and weight loss was determined gravimetrically.

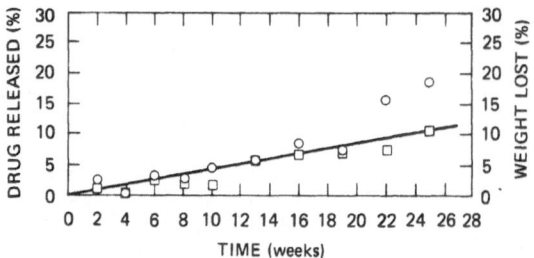

Fig. 8   In vivo cumulative weight loss (□) and cumulative release of
          levonorgestrel (○) from crosslinked poly(ortho ester) rods,
          2.4 x 20 mm, containing 30 wt% levonorgestrel and 7.1 wt%
          Mg(OH)$_2$. Total drug content 32.0 mg. Devices implanted
          subcutaneously into rabbits.
          (Reprinted from J. Heller, B. K. Fritzinger, S. Y. Ng, and
          D.W.H. Penhale, In vitro and in vivo release of levonorgestrel
          from poly(ortho esters) II crosslinked polymers, J. of
          Controlled Release, 1 (1984) 233-238. With permission).

Even though the data points that were based on single devices show
considerable scatter, polymer erosion and drug release occur
concomitantly for at least 20 weeks, after which drug release may
accelerate even though polymer erosion remains constant. However,
because the experiment was discontinued after 25 weeks, it is not
certain whether the values for percent drug release at weeks 22 and 25
are real, or more likely experimental error. Also, all ingredients were
mixed into the viscous prepolymer manually so that the distribution of
drug and excipient in the device was quite inhomogeneous.

Further evidence for surface erosion has been obtained by scanning
electron microscopy (SEM) examination of explanted devices after 2 and
25 weeks in rabbits, as shown in Figure 9. In these photographs the 2-
week device is essentially unchanged, but the 25-week device shows
definite signs of erosion around the periphery and a small reduction in
diameter. Most importantly, the interior of the device remains
unchanged.

Levonorgestrel blood plasma levels achieved when 1 and 3 rods are
implanted subcutaneously into rabbits is shown in Figure 10 (9). After
an initial burst, constant blood levels for about 9 months have been
demonstrated. However, these blood levels are quite low and a more

rapidly eroding polymer was needed. To achieve this, we prepared a crosslinked polymer containing 1 mole% of copolymerized 9,10-dihydroxystearic acid and fabricated devices containing 30 wt% levonorgestrel and 7 wt% Mg(OH)$_2$. We have previously shown that the copolymerization of 9,10-dihydroxystearic acid significantly increases rate of polymer erosion (10). These devices were also implanted in rabbits and explanted at 2-week intervals.

Figure 11 shows SEM photography of devices explanted after 6, 9, 12, and 16 weeks in rabbits. Clearly, as also found in vitro, devices that contain copolymerized dihydroxystearic acid erode at significantly higher rates than similar devices without copolymerized dihydroxystearic acid.

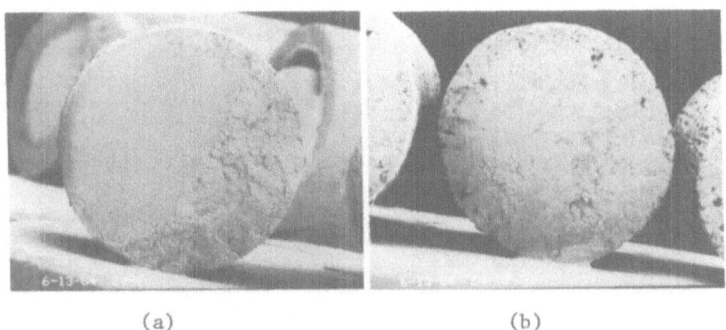

(a)   (b)

Fig. 9   Scanning electron micrograph of crosslinked poly(ortho ester) rods, 2.4 x 20 mm, containing 30 wt% levonorgestrel and 7.1 wt% Mg(OH)$_2$.
(a)   After 2 weeks in rabbit, 30X
(b)   After 25 weeks in rabbit, 30X
(Reprinted from J. Heller, B. K. Fritzinger, S. Y. Ng, and D.W.H. Penhale, In vitro and in vivo release of levonorgestrel from poly(ortho esters) II crosslinked polymers, J. of Controlled Release, 1 (1984) 233-238. With permission).

Because all ingredients were manually mixed into the viscous prepolymer, the distribution of drug and excipient in the device is very inhomogeneous and it is likely that the voids shown in the erosion zone are largely due to this uneven distribution. Nevertheless, the devices show a progressive diminution of a central, uneroded zone.

Levonorgestrel blood plasma levels achieved when 1 and 2 rods are implanted subcutaneously into rabbits is shown in Figure 12 (9). Results of this ongoing study show that the increased rate of polymer erosion resulted in a significant increase in levonorgestrel blood plasma level.

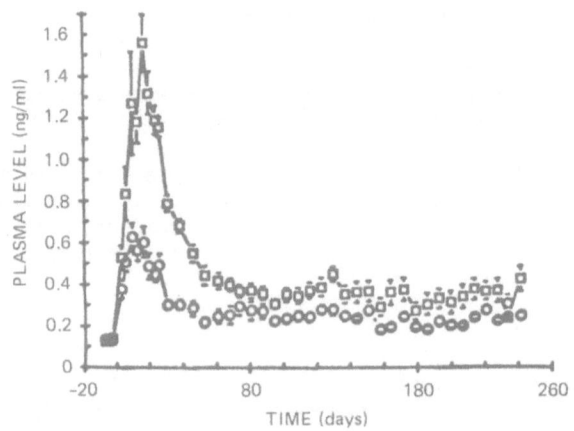

Fig. 10   Daily rabbit blood plasma levels of levonorgestrel from
crosslinked poly(ortho ester) rods 2.4 x 20 mm containing 30
wt% levonorgestrel and 7.1 wt% Mg(OH)$_2$
○ 1 device per rabbit   □ 3 devices per rabbit.

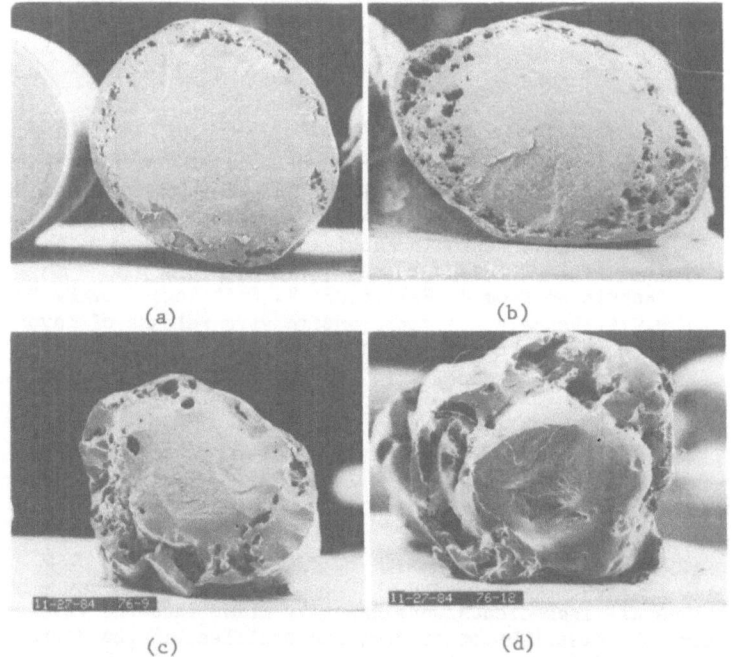

(a)   (b)

(c)   (d)

Fig. 11   Scanning electron micrograph of crosslinked poly(ortho ester)
rods 2.4 x 20 mm containing 30 wt% levonorgestrel, 7.1 wt%
Mg(OH)$_2$, and 1 mole% copolymerized dihydroxystearic acid.
(a) After 6 weeks in rabbit, 30X
(b) After 9 weeks in rabbit, 30X
(c) After 12 weeks in rabbit, 25X
(d) After 16 weeks in rabbit, 25X

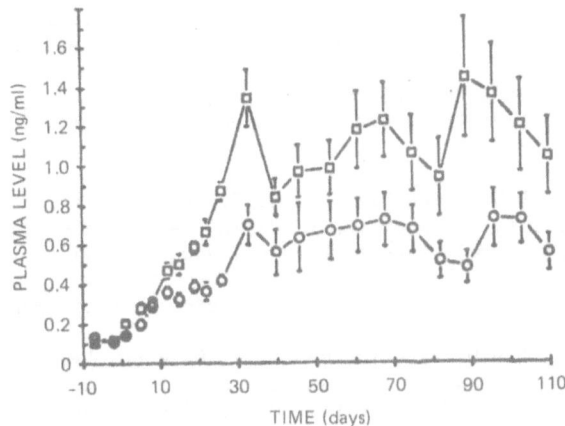

Fig. 12 Daily rabbit blood plasma levels of levonorgestrel from
crosslinked poly(ortho ester) rods 2.4 x 20 mm containing 30
wt% levonorgestrel, 7.1 wt% Mg(OH)$_2$, and 1 mole% copolymerized
dihydroxystearic acid
o 1 device per rabbit □ 2 devices per rabbit.

Additional studies with homogeneous devices are now in progress.
In these experiments drug and excipient are blended into the prepolymer
by means of a 2CV Atlantic Research Vertical Cone Micromixer, and more
soluble basic excipients are being used.

ACKNOWLEDGEMENT

Work on short-term release devices was conducted at the Interx
Laboratories of Merck Sharp and Dome by R. V. Sparer, C. Shi, C. D.
Ringeisen, and K. J. Himmelstein using polymers prepared at SRI
International.

Work on long-term devices was performed at SRI International under
sponsorship of the Contraceptive Development Branch of the National
Institute of Child Health and Human Development under Contract No. N01-
HD-7-2826 by B. K. Fritzinger, S. Y. Ng, and J. Heller. Implantation
studies were carried out by Dr. L. T. Juhos and SEM studies by Mr. J. C.
Terry.

REFERENCES

1. J. Heller, Controlled release of biologically active compounds from bioerodible polymers, Biomaterials, 1 (1980) 51-57.

2. J. Heller, Biodegradable polymers in controlled drug delivery, CRC Critical Reviews in Therapeutic Drug Carrier Systems 1 (1984) 39-90.

3. J. Heller, D.W.H. Penhale, and R. F. Helwing, Preparation of polyacetals by the reaction of divinyl ethers and polyols, J. Polymer Sci., Polym. Lett. Ed. 18 (1980) 293-297.

4. J. Heller, S. Y. Ng, and S. H. Pangburn, work in progress.

5. J. Heller, D.W.H. Penhale, and R. F. Helwing, Preparation of poly(ortho esters) by the reaction of ketene acetals and polyols, J. Polymer Sci., Polym. Lett. Ed. 18 (1980) 619-624.

6. J. Heller, B. K. Fritzinger, S. Y. Ng, and D.W.H. Penhale, In vitro and in vivo release of levonorgestrel from poly(ortho esters): I Linear Polymers, J. Controlled Release, 1 (1984) 225-232.

7. J. Heller, B. K. Fritzinger, S. Y. Ng, and D.W.H. Penhale, In vitro and in vivo release levonorgestrel from poly(ortho esters): II Crosslinked Polymers, J. Controlled Release, 1 (1984) 233-238.

8. R. V. Sparer, C. Shih, C. D. Ringeisen, and K. J. Himmelstein, Controlled release from erodible poly(ortho ester) drug delivery systems, J. Controlled Release 1 (1984) 23-32.

9. J. Heller, B. K. Fritzinger, and S. Y. Ng, work in progress.

10. J. Heller, D.W.H. Penhale, B. K. Fritzinger, and S. Y. Ng, Controlled release of contraceptive agents from poly(ortho esters) in G. I. Zatuchni, A. Goldsmith, J. D. Shelton, and J. J. Sciarra (Eds.) Long-Acting Contraceptive Delivery Systems, Harper and Row, New York (1984), pp. 113-128.

A NEW INDOMETHACIN LYSINATE MODIFIED RELEASE SYSTEM

Paolo Colombo, Ubaldo Conte, Andrea Gazzaniga,
Carla Caramella and Aldo La Manna

Dipartimento Chimica Farmaceutica - Università di Pavia
Via Taramelli 12 - 27100 Pavia, Italy

SUMMARY

Considering some disadvantages shown by indomethacin osmotic devices, a new drug delivery system capable of releasing indomethacin lysinate at a constant rate has been prepared.

The system mainly consists of:

i - a monolithic core, obtained by compression of a water insoluble polymer (CAP) mixed with the drug, an alkaline buffer agent and a polyalcohol having negative dissolution heat.

ii - a polymeric film insoluble in biological fluids, obtained by spray-coating the monolithic surface, acting as release controlling membrane.

In order to determine the optimal delivery rate capable of maintaining indomethacin therapeutic plasma levels, dissolution tests (U.S.P. XX paddle) data were compared with in vivo (humans) plasma levels.

A good correlation between in vitro and in vivo data was obtained thus demonstrating the reliability of the system.

This new drug delivery system offers the following advantages:

a - the abrupt drug release from the core is hindered by its monolithic structure

b - the drug release is independent of the environmental pH value

c - a zero order release kinetics is maintained until 70% of drug is released

d - the geometric characteristics of the new drug delivery system assure a high bioavailability.

INTRODUCTION

Recently many modified release indomethacin preparations have been introduced to patient treatment thus reducing the incidence of adverse reactions, which often limit conventional therapy (1).

An osmotically driven zero-order release indomethacin oral system introduced to the market a few years ago (1982) was withdrawn (1983) following reports of severe adverse reactions; those were probably due to the release characteristics of the system (2), which exposes the mucosa to high flux of indomethacin saturated solution.

Other than with osmotic devices, the costant release of drug can be achieved using a reservoir coated with a polymeric diffusive membrane; in this way the flux of the drug solution is slower, being distributed over the entire diffusive membrane surface.

Film coating represents a reliable method for preparing such modified release systems (MRS), for the additional reasons that it facilitates modulation of the release rate (through appropriate selection of film composition and thickness) and it involves well consolidated preparative technology and equipment (3).

Nevertheless, certain disadvantages should be considered the possibility, for example, of dose dumping due to film rupture and of a delay in the release process, due to hydration of the diffusive membrane.

In the present paper we describe the design and realization of a novel indomethacin modified release system (4) (INDL-MRS), which is composed of an active-principle reservoir coated with a diffusive membrane and capable of releasing 70% of active ingredient contents at a constant rate.

We dealt with such problems as the poor water solubility of the drug, the need to avoid dose dumping and to reduce release time lag.

These problems were resolved, respectively, by using a water soluble derivative of indomethacin (indomethacin lysinate) (INDL) (5) by preparing a matrix type core and by adding an immediate dose of active principle to a water soluble layer which is applied externally to the diffusive membrane.

Moreover, an osmotic agent and/or a buffering agent were added to the core composition to facilitate water penetration and to reduce the influence of external pH on the release; an appropriate selection of the shape and dimensions of the core allowed us to obtain the desired release rate without affecting the film's strength.

Different types of system were prepared and tested for in vitro release and in vivo performance in a single subject to see if a correlation could be established between in vitro and in vivo release rate.

MATERIALS AND METHODS

1. Materials

a. Polymers. An insoluble grade of cellulose acetate propionate (CAP 482-20; Eastman-Kodak, USA) with hydroxyl contents of 2,1% and Tg=147°C was employed as matrix former.
As film formers, we used a freely water-permeable and a slightly water permeable acrylic methacrylic acid ester copolymer (Eudragit RL and Eudragit RS respectively; Rohm Pharma, Germany) with M_w=150,000, mixed in appropriate proportions, ranging from 4:1 to 7:1.

370

A hydroxypropylmethylcellulose grade (Pharmacoat 606 (Shin-Etsu Chemical, Japan)) was used for immediate dose layer preparation.

b. Drug and fillers. The drug was indomethacin lysinate (MW =503,9, m.p.=204-207°C, water solubility = 1,2% w/v (20°C). Mannitol (FU), sorbitol (FU), talc (FU), were used as soluble or insoluble fillers and Na$_2$HPO$_4$ (FU) was used as an alkaline buffer.

2. Methods

a. Preparation of modified release systems (INDL-MRS). The composition of the matrix-core varied within the following ranges:

indomethacin lysinate	30-40%
Na$_2$HPO$_4$	20-25%
talc	15-20%
mannitol or sorbitol	15-30%
CAP	2-5 %

The drug, mixed with the insoluble and the soluble fillers and with the buffering agent in a sigma mixer, was granulated by massing it with binding solution (10% CAP in acetone:isopropanol (1:1)), and by passing it through a n. 25 ASTM sieve and drying it overnight at 40° C.
The granulation, lubricated with 1% magnesium stearate, was compressed at 300 MPa with an instrumented reciprocating tableting machine employing two concave punch sets with respective diameters of 9 mm (curvature radius 10 mm) and 12 mm (curvature radius 14 mm). The matrix cores were pan-coated with an Eudragit RL/Eudragit RS solution (6% w/v in acetone:isopropanol 1:1; RL/RS ratio 7:1 or 4:1) containing 1% w/v of castor oil, applying up to 8-12 mg of dry polymer per core. The finished systems were dried overnight at 40° C.
The immediate dose (10 mg), when present, was applied by spraying the dried systems with the following mixture (for 100 systems): indomethacin lysinate 1 g, MgCO$_3$ 0.4 g, Pharmacoat 606 1.2 g, isopropanol 15 ml, dichloromethane 15 ml.

b. Release experiments. The release tests (6 replicates) were performed in water (1,000 ml) using a USP XXI apparatus II, (paddle rotation speed = 100 rpm). Drug concentrations were continuously assayed by spectrophotometry (280 nm).
The release data were fitted according to the Weibull distribution function from which overall time parameter (td) and mean dissolution times (MDT) were calculated (6).
The effect of film integrity on the release of active principle was assessed with the following experiment: six tablets were submitted to the dissolution test; after one hour of release, three of the six tablets were withdrawn from the dissolution vessel and, after radial cutting of part of their diffusive film, they were again put into the vessel for dissolution to continue. At the end of the experiment, the two dissolution curves were drawn, fitted with the Weibull distribution function and the td parameter confidence limits (P=0.95) were calculated.

c. In vivo experiments. In vivo drug plasma levels were determined in a single volunteer after administration of, respectively, a 100 mg INDL plain capsule, a matrix core, and differing modified release systems; each administration was separated by an appropriate wash-out period.

The subject fasted overnight, received administration at 9:00 a.m. and fasted till noon. Blood samples were taken before administration and at opportunely spaced time intervals until 12-24 h after administration. The indometacin serum levels were determined by HPLC (7).

The AUC (0 - ∞) was calculated by the trapezoidal rule. Since an extrapolation from the last datum onward was appropriate, the terminal slope was estimated from concentration time data when administering a plain capsule (8). Bioavailability (F) of the INDL-MRS relative to the plain capsule was determined from AUC values. The mean residence time MRT (0 -∞) was calculated accordingly (9).

RESULTS AND DISCUSSION

System description

A cross-section of the systems prepared is shown in Fig. 1.

In vitro release patterns

The analysis of the release kinetics of an INDL- MRS during the development steps (Fig. 2) shows that the release kinetics of the matrix core obeys the square root law (n=0.5 in the log-log fitting of data according to the equation $y=kx^{n}$ (10)); it becomes zero-order (until 70% of

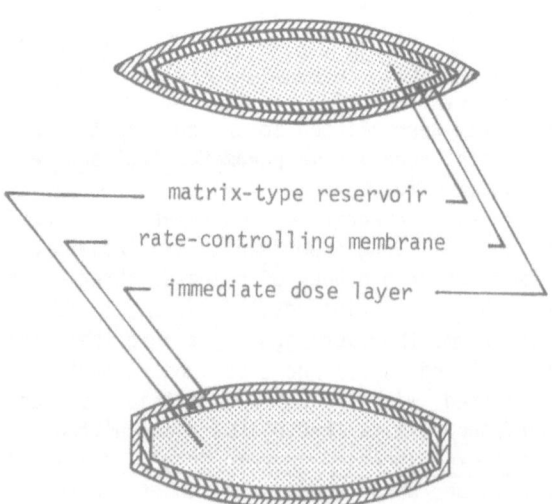

Figure 1. Cross-section of indomethacin lysinate - MRS.

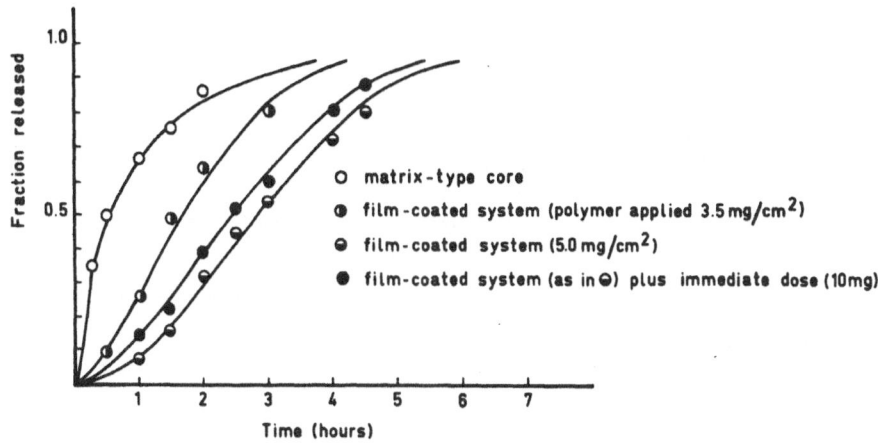

Figure 2. Evolution of the release kinetics of an indomethacin - MRS during the development steps.

drug is released) (n=1) in the coated systems where the polymeric film acts as a rate-controlling membrane. In coated systems, the release rate decreases with increase in film thickness, while film hydration times tend to increase. When an immediate dose is added to the outer layer, the kinetics does not change, but the time-lag is reduced.

The results of the release from an unbroken system compared with that of a system whose coating had been partially cut show that (as expected, on the basis of the monolithic structure of the core) the release in the case of cut films seems faster than that of unbroken films, but not dramatically different (Fig. 3).

In vivo experiments

In table I we give the composition of the INDL-MRS tested in vivo. Differences in core compositions were minimal, whereas the finished systems differed in film composition, shape (on which surface area depended) and in the presence (or not) of an immediate dose.

In particular, system A had the less permeable film, a biconvex shape and carried no immediate dose. Formulation B had the most permeable film, was lenticular in shape and contained an immediate dose, whereas formulation C differed from A only in surface area. From the many systems prepared and tested in vitro, those described in Table I were chosen for in vivo experiments since they were thought to be representative in terms of modifications in formulation factors (film composition and area, immediate dose loading, ...).

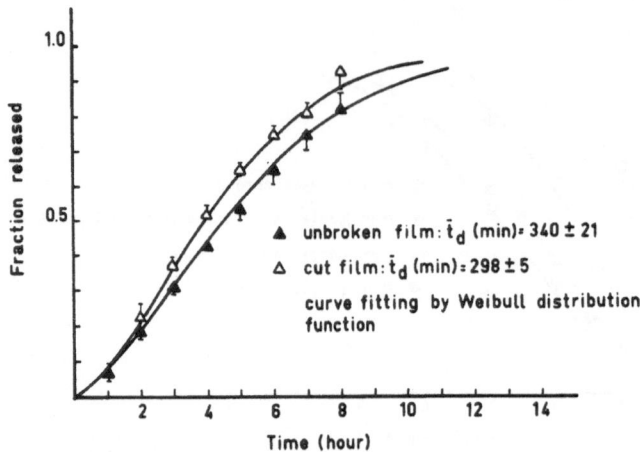

Figure 3. Influence of film integrity on the release rate of an indomethacin lysinate - MRS.

Table 1. Composition and main features of indomethacin lysinate - MRS in vivo.

SYSTEM	A	B	C
CORE			
INDL	100	120	100
Na_2HPO_4	80	80	80
MANNITOL	90	--	70
SORBITOL	--	30	--
TALC	40	40	60
CAP	10	10	10
MEMBRANE			
EUDRAGIT RL/RS 4:1	5 mg/cm^2	--	5 mg/cm^2
EUDRAGIT RL/RS 7:1	--	5 mg/cm^2	--
IMMEDIATE DOSE	--	10 mg	--
SHAPE (surface area, cm^2)	Biconvex (1.90)	Lenticular (2.50)	Lenticular (2.50)

Plasma level curves obtained for these systems are presented in Fig. 4 in comparison with the standard formulation curve (100 mg INDL plain capsule), together with the corresponding in vitro release curves (in the insert). The td values (mean and confidence limits) were, respectively, 332+21; 199+12 and 274+17 for formulation A, B and C.

System A produced fairly constant plasma concentrations throughout its lifetime, which, on the basis of its in vitro release kinetics, was predictable. However, plasma levels were in the subtherapeutic range (7) and relative bioavailability was low (48%). This was probably due to an inadequate in vivo release, possibly caused by inappropriate film composition and/or system geometry.

The plasma concentration curve obtained with system B showed a rapid attainment of the therapeutic level and its maintainance for a reasonable timespan. This in vivo pattern could be attributed to the presence of an immediate dose and to the increased film permeation rate and was in line with in vitro release (which was faster than that of A and did not show any time lag). The prolonged lifetime of system B was attributable also to the slightly higher drug contents of the core. The maximum plasma levels obtained ($\cong$3 mcg/ml) were considered to be too high, even after subtracting the area contributed by the immediate dose from total AUC.

The in vivo release rate of system C matched therapeutic needs : the relative bioavailability was complete, and therapeutic plasma levels were

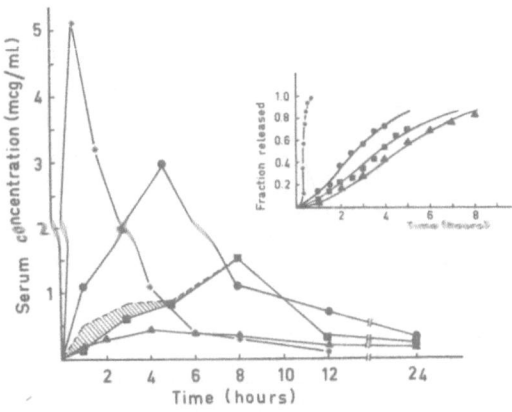

Figure 4. Indomethacin serum levels following administration and in vitro release curves (in the insert) of: 100mg INDL powder(*); INDL - MRS: A (▲), B (●), C (■).

reached and maintained for a prolonged time. This was consistent with the in vitro results and attributable to an appropriate selection of the factors influencing the release rate. A certain delay in the onset of the therapeutic plasma levels was still evident, which could be easily obviated by the addition of an immediate dose (see simulated dashed portion of the AUC).

When analogous parameters , such as the mean residence time (in vivo) and the mean dissolution time (in vitro) are considered, a good vivo-vitro correlation is found for the capsule, the matrix-core and the three INDL-MRS examined (Fig. 5) This confirms the predictability of in vitro data with regard to in vivo performance.

Since the correlation is good also at the highest dissolution rates, it seems conceivable that in this case dissolution limits absorption rate (11). If we consider the in vitro and in vivo mean release times (the latter calculated according to (12) as the difference between the mean residence time of the MRS and the mean residence time of the standard preparation), a time-scaling factor (8) approximately equal to 2.5 can be calculated for these systems.

CONCLUSIONS

Those sytems prepared with a lenticular shape and composed of a matrix-core and a diffusive polymeric membrane allowed us to obtain an in

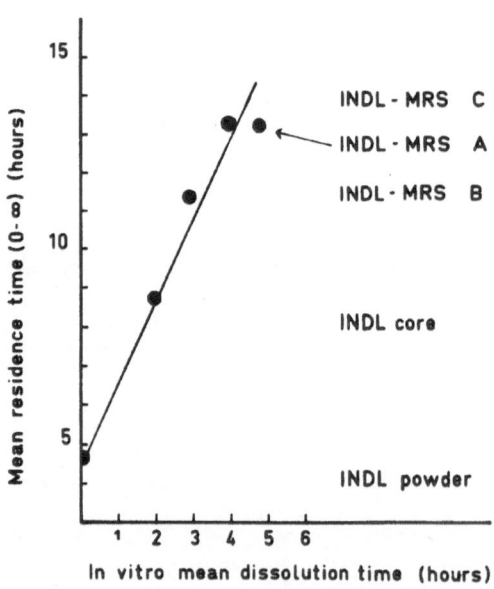

Figure 5. Linear regression between mean residence time and in vitro mean dissolution time (r=0.97).

vitro zero-order release of up to =70% of the amount released. The safety of the systems is assured mainly by the matrix-like structure of the core.

In vivo studies conducted in the early stages of development allowed an interactive correlation to be established between in vitro and in vivo data. In particular, in vitro mean dissolution times of about 3-4 hrs could allow us to obtain in vivo mean release times of 8-10 hrs, which are consistent, in the examined cases, with therapeutic needs.

These results illustrate the feasibility of the vivo-vitro correlation approach in the development of a modified release system.

Acknowledgements

The authors wish to thank Sigma-Tau (Pomezia - Italy) for providing support for this research.

REFERENCES

1) Medical tribune 4, 8; 1985
2) Scrip 1983, n.821, 825, 826
3) Rowe, R.C.; Pharm. Int. 6,14; 1985
4) Colombo, P.; Conte, U.; Reiner, A.; Italian Patent Application 48632 A84 (July 26th, 1984)
5) Reiner, A.; Italian Patent Application 48807 A84 (September 6th, 1984)
6) Langenbucher, F.; Moller, H.; Pharm. Ind. 45, 629; 1983
7) Segre,G.; personal communication
8) Brockmeier, D.; Arzneim.-Forsch./Drug res., 34,1604; 1984
9) Kiyoshi Yamaoka; Terumichi Nakagawa; Toyozo Uno; J. Pharmacokin. Biopharm., 6, 547; 1978
10) Korsmeyer, R.W.; Gurny, R.; Doelker, E.; Buri, P.; Peppas, N.A.; Int. J. Pharm., 15, 25; 1983
11) Erni, W.; Eckert, M.; Acta Pharm. Technol. 29, 281; 1983
12) Langenbucher, F.; Moller, H.; Pharm. Ind. 45, 623; 1983

A HYBRID SYSTEM FOR ZERO ORDER DRUG DELIVERY REGIMEN

Y. Yaacobi, N. Lotan and S. Sideman

Biomedical Engineering Department
Technion-Israel Institute of Technology
Haifa 32000, Israel

INTRODUCTION

The disadvantages of intermittent administration of drugs, as encountered during various therapeutic treatments of chronic patients, brought about the development of controlled drug delivery systems[1-11] which are effective for various, relatively short, durations.

Several methods have been proposed[12] for achieving a long term continuous supply of drugs. An interesting method amongst these is one in which the polymeric matrix, which contains the drug, undergoes a gradual dissolution[13]. The rate of drug release is thus a function of the polymeric matrix dissolution rate. In another quite different procedure the drug is released from a reservoir via diffusion controlling polymeric membrane . In this case, however, the rate of drug release decreases as the concentration of drug within the reservoir decreases with time. As constant rate of drug supply is, however, required in many instances some approaches which are used for attaining this goal are summarized in Table 1.

Our studies[14-16] deal with drug release devices based on biochemical principles. These may be implemented either as a single enzyme-or a multiple enzyme system. They are schematically depicted in Fig. 1. The drug is represented by the released low molecular weight monomeric unit. The single enzyme unit usually contains an exo-type enzyme, whereas the multiple enzyme unit includes both the exo- and endo-type enzymes. The basic features of endoenzyme-polymer systems, corresponding to step I of the multiple enzyme system, as well as the characteristic degradation pathways and the kinetic course achieved, were reported earlier[14-16]. Obviously, step I by itself cannot be considered the basis for a constant rate drug delivery device, and requires the complementary action of an exo-type enzyme.

In the present report we limit ourselves to the exo-enzymatic degradation of a linear polymer. The constant rate drug delivery characteristics are achieved by an implanted, hybrid-mode device in which the enzyme cleaves the polymer, and progressively releases the low molecular weight drug through a properly chosen membrane. The latter allows the selective escape of the drug into the organism. The first part of this report presents the model employed and the theoretical considerations related to

Table 1. Constant Rate "Zero Order" Drug Release Systems

1.	Mechanical	- Infusion Pumps	Ref (12)
2.	Mechano-Chemical	- Osmotically driven "Squeeze" devices	"
3.	Chemical	- Hydrolysis of entire carrier	"
		- Hydrolysis of drug-carrier bond	"
4.	Physico-Chemical	- Saturated solution, excess solid drug	"
5.	Physical	- Appropriately designed geometry	"
6.	Biochemical	- Enzyme-based devices	This study

its function. The second section describes the experimental determination of the diffusion characteristics of the drug Melphalan through an appropriate membrane. The last section presents the numerical simulation of the performance of the hybrid device, including its application when a specific drug delivery regimen is required.

THEORETICAL

The System

The basic element in the proposed drug delivery model system is a subcutaneously implanted cell, with a prickable rubber seal in the wall facing the skin. This allows repeated penetrations with a hypodermic needle, in order to refill the reservoir. The opposite wall of the cell is made of a selective membrane of appropriate permeability and thickness, which retains macromolecules while enabling the diffusion of the enzymically-released low molecular weight drug molecules.

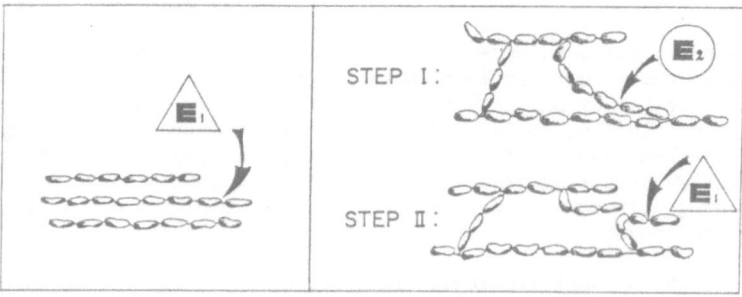

Fig. 1. Enzyme-based systems for zero order drug release. E_1 releases chain-end units in consecutive steps. E_2 acts randomly, and produces new chain-ends.

The cell is filled with a solution containing the enzyme and a polymeric drug, each at a predetermined, required concentration. The enzyme is of the exo-acting type and its kinetic parameters are well defined. The polymer is linear, and may be a homopolymer (i.e. polydrug) or a copolymer in which the drug residues and the "diluent" residues are covalently bonded and randomly distributed along the chain. All the residues in the polymer are prone to hydrolytic release by the exoenzyme.

The substrate moieties to be acted upon by the enzyme throughout the process are the chemical bonds joining the polymer residues. The concentration (R) of these bonds continuously decreases as the polymer degradation process progresses. However, in the particular, exo-type mode of action of the enzyme considered here, the substrate units which are actually presented to the enzyme, at any given time during the process, are only the end-bonds holding the terminal residues of the polymeric chains. Unlike R, the concentration (R_{end}) of these end-bonds remains constant throughout the lifetime of the system. Obviously R_{end} is fixed at the beginning of the process by the choice of the total concentration of polymer and its average degree of polymerization. The constant R_{end} and the exo-type acting enzyme thus provide the basis for the proposed constant rate drug release process. Consequently, the constant rate release regimen will prevail as long as the endwise degradation of the polymer is operative, and inspite of the progressive reduction in R.

The Drug Transfer Rate

For the system considered, a steady state operation involving a constant rate drug delivery is reached when, at any moment, the amount of drug produced within the device is identical to that crossing the membrane. This situation is described by Eq. (1) which relates the

$$CV = JA \tag{1}$$

volumetric rate of the enzyme reaction (C), the device volume (V) and the drug flux (J) through the membrane of given area (A). We consider that the enzymic reaction follows the Michaelis-Menten kinetics, and that the drug transfer across the membrane is a diffusion controlled process. For such a case, the quantities C and J are given by Eqs. (2) and (3) respectively,

$$C = K_m S_o / K_m + S_o) \tag{2}$$

$$J = D(P_{in} - P_{out})/z \tag{3}$$

where K_m is the Michaelis-Menten constant, S_o is the concentration of polymeric chain ends in the device, D is the diffusion constant of the drug through the membrane, P_{in} and P_{out} are the instantaneous drug concentrations within and outside the device, respectively and z is the thickness of the membrane.

Considering that the implanted device releases the drug into the whole body, the volume $V_{body} \gg V$ and, consequently, $P_{out} \ll P_{in}$. Eq. (3) then reduces to

$$J = DP/z = k_L P \; ; \; P = P_{in} \tag{4}$$

where k_L is the mass transfer coefficient of the drug across the membrane. Thus, the constant rate drug delivery from such a device is described by substituting Eqs. (2) and (4) into Eq. (1).

The characteristic behavior of the system is, however, somewhat different when the drug release process is actuated by mixing the enzyme and polymeric drug solution right at the time of implantation of the device. In this case, a transient, pre-steady state period exists. As described by Eq. (5), the amount of drug accumulated within the device,

$$V(dP/dt) = CV - DPA/z = CV - k_L AP \qquad (5)$$

$V(dP/dt)$, equals the difference between the drug produced by the enzymic reaction and that diffused through the membrane. Upon rearrangement and integration one obtains

$$P = (CV/k_L A) \{1-\exp(-k_L At/V)\} \qquad (6)$$

$$t = (V/k_K A) \ln\{1/(1-k_L AP/CV)\} \qquad (7)$$

Note that the instantaneous rate of drug transfer into the body (in units of mass/unit time) is given by multiplying Eq. (6) by $k_L A$.

It is evident from Eq. (6) that the system will approach steady state only assymptotically. Therefore, we define an operational quasi steady state (QSS) as the condition in which P, the instantaneous concentration, reaches 95% of its steady state value, i.e.:

$$P_{QSS} = 0.95 \ CV/k_L A \qquad (8)$$

The time needed for the system to reach this quasi steady state condition is obtained by Eq. (9). As seen, t_{QSS} depends only on the

$$t_{QSS} = 3V/k_L A \qquad (9)$$

geometry of the reservoir and on the transport characteristics of the membrane.

This model also allows us to consider a more complex drug delivery regimen. For instance, consider a therapeutic schedule where a high dose of drug, which is n times larger than the desired maintenance dose, is required at the onset of the treatment, and a constant maintenance dose is then required for an extended period of time. Introducing the new initial conditions (t=0, $P=nCV/k_L A$) yields a modification of Eq. (6) which describes the desired operating mode. Thus,

$$P = (CV/k_L A) \{1+(n-1)\exp(-k_L At/V)\} \qquad (10)$$

and the instantaneous rate of drug delivery from the device is given by $(k_L A)P$ (in units of mass per time).

EXPERIMENTAL: THE DIFFUSION COEFFICIENT

The application of Eqs. (6) and (10) requires the knowledge of the value of D. The diffusion coefficient of Melphalan, a model drug used here, through a cellulose-type membrane was experimentally determined by using a two compartment diffusion cell. One compartment was filled with a solution of a known concentration of Melphalan, and the drug was allowed to diffuse freely into a second compartment. The solutions in both compartments were continuously recirculated. The drug concentration in the down-

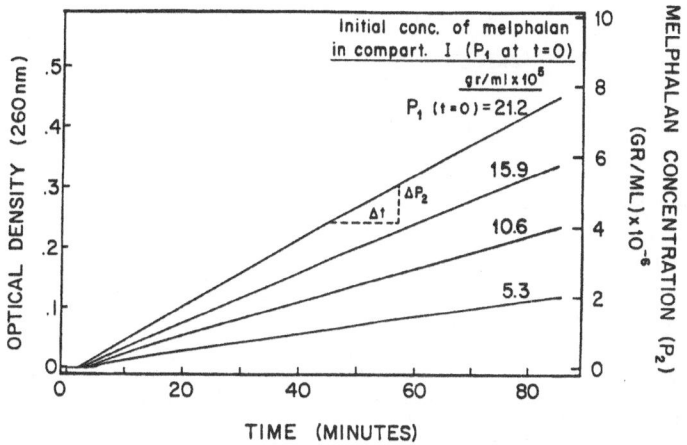

Fig. 2. Determination of diffusion coefficient of Melphalan through a cellulosic membrane. The initial concentrations of drug in the upstream compartment are indicated.

stream compartment was continuously monitored spectrophotometrically, and recorded. The results thus obtained are shown in Fig. 2.

The diffusion coefficient, D, was calculated at some arbitrary points of the lines by

$$D = (V_2 z/A) \cdot (\Delta P_2/\Delta t) \cdot \{1/(P_1-P_2)\} \tag{11}$$

where V_2 is the volume of the downstream compartment, P_1 and P_2 are the drug concentrations in the upstream and downstream compartments respectively, and $(\Delta P_2/\Delta t)$ is the slope of the experimental P_2 vs time curve. In our experiments $V_2 = 20$ ml, A = 9.62 cm^2, z = 0.041 mm. The average value of D = 7×10^{-10} cm^2/sec was obtained, and used in the subsequent numerical simulation.

POTENTIAL IMPLEMENTATION

The performance characteristics of the drug delivery system considered here is presented in Fig. 3 for representative cases of therapy without initial overdose. Calculations were performed by using Eq. (6), and the system parameters listed in Table 2. As seen, the quasi steady state is reached at $t_{QSS} \approx 50$ hrs for all doses considered, followed by an essentially constant rate delivery regimen. It is noteworthy that the t_{QSS} value obtained here is negligible when compared to the effective life time of up to 2.5 years of the system.

Calculations were also carried out for the cases wherein the required initial dose is n times higher than the maintenance dose. Fig. 4 depicts the drug release characteristics when the same basic system defined in Table 2 is used with n=5. The initial overdose period in these cases extends for approximately 50 hrs, followed by the requested constant rate release regimen.

The utilization of the suggested system, as specified in Table 2, assumes that the therapeutic requirements are specified by the attending physcian. Assume that the required doses are either 0.25, 0.50, 0.75,

383

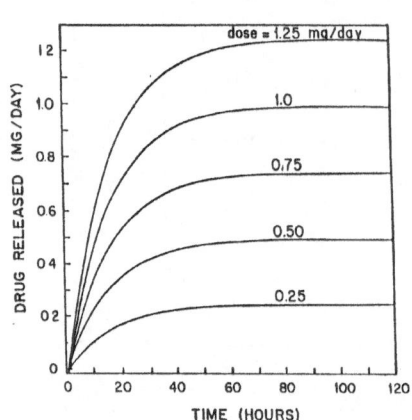

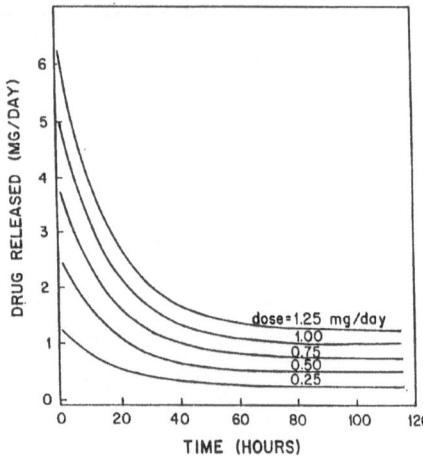

Fig. 3. Drug release characteristics for regular-type therapy. The daily maintenance doses are indicated.

Fig. 4. Drug release characteristics involving initial doses of five times the daily maintenance doses.

1.00 or 1.25 mg drug per day, each being administered for either 0.5, 1.0, 1.5, 2.0 or 2.5 years. The required polymer concentration in the device is obtained from Fig. 5. This value is now used in conjunction with Fig. 6, and the corresponding design enzyme concentration is obtained. Where these concentrations are not appropriate, the characteristics of the system (e.g. volume, membrane characteristics, polymer characteristics) are to be modified accordingly.

Table 2. Characteristics of the hybrid drug delivery device

Device	– Volume	78.5	ml
Membrane	– Thickness (z)	0.041	mm
	– Molecular weight cut-off	12000	
	– Surface area (A)	78.5	cm^2
Polymer	– Polymelphalan		
	– Average molecular weight	40000	
Enzyme*	– Exoacting type		
	– Molecular weight	35000	
	– Michaelis-Menten constant (K_M)	1.6	M
	– Catalytic constant (k_2)	120	min^{-1}
Drug	– Melphalan		
	– Molecular weight	305	
	– Diffusion coefficient (D)	7×10^{-10}	cm^2/sec**
	– Dose	0.25–1.25	mg/day
	– Release duration	0.5 –2.5	years

    * Data are representative for Carboxypeptidase A
   ** Experimentally determined in this study

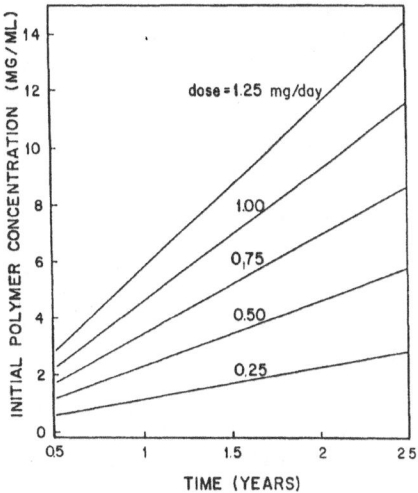

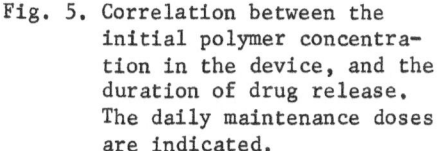

Fig. 5. Correlation between the
initial polymer concentration in the device, and the
duration of drug release.
The daily maintenance doses
are indicated.

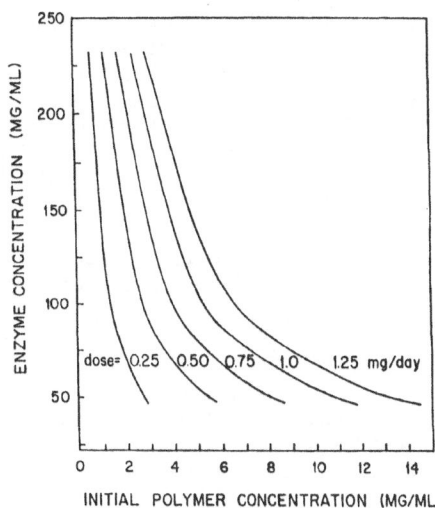

Fig. 6. Correlation between the enzyme
concentration and initial polymer concentration in the device,
for the indicated daily
maintenance doses.

Obviously, the analysis presented relates to an ideal system which
is free of mass transfer limitations such as polarization and concentration gradients on both sides of the membrane, or of potential clogging of
the membrane. Also neglected is the possible loss of enzyme activity
throughout the life time of the system. These aspects, however, must be
taken into account when designing the particular system of interest. It
is, however, noted that various technical approaches to overcome some of
these anticipated difficulties are already available. For instance, one
can envisage using a "self cleaning" system obtained by immobilising
appropriate enzymes on the surface of the membrane.

CONCLUSIONS

It is suggested that enzyme-mediated drug delivery systems are most
attractive to reach a constant-rate, concentration independent, zero-
order release of the drug for prolonged periods of time.

An analytical model describing a novel hybrid approach for achieving
zero order drug delivery is presented. An implantable device which
contains a solution of a polymeric drug together with an exo-type degrading
enzyme is considered. The device has a drug escaping port, sealed by a
selective membrane which is permeable only to low molecular weight molecules. The release of the drug is achieved by progressive enzymic degradation of the polymeric drug within the reservoir, and the subsequent escape
of the active, low molecular weight species through the transport-
controlling membrane at the exit port.

The behavior of the system is studied in terms of the Michaelis-
Menten kinetic mechanism using the experimentally determined diffusion
coefficient of Melphalan through a cellulose-type membrane. A mathe-

matical description of the performance of the system has been carried out. The results indicate that the appropriate choice of polymer, enzyme and membrane characteristics, allows for a wide range of therapeutic regimes to be achieved. For example: an 80 ml device which contains a solution of polymeric drug (with a degree of polymerization of 130, at 4.6 mg/ml) and a degrading enzyme (at 0.06 mg/ml with a catalytic constant of $k_2=120$ min^{-1} and Michaelis-Menten constant of $K_m=1.6M$), will release a low molecular weight drug for a period of two years, at a constant dose of 0.5 mg/day. Obviously, practical difficulties associated with the implementation of this procedure must be considered once a particular system is designed.

ACKNOWLEDGEMENT

This study was supported in part by a fellowship from the Foulkes Foundation (to Y. Yaacobi) and by the MEP Group, Women's Division, American Technion Society, N.Y. Both are gratefully acknowledged.

REFERENCES

1. S. D. Brick, "Controlled Drug Delivery", CRC, Boca Raton (1983).
2. D. H. Lewis, "Controlled Release of Pesticides and Pharmaceuticals", Plenum, New York (1981).
3. R. Baker, "Controlled Release of Bioactive Materials", Academic, New York (1980).
4. J. Libman, "New Drug Delivery Systems", National Council for Research and Development, Jerusalem, Israel (1980).
5. J. R. Robinson, "Sustained and Controlled Release Drug Delivery Systems", Dekker, New York (1978).
6. R. J. Kostelnic, "Polymeric Delivery Systems", Gordon & Breach, New York (1978).
7. D. R. Paul and F. W. Harris, "Controlled Release Polymeric Formulations", American Chemical Society, Washington (1976).
8. S. K. Chandrasekaran and J. E. Shaw, Design of Transdermal Therapeutic systems, in "Contemporary Topics in Polymer Science", E. M. Pearce and J. R. Schaefgen, eds. Vol. 2, Plenum, New York (1977).
9. R. Langer, Implantable Controlled Release Systems, Pharmac. Ther. 21:35 (1983).
10. M. T. Olivari and J. N. Cohn, Cutaneous Administration of Nitroglycerin: A Review, Pharmacotherapy, 3:149 (1983).
11. R. Langer, Polymeric Delivery Systems for Controlled Drug Release, Chem. Eng. Commun. 6:1 (1980).
12. A. F. Kydonious, "Controlled Release Technologies: Methods, Theory and Applications", CRC, Boca Raton (1980).
13. J. Heller, R. W. Baker and J. O. Rodin, Controlled Drug Release by Polymer Dissolution. I. Partial Esters of Maleic Anhydride Copolymers - Properties and Theory, J. Appl. Pol. Sci. 22:1991 (1978).
14. N. Lotan, Y. Yaacobi, J. Finberg, R. Lamed and S. Sideman, Model systems for the enzymic degradation of polymeric drugs, Abst. 10th Congr. Europ. Soc. Artif. Organs. Bologna, Italy (1983).
15. S. Livny, "Biopolymers as Carriers in Controlled Drug Release Systems", M.Sc. Thesis, Technion, Haifa, Israel (1984).
16. Y. Yaacobi, S. Sideman and N. Lotan, Enzymic Degradation of Synthetic Biopolymers: Chain Length Dependent Kinetics, Life Support Systems (1985) in press.

ULTRASONIC MODULATED DRUG DELIVERY SYSTEMS

Joseph Kost,[1,2] Kam Leong,[2] and Robert Langer[2]

[1]Department of Chemical Engineering
Ben-Gurion University of the Negev
Beer-Sheva, Israel 84105
[2]Department of Applied Biological Sciences
Massachusetts Institute of Technology
Cambridge, MA 02139

Introduction

Considerable research efforts have been directed towards developing polymeric controlled drug delivery systems.[1,2] However, relatively little attention has been paid to designing systems where the rate of delivery can be regulated externally. Nevertheless, there are a number of situations where augmented delivery could be beneficial (Table 1).

Table 1: Examples of Clinical Situations where
Demand Delivery is Desirable

Metabolic, Endocrine	- Insulin (diabetes)
Gastrointestinal	- Gastric acid inhibitors (ulcer control)
Cardiovascular	- Anti-anginals
	- Anti-arrythmics
Respiratory	- Epinephrine

The systems where external or feedback control have been studied are still largely experimental. They consist of pumps that can be activated to provide different rates,[3] polymers responding to pH stimuli,[4] non-erodible polymers containing enzymes that cause the polymer to swell and regulate the rate of delivery in response to external stimuli,[5,6] pH-sensitive erodible polymers containing enzymes in hydrogels that degrade more rapidly in response to external stimuli,[7] lectin drug - systems that release additional drug due to the affinity of an external molecule for the lectin,[8,9] and polymer drug-magnetic systems that release additional drug in response to an oscillating external magnetic field.[10,11] Most of these systems will ultimately require removal of the drug-carrier system from the patient and require fairly complex device formulation compared to conventional drug dosage forms.

In this paper we propose a novel system in which release rates of substances from polymeric matrices can be repeatedly modulated at will from a position external to the delivery system by ultrasonic energy. This system is applicable to bioerodible as well as non-erodible polymers. The system requires no special device formulation and can be used with a variety of drugs.

Experimental

Both bioerodible and non-erodible polymers were used as the drug carrier matrices. The bioerodibles were poly[bis(p-carboxy phenoxy) alkane anhydride], synthesized by following the method described by Conix.[12] The non erodible polymer was ethylene-vinyl acetate copolymer (EVA) (Dupont Chemical Co. Elvax, 40P).

Drug incorporated polyanhydride matrices were formulated by either compression or injection molding. A mixture of finely ground and sieved (90-150um) polymer and 10% of p-nitroaniline as a model drug was pressed into circular disks in a Carver test cylinder outfit, at 30 Kpsi and 5°C above the Tg for 10 minutes. Injection molding was performed in the ASCI Mini Max Injection Molder at temperatures of 10°C above the Tm of the polymer.[13]

Bovine zinc insulin (Eli Lilly) or bovine serum albumin (BSA Sigma) was mixed with 10% wt/vol. of EVA in methylene chloride. The suspension was quickly poured into a glass mold which had been previously cooled on dry ice. When the solution was solidified (10 min), the sample was removed from the mold and placed consecutively in a freezer under vacuum at -20°C for 48 hours and then an additional 48 hours at 20°C.[14]

The triggering device was a RAI Research Corporation ultrasonic cleaner model 250 which generated an ultrasonic frequency of 75 KHz.

Drug incorporated polymeric matrices were placed in a jacketed vial filled with 0.1M phosphate buffer pH 7.4 at 37°C (Figure 1) and were exposed to alternating periods of triggering and non-triggering. After each period the buffer in the vial was replaced and assayed in a UV spectrophotometer for drug concentration. The concentration of p-nitroaniline was determined at 381 nm and the polymer degradation products at 250 nm. The BSA and insulin concentrations were assayed at 220 nm. Experiments had also been conducted where the drug concentration was detected continually; the buffer was circulated by a peristaltic pump to a UV spectrophotometer in a closed-loop manner at 9 ml/min.

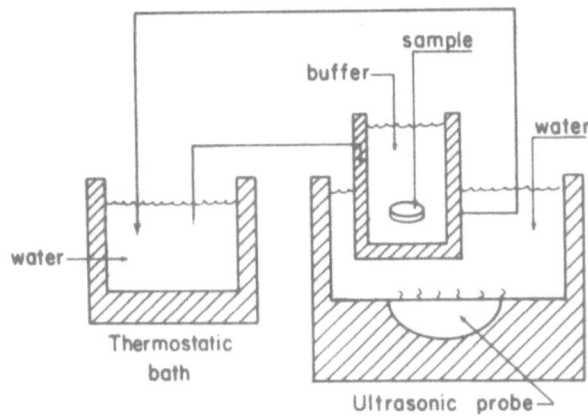

Figure 1.   Schematic diagram of the experimental setup used to determine the effect of ultrasound on degradation and release rates.

Results and Discussion

The effect of the ultrasonic triggering on degradation and release rates of injection molded poly[bis(p-carboxyphenoxy)methane] (PCPM) samples is shown on figure 2. Pronounced enhancement of polymer degradation and release rates while exposed to ultrasound can be observed. The close correlation between the two rates suggests that the increase in release rates during the triggering period is in part due to the enhanced erosion of the polymeric matrix.

The extent of enhancement is more clearly expressed as the ratio of the rate of release (or degradation) at a given period of ultrasound exposure compared to the average of the rates immediately preceeding and following this exposure (Figure 3).

To determine the kinetics of modulation, several experiments were performed in a closed-loop on-line detection system in which the concentration of the releasing medium was monitored continuously. The release rates (Figure 4) are represented by the slope. As can be seen the release rates at different periods are constant with time and the response time is very short when the ultrasound is turned on and is almost instantaneous when the ultrasound is turned off.

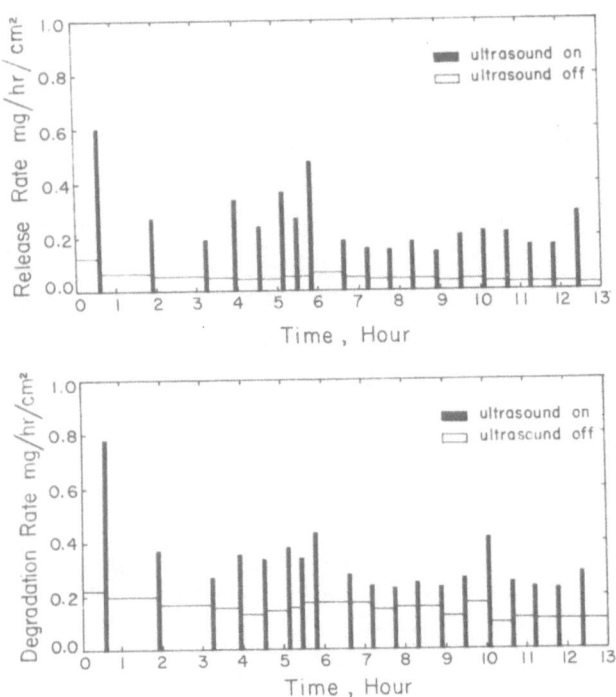

Figure 2. The effect of ultrasound on degradation and release rates of p-nitroaniline from poly[bis(p-carboxy phenoxy)methane] as a function of time.

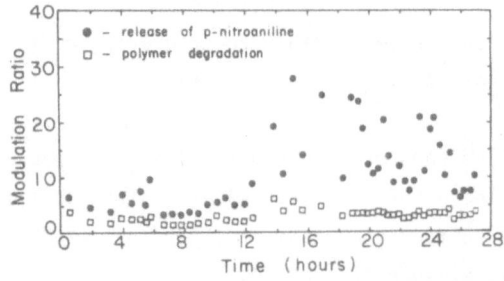

Figure 3. Modulation ratio of the rate of release on degradation at a given period of ultrasound exposure to the average of the rates immediately preceeding and following this exposure.

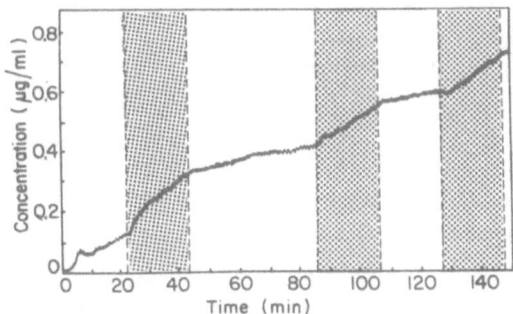

Figure 4. Release of p-nitroaniline from poly[bis-
(p-carboxyphenoxy) hexane] with on-line
detection. The shaded areas are the periods
in which the device was exposed to ultrasound.

Enhancement of release rates while exposed to ultrasound was also
observed in non-erodible polymers (Figure 5). This suggests that the
ultrasound can affect not only the biodegradability but also the
permeability of the polymeric matrices.

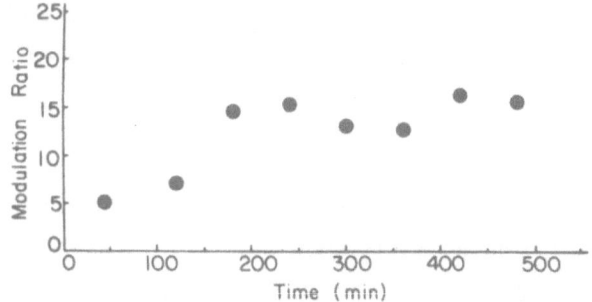

Figure 5. Modulation ratio vs. time for zinc
bovine insulin release from ethylene
vinyl acetate copolymer exposed to
alternating triggering periods of 30
minutes. The insulin powder was sieved
to less than 75 um, and the loading
level was 40%.

It has also been demonstrated that the extent of enhancement can be
regulated (Figure 6). By varying the intensity of the ultrasound, the
enhancement ratios for both the polymer degradation and drug release
could be altered.

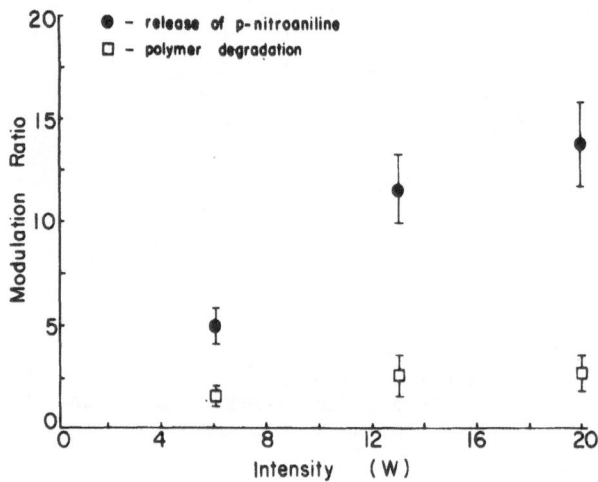

Figure 6. The effect of ultrasound intensity on modulation ratio for the release of p-nitroaniline from poly[bis(p-carboxyphenoxy) methane], with a loading level of 10%.

The effect of the ultrasound on the surface morphology of the polyanhydride polymers can be seen on the scanning electron micrographs (Figure 7), which illustrate the mechanical erosion effect of ultrasound on the solid polymer. Samples exposed to ultrasound exhibited a cracking pattern far more extensive than non-triggered samples which had a similar extent of degradation and release.

Several control experiments were performed to investigate the enhancement mechanism. In particular, the factors examined were: temperature, mixing and cavitation.

Exposure of the samples to ultrasound would increase the temperature inside the polymer. Figure 8a shows the temperature increase of the device while exposed to ultrasound. Figure 8b presents the effect of such temperature increase on release rates. These results suggest that the enhanced release due to ultrasound can not be due to a temperature increase.

To examine the effect of ultrasound on the diffusion boundary layer, release experiments were performed under vigorous shaking (Figure 9). As can be seen the effect of the ultrasound on the augmented release can not be due to mixing. In release experiments performed under vigorous shaking the increase in release rates is minimal. Even though the ultrasound

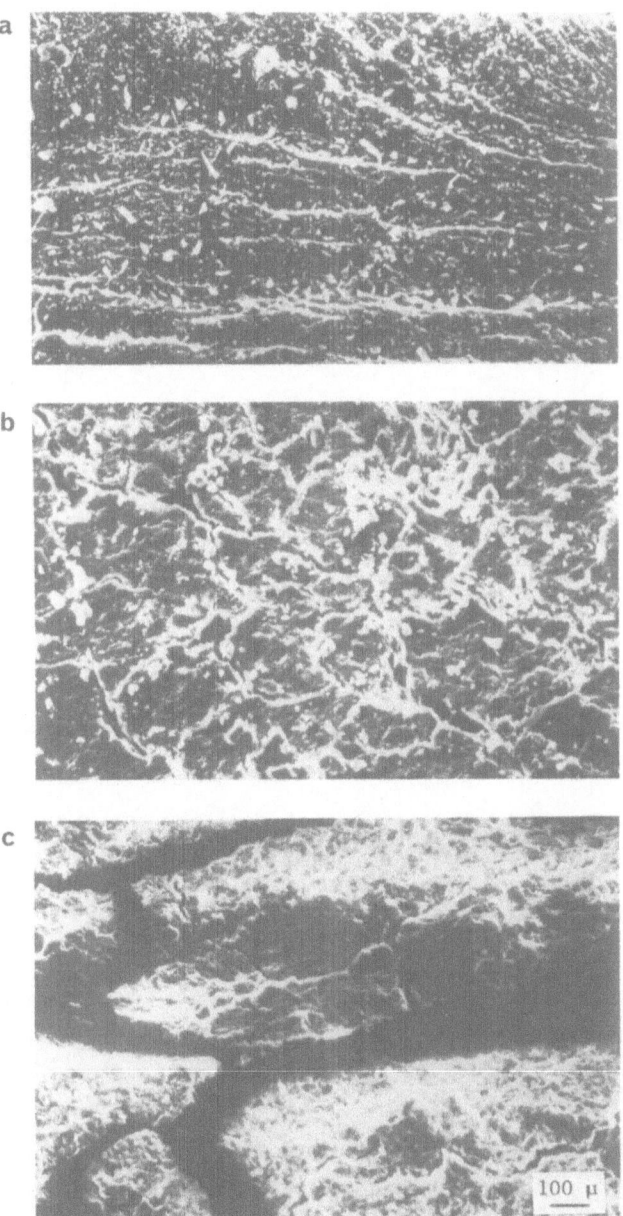

Figure 7. Scanning electron micrographs of poly[bis-(p-carboxyphenoxy) methane] matrices loaded with 10% p-nitroaniline: (a) Samples unleached and not exposed to ultrasound. (b) Samples with approximately 10% of degradation and 25% of release, without exposure to ultrasound. (c) Samples with extent of degradation and release similar to (b), with a cumulative ultrasonic exposure of two hours.

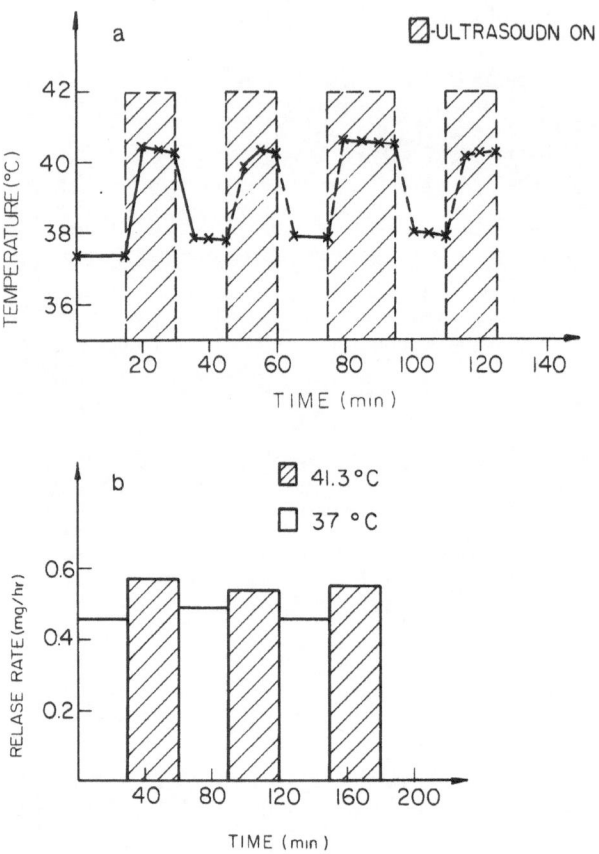

Figure 8. (a) Temperature profiles of samples exposed and unexposed to ultrasound. (b) The effect of temperature on the release rates of bovine serum albumin from EVA copolymer.

would eliminate or diminish the diffusion boundary layer, that effect cannot be held responsible for the ten to thirty fold increase in release.

Cavitation is presumably the leading cause for the augmented release. This conclusion comes from experiments which show that the effect of the ultrasound on degradation and release rates is much smaller in the degassed buffer where the cavitation phenomenom is less pronounced (Figure 10).

Conclusion

The development of a demand responsive drug delivery systems offers an alternative to other cumbersome and less efficient means of medical therapeutics. The ultrasonic system possesses advantages over other systems in that the polymer can be injected, and since it is bioerodible there is no need for surgical removal.

No special formulation procedures such as adding magnetic beads are required. The extent of modulation can readily be controlled externally over a range of ultrasonic frequencies and intensities.

The application of ultrasound in humans, both for diagnostic and therapeutic purposes, has been extensively studied[15] and is considered a safe practice. The challenge remains to explore the conditions under which the system will be effective in vivo.

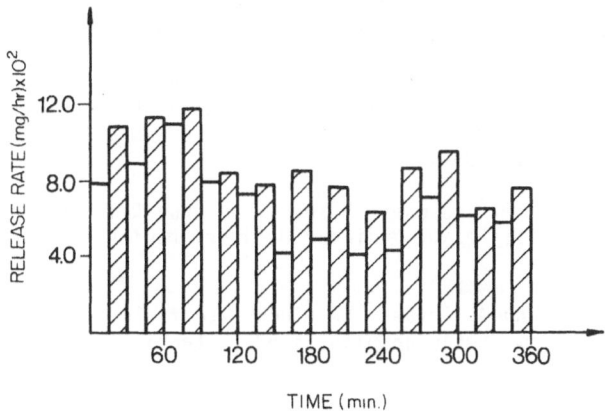

Figure 9. The effect of vigorous shaking on release rates of bovine serum albumin from ethylene vinyl acetate copolymer (30% loading). The shaded areas are release rates during the shaken periods.

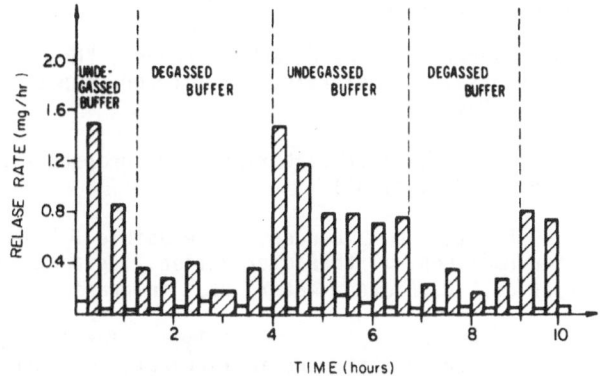

Figure 10. Release rates of bovine serum albumin vs. time (30% loading). The sample was exposed to ultrasound alternately in degassed and undegassed buffer.

References

1. R. Langer and N. Peppas, "Chemical and Physical Structure of Polymers as Carriers for Controlled Release of Bioactive Agents: A Review," J. Macromol. Sci. Reviews, Macromol. Chem. Phys. C23, 61, (1983).

2. J. Kost and R. Langer, "Controlled Release of Bioactive Agents", Trends in Biotechnology, 2, 47, (1984).

3. M. Sefton, "Implantable Pumps," in Medical Application of Controlled Release", Langer, R. and Wise, D., eds., CRC press, 129, (1984).

4. F. Alhaique, M. Marchetti, F. Riccieri, and E. Santucci, "A Polymeric Film Responding in Diffusion Properties to Environmental pH Stimuli: A Model for a Self-Regulating Drug Delivery System," J. Pharmacol. 33, 413, (1981).

5. J. Kost, T. Horbett, B. Ratner, and M. Singh, "Glucose Sensitive Membranes Containing Glucose Oxidase: Activity, Swelling and Permeability Studies," J. Biomed. Mater. Res. in press

6. K. Ishihara, M. Kobayashi, N. Ishimanu, and I. Shinohara, "Glucose Induced Permeation Control of Insulin through a Complex Membrane Consisting of Immobilized Glucose Oxidase and a Poly(amine)", Polymer. J. 16, 8, 625, (1984).

7. J. Heller and P. Trescony, "Controlled Drug Release by Polymer Dissolution II: Enzyme-Mediated Delivery Device," J. Pharm. Sci. 68, 919, (1979).

8. M. Brownlee and A. Cerami, "A Glucose-Controlled Insulin Delivery System: Semisynthetic Insulin Bound to Lectin," Science 206, 1190, (1979).

9. S. Kim, S. Jeong, S. Sato, J. McRea, and J. Feijen, "Self Regulating Insulin Delivery System-A Chemical Approach," in Recent Advances in Drug Delivery Systems, Anderson, J. and Kim, S., eds., Plenum Press, 123, (1984).

10. E. Edelman, J. Kost, H. Bobeck, and R. Langer, "Regulation of Drug Release from Polymer Matrices by Oscillating Magnetic Fields," J. Biomed. Mater. Res. in press

11. J. Kost, R. Noekker, E. Kunica, and R. Langer, "Magnetically Controlled Release Systems: Effect of Polymer Composition," J. Biomed. Mater. Res. in press

12. A. Conix, "Poly [1,3,-bis(p-carboxyphenoxy) propane anhydride]," Macro. Synth., 2, 95, (1966).

13. K. Leong, B. Brott, and R. Langer, "Bioerodible Polyanhydrides as Drug-Carrier Matrices. I: Characterization, Degradation and Release Characteristics," J. Biomed. Mater. Res. in press

14. W. Rhine, D. Hsieh, and R. Langer, "Polymers for Sustained Macromolecule Release: Procedure to Fabricate Reproducible Delivery Systems and Control Release Kinetics," J. Pharm. Sci. 69, 3, 265 (1980).

15. F. Fry, Ultrasound: Its Applications in Medicine and Biology, Elsevier Scientific Publishing Company, (1979).

CHARACTERIZATION OF DRUG LOADING INTO SWELLABLE POLYMERS BY SURFACE ANALYSIS

Fabio Carli°, Italo Colombo°, Mara Lovrecich*,
Fulvio Rubessa*, and Clara Torricelli°

° Physical Pharmacy Laboratory, Pharmaceutical Research and
  Development, Farmitalia Carlo Erba, Milano, Italy
* Istituto di Chimica Farmaceutica, Università di Trieste
  Italy

INTRODUCTION

Swellable polymers can be used in the pharmaceutical dosage forms either to control the release process or to improve the solubility and the dissolution rate of slightly soluble drugs. In the former case slowly swellable polymers are used, such as crosslinked polyvinyl alcohol[1], ethylene/vinylacetate copolymers[2], crosslinked hydroxyethylmetacrylate[3]; the swelling mechanism of these polymers largely influences the drug release process[4]. In the later case, polymers with a high swelling rate are used, such as crosslinked polyvinyl pyrrolidone[5-7], crosslinked casein and crosslinked sodium carboxymethylcellulose[8].

Drug location in the slowly swellable polymeric particles has been proved to play a major role in the overall drug release process. Lee[9] showed that zero-order release rate from PHEMA hydrogel beads can be obtained if specific drug concentration profiles inside the particles are established; the drug location was checked by Scanning Electron Microscopy (SEM) X-Ray microanalysis.
On the other hand, no specific attempt to identify the drug location in the fast swelling polymers has been carried out.

It is object of this work to test the ability of X-ray photoelectron spectroscopy (XPS or ESCA) to map the drug molecules loaded both in the slowly and fast swelling polymers. Furthermore contact angle measurements are carried out and compared with the XPS technique.

Two different drug-polymer systems were prepared by swelling the polymeric particles with the drug solution and subsequent drying: crosslinked polyvinylpyrrolidone was used to improve the solubility properties of griseofulvin, whereas crosslinked polystyrene to control the release process of theophylline.

THEORY

XPS

X-ray photoelectron spectroscopy is a technique based on the photo -
electric effect[10]; when an atom is irradiated by a beam of X-rays of $h\nu$
energy, electrons at originary energy level, E_B, are emitted with a kinetic
energy, E_K,:

$$E_B = h\nu - E_K \tag{1}$$

Only electrons close to the surface (20-50 $\overset{o}{A}$ for polymers) can escape, due
to the recombination or decay effect. Because each E_B is related to a specific atomic level, it is possible to identify, from the energies of the
emission peaks, the atom on the surface of the solid sample. Furthermore,
being the intensity of each photoemission peak proportional to the surface
concentration of that atom, one is able to derive a quantitative surface
analysis, i.e. the percentage atomic composition.

Solid/liquid Contact Angle

The measurement of contact angle of liquid drops on the surface of solid
samples is a widely employed technique to determine the surface free energy
or polarity of solids[11]. For binary systems it is possible to use the Cassie-
Baxter equation[12], which allows the determination of the surface fraction
of each component:

$$\cos\vartheta = f_1\cos\vartheta_1 + (1-f_1)\cos\vartheta_2 \tag{2}$$

where ϑ is the contact angle, with a given liquid, of the binary system
(e.g., the drug/polymer system), ϑ_1 and ϑ_2 are the contact angles of the
pure drug and pure polymer respectively and f_1 is the surface fraction
of the pure drug.

EXPERIMENTAL

Materials

The chemically crosslinked polyvinylpyrrolidone (special sample from
BASF, Germany), the popcorn polyvinylpyrrolidone (Kollidon, BASF, Germany),
and the polystyrene crosslinked with 1% divinylbenzene (BIO-BEADS SX1, Biorad,
Italy) were used as received. Griseofulvin (Glaxo Group Ltd, England), and
theophylline (Farmitalia-Carlo Erba, Italy), were used as model drug.

Methods

Drug loading. A) The powdered polymer is suspended in an excess of the
drug solution, stirred 24 hours and filtered; the resultant systems were
dried at 60° (theophylline) and 120° (griseofulvin) for 1-2 hours and after
trituration passed through a 75 μm sieve. B) An appropriate volume (smaller
than the maximum swelling volume of the polymer) of the drug solution is
slowly added to the powdered polymer under mixing and treated as in case A.

Determination of Drug Content   An accurately weighed amount of the drug/
polymer powder is placed in a Soxhlet apparatus and the drug extracted with
an appropriate solvent and assayed spectrophotometrically (Perkin Elmer 559
spectrophotometer).

XPS analysis   The sample has been prepared by pressing a suitable amount
of powder onto a foil of pure indium (Goodfellow, 99,999% purity, England)
and placed in an XPS-AES instrument (Physical Electronics Inc. Phi Mod. 548,
USA) using Al- $K\alpha$(1448,6 eV) as anode material. A base pressure of $3x10^{-7}$
Pa is obtained in the analysis chamber. After smoothing and background sub-
traction the peak intensities were calculated by digital integration. These
values were used to determine the surface concentration of the drug.

Solid/water Contact Angle   Solid/water contact angles were measured
with a wettability tester (Lorentzen- Wettre Sweden). Small drops of demine-
ralized water were placed on the surface compacts by a microsyringe and,
after stabilization, the magnified image of drops were projected onto a
screen. The contact angle values were derived from the heigth and length
of the drop image via a trigonometric relationship. At least six replicates
were carried out.

Griseofulvin Solubility Measurement   The solubility of loaded crosspo-
vidones was measured by placing an excess amount of the powdered systems
in a flow cell containing 50 ml of pH 7.5 buffer solution, at 37°C, under
constant magnetic stirring. The solution was filtered and pumped directly
to a spectrophotometer cell (SP-8-100, Pye Unicam, U.K.).

Theophylline Dissolution Rate Measurements   The theophylline release
test was carried out in a rotating paddle system (USP XX) at 150 rpm in
900 ml of a buffer solution, pH 7.5, at 37° C. An accurately weighed sample
of polystyrene systems was placed in a capsule and introduced at the bottom
of the vessel.

RESULTS AND DISCUSSION

XPS Analysis

Typical XPS spectra are reported in Figure 1: for griseofulvin/crosspo-
vidone systems the oxigen, nitrogen, carbon and chlorine peaks are clearly
detectable whereas for the polystyrene/theophylline systems only the first
three atoms can be identified; this is due to the different chemical structure of the components (Figure 2).

Griseofulvin/crosspovidone system   In table I the XPS surface elemental
compositions of griseofulvin/crosspovidone system are reported. In this case
the chlorine peak can be univocally attributed to the drug molecule, whereas
the nitrogen peak is only due to the PVP monomer. Thus, by comparing the
experimental Cl/N values with the theoretical ones calculated for an homogeneous intramacromolecular drug distribution, one should be able to identify
the drug location in the crosspovidone particles: Cl/N values higher than
the theoretical ones suggest a drug excess in the surface layers, whereas
lower Cl/N values indicate a higher concentration of the drug in the inner
core of the polymer particles.

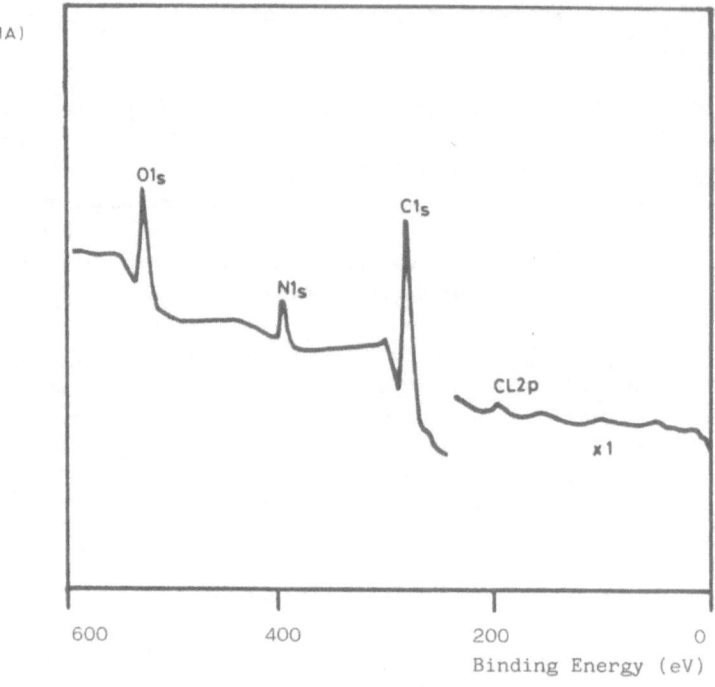

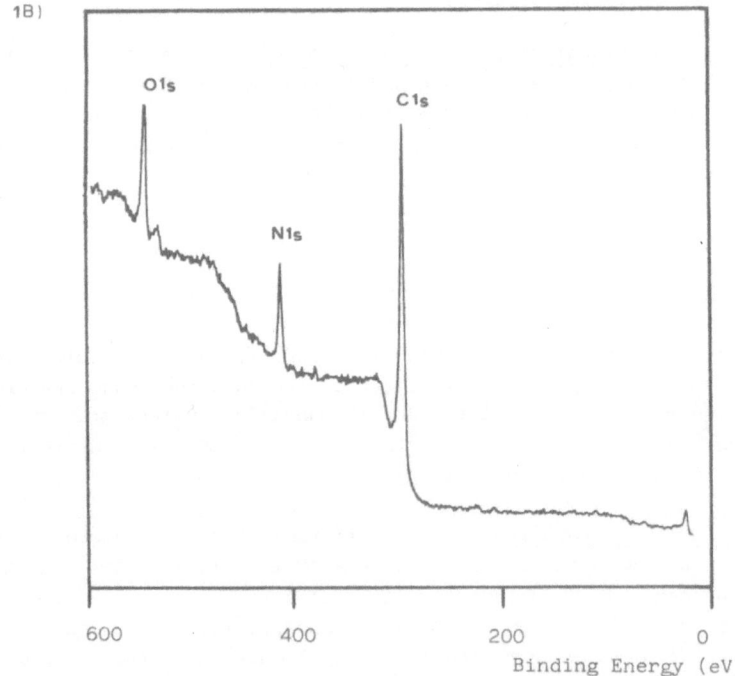

Fig.1. Typical X-ray Photoelectron Spectra of griseofulvin/crosspovidone
system (1A) and theophylline/polystyrene system (1B).

Table I. Surface Elemental Composition Of Griseofulvin/Crosspovidone Systems prepared by direct loading technique.

System	PERCENTAGE ELEMENTAL COMPOSITION					
	O	C	N	Cl	S	Cl/N
Crosspovidone (Chem.cl)	17.7	70.8	11.5	–	–	
Crosspovidone (Popcorn)	12.0	76.5	11.5	–	–	
Griseofulvin	25.4	69.9	–	4.7	–	
Drug/Polymer 1:5 w/w (chem. cl.)	17.3	70.4	11.3	0.5	0.3	0.055
Drug/Polymer 1:7 w/w (chem. cl.)	16.8	73.9	8.7	0.3	0.3	0.036
Drug/polymer 1:10 w/w (chem. cl.)	18.9	72.3	8.2	0.3	0.3	0.038
Drug/Polymer 1:5 w/w (popcorn)	13.8	77.0	8.4	0.8	–	0.095

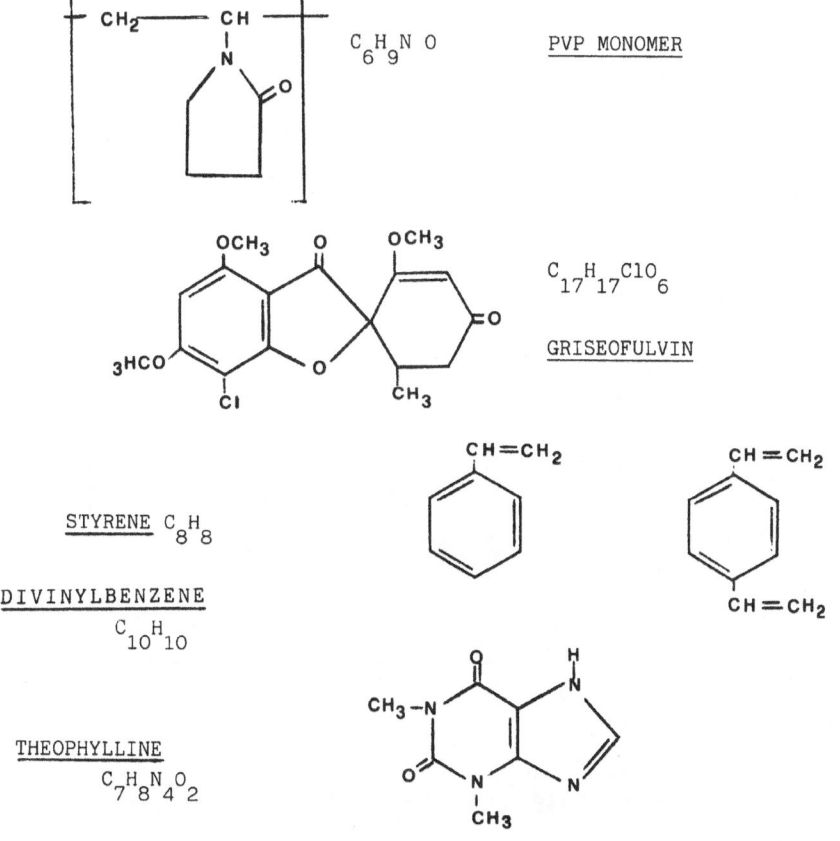

C_6H_9NO PVP MONOMER

$C_{17}H_{17}ClO_6$

GRISEOFULVIN

STYRENE C_8H_8

DIVINYLBENZENE
$C_{10}H_{10}$

THEOPHYLLINE
$C_7H_8N_4O_2$

Fig.2. Chemical structures of Polymers and Drugs used.

As shown in Figure 3, all the three systems prepared by loading the chemically crosslinked polyvinylpyrrolidone with different amounts of griseofulvin present Cl/N values very close to the theoretical ratios, indicating a good and homogeneous intramacromolecular griseofulvin distribution at all the ratios of drug loading. On the contrary, the popcorn crosspovidone system presents a Cl/N value higher than the theoretical one, suggesting a surface excess of griseofulvin.

Theophylline/crosslinked polystyrene systems The surface elemental composition of theophylline/polystirene is reported in table II. In this case only the nitrogen peak can be univocally attributed to the drug molecule; thus the N/C ratio is chosen to identify the theophylline location. Also in this case experimental N/C values are higher than the theoretical N/C ratios suggesting a drug excess in the surface layers. As shown in Figure 4, both the two polystyrene systems presented N/C ratios dramatically higher than the theoretical ones for an homogeneous distribution indicating a large excess of theophylline over the surface of the polystyrene beads.

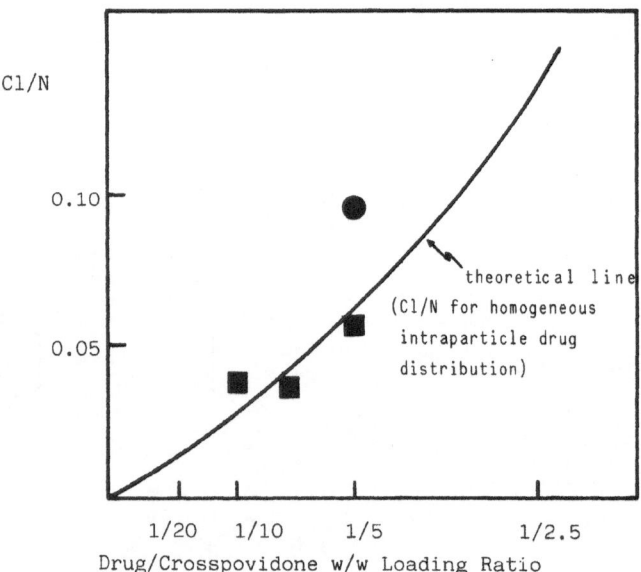

Fig.3. Cl/N ratios (derived from XPS) of griseofulvin/chemically crosslinked (■) and popcorn (●) povidone systems.

Table II. Surface Elemental Composition of Theophylline/polystyrene
Systems

System	PERCENTAGE ELEMENTAL COMPOSITION				
	O	C	N	N/C	N/C (theoretical) homogeneous distribution
Crosslinked polystyrene	1.9	98.1			
Monohydrate theophylline	14.7	54.0	30.8		
System A (28mg/g) (suspension loading)	5.5	88.0	5.9	0.067	0.008
System B (40mg/g) (direct swelling)	5.5	90.9	3.6	0.040	0.011

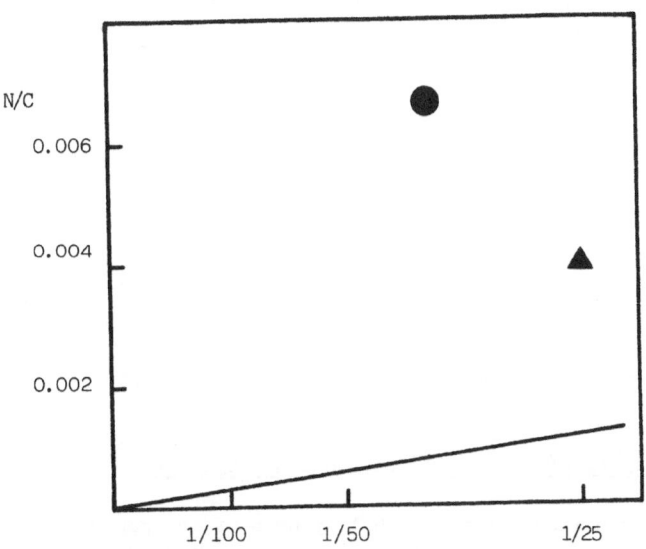

Fig.4. N/C ratios (derived from XPS) of theophylline /polystyrene system
A (●) and system B (▲)

Contact Angle Measurements

Water contact angle measurements have been carried out only on polystyrene systems; in fact the crosspovidone is so highly and fast swellable that drop dimensions can not be measured.

Contact angle data of theophylline/polystyrene systems are reported in table III: the drug presents a water contact angle much lower than that of the polymer; thus it is possible to apply the equation 2 to calculate the surface fraction of the polymeric particles coated by the drug itself. Both the direct and the suspension loading technique result in a high percentage of polystyrene surface coated by theophylline: these data are in agreement with the XPS data, indicating a large excess of the drug in the surface layers.

Table III. Surface Wettability of Theophylline/
Crosslinked Polystyrene systems

System	Solid/water contact angle	Polymer Surface fraction coated by the drug
Theophylline monohydrate	44°	
Recrystallized (CHCl$_3$) theophylline	45°	
Crosslinked polystyrene	91°	
Polystyrene loaded by direct swelling	78°	0.31
Polystyrene loaded by suspension swelling	76°	0.35

Solubility of Griseofulvin/crosspovidone Systems

The continuous flow solubility data of the popcorn crosspovidone and chemically crosslinked povidone loaded with 20% w/w of griseofulvin are reported in Figure 5: the later polymer system presents an oversaturation concentration much higher than the popcorn povidone system; this fact can be reasonably attributed to the higher dispersion of griseofulvin molecules in the inner core of the chemically crosslinked povidone, as indicated by XPS data.

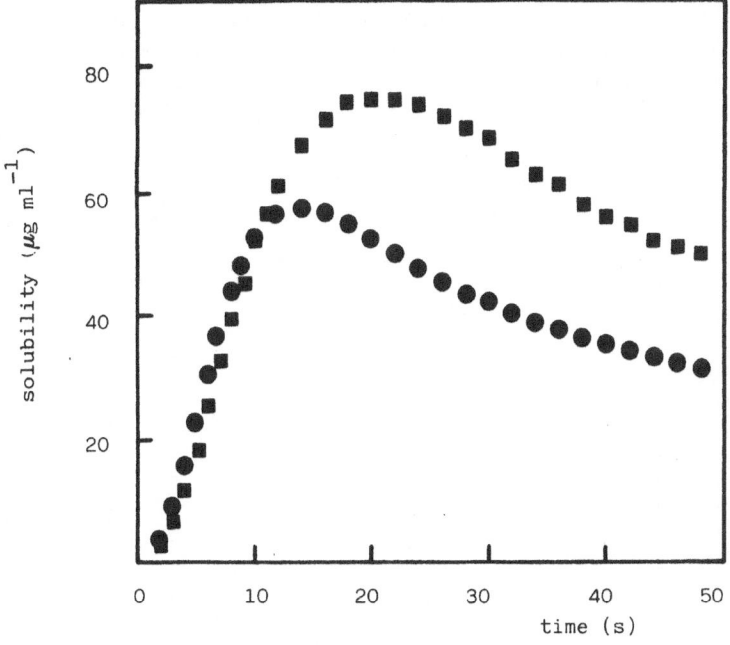

Fig.5. Solubility data of griseofulvin/popcorn (●) and chemically
crosslinked (■) povidone 1/5 w/w systems.

Theophylline release rate from crosslinked polystyrene systems

Theophylline release rate data from the two crosslinked polystyrene
systems are presented on a double logarithmic plot (Figure 6) as suggested
by Korsmeyer et al.[13] : two different slopes are clearly evidentiated; the
first one is due to the dissolution of the drug surface deposited crystalline
particles, the second one due to the drug diffusion from the core of the polymeric particles. This interpretation is further substantiated by considering
that the first stage release data fit the Hixson-Crowell equation[14] , whereas
the second stage data fit the diffusion square root of time equation[15] . The
former equation is usually applied to the dissolution of free drug particles,
the latter to drug diffusion through a polymer matrix[16].

CONCLUSIONS

X-Ray photoelectron spectroscopy proves to be a very sensitive analytical
tool to identify the drug location in the polymeric particles: homogeneous
intrapolymer drug distribution is detected in the case of crosspovidone systems
whereas drug excess on the particle surface is found in the case of crosslinked polystyrene systems. In the latter case, the XPS data are substantiated by contact angle measurements.

It has been also shown that this drug location characteristic strongly influences the drug release process , as indicated by the high solubility of the homogeneously griseofulvin-loaded crosspovidone and the two step release process of the heterogeneously theophylline distributed polystyrene systems.

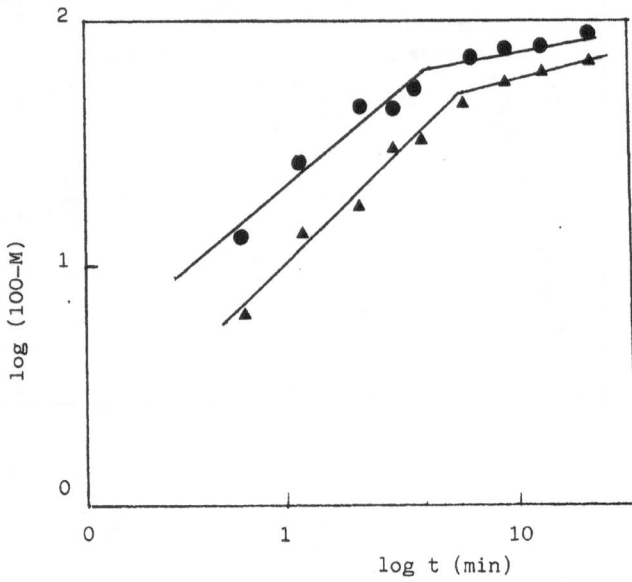

Fig.6. Double logarithmic plot of theophylline fraction released (M=percentage of unreleased drug) from polystyrene system A (●) and system B (▲)

ACKNOWLEDGEMENTS

This research was partially supported by CNR (grant to Farmitalia Carlo Erba), progetto finalizzato Chimica Fine e Secondaria.
The assistance of Mrs. R.Fiannaca is acknowledged.

REFERENCES

1. R.W. Korsmeyer and N. Peppas, J. Membrane Sci., 9:211 (1981)
2. M.V. Stefton, L.R. Brown and R.S. Langer, J.Pharm.Sci., 73:1859 (1984)
3. P. Gyselinck, E. Schacht, R. Van Severen, and P. Braechman, Acta Pharm. Technol., 29:9 (1983)
4. N.A. Peppas and N.M. Franson, J. Appl. Polym. Sci., 21:983 (1983)
5. B.C. Lippold and R. Lütschg, Pharm. Ind., 40:541 (1978)
6. F. Carli, I. Colombo, L. Magarotto and A. Motta, 4th Pharm. Technol. Conf., Edimburgh, U.K. (1984)
7. F. Carli, 12th Int. Symp. Controlled Release Bioactive Materials, Geneva Switzerland (1985)
8. H. Mahmoud, 12th Int. Symp. Controlled Release Bioactive Materials, Geneva, Switzerland (1985)
9. P.I. Lee, 11th Int. Symp. Controlled Release Bioactive Materials, Ft. Lauderdale,USA (1984)

10. K. Siegbahn et al. in "ESCA", Almquist and Wigsells Eds, Uppsala (1967)

11. S. Wu, J. Polymer. Sci. part C, 34:19 (1971)

12. A.B.O. Cassie and S. Baxter, Trans. Faraday Soc.,40:546 (1944)

13. R.W. Korsmeyer, R. Gurny, E. Doelker, P. Buri, and N. Peppas, Int. J. Pharm .,15:25 (1983)

14. A. Martin, J. Swarbrick, and A. Cammarata,Physical Pharmacy,p.411, Lea and Febiger Publishers, Philadelphia, (1983)

15. T. Higuchi, J.Pharm. Sci., 52:1145 (1963)

16. F. Carli, G. Capone, I. Colombo, L. Magarotto, and A. Motta, Int. J. Pharm., 21:317 (1984)

POLYMERIC INSERTS FOR SUSTAINED OCULAR DELIVERY OF PILOCARPINE

Marco F. Saettone, Boris Giannaccini, Giacomo Marchesini, Giancarlo Galli* and Emo Chiellini*

Istituto di Chimica Farmaceutica, Università di Pisa, Via Bonanno 6, 56100 Pisa, Italy, and *Dipartimento di Chimica e Chimica Industriale, Università di Pisa, Via Risorgimento 35, 56100 Pisa, Italy

INTRODUCTION

A well known disadvantage of liquid ophthalmic medications (eyedrops) is their rapid elimination from the eye by blinking, induced lacrimation, tear turnover, etc. As a consequence, only a small fraction of the applied drug dose penetrates into the anterior chamber of the eye, and the drug concentration in the biophase is far from being constant, but decreases rapidly after an initial peak (pulse entry mechanism). The ocular bioavailability of topically applied drugs can be moderately increased by modifying the vehicle viscosity, consistency and/or structure. Viscous liquid and semisolid preparations (either paraffin ointment or hydrogels), however, produce at most a prolonged-pulse type of delivery, and do not adequately control release of most medicaments.

These considerations have stimulated much research on alternative therapeutic systems, ideally capable of being retained in the eye for a long time, while releasing the drug to the tear fluid at a slow, controlled rate. The still widely used anti-glaucoma drug, pilocarpine, which necessitates frequent (up to 12 daily) instillations when administered as eyedrops, is a typical example of therapeutic agent that would greatly benefit from a dosage form of this kind. A device releasing pilocarpine to the tear fluid at a controlled rate would: a) provide a constant control of elevated intraocular pressure (IOP); b) reduce the severity of undesired side-effects (myopia, miosis, etc.); c) allow a reduction of the total administered dose, thus minimizing the risk of systemic toxicity resulting from trans-mucosal absorption; and d) improve the patient compliance.

Administration of the drug in solid polymeric devices (inserts) is nowadays actively investigated as a possible approach to controlled delivery. According to their mode of release, inserts can be classified as:

A) devices displaying zero-order release;

B) devices displaying prolonged or sustained (but not zero order) release.

A typical example of the first (A) category is the well-known OCUSERT therapeutic system. In spite of its ingenuity and efficacy, however, this insoluble device has not been widely accepted, probably on account of its high price and poor retention in the eye. The possibility of obtaining a constant rate of pilocarpine release with simpler, monolithic soluble or erodible inserts does not appear as an easily attainable goal. Promising results have been recently reported with erodible matrices of butyl half ester of poly(methyl vinyl ether-_alt_-maleic anhydride), that showed "in vitro" a zero-order release of pilocarpine for over 1 hr (1).

The second (B) category has also been the object of active investigation. Inserts of this type are usually monolithic, soluble or bioerodible, and after introduction into the conjunctival sac may release the drug by different mechanisms (diffusion, surface erosion, etc.). The so-called S.O.D.I. (Soluble Ophthalmic Drug Inserts), introduced in the ophthalmic practice in the U.S.S.R., may be mentioned as an example (2). In spite of their relatively simple structure and composition, these devices may prove far superior respect to the traditional eyedrops. The present authors have described in previous reports (3,4) the preparation and preliminary pharmacological testing of polymeric inserts containing pilocarpine (Pi). The matrices were based on commercially available polymers, such as poly(vinyl alcohol), PVA, or hydroxypropylcellulose, HPC, and contained the drug as the nitrate, or as the "salt" of poly(acrylic acid), PAA.

When tested for miotic activity on rabbits, the PVA inserts containing the Pi/PAA ionic complex increased significantly the drug bioavailability with respect to both an aqueous solution of Pi nitrate and PVA or HPC inserts containing the same salt. Although all devices released the drug "in vitro" by "square root", diffusive kinetics, and produced in rabbits an activity pattern corresponding to a prolonged-pulse release, they gave in humans particularly interesting results. A 24-hr control of IOP in glaucoma patients could be obtained with a single PVA/PAA insert containing 0.77 mg Pi base, while PVA (or HPC) inserts containing Pi nitrate were less effective (5). The hypothesis was advanced that the greater bioavailability, both in humans and in rabbits, of the devices containing the Pi/PAA complex might be due to the restraint imposed by the polyanionic structure to the diffusion of the drug base from the slowly dissolving,hydrated polymeric matrices.

These results have stimulated further research, aimed at the synthesis of new polymeric materials, specifically designed as insert formers, and possibly incorporating the features and advantages of the PVA/PAA matrices. Some desired characteristics were: a) the presence of carboxyl groups, functioning as anchoring sites for the drug base and restraining its release; b) sufficient

410

water solubility, and plasticity of the polymeric films cast from water solutions; and c) absolute lack of irritancy for the ocular tissues.

The present report is concerned with the synthesis of a series of new functional co- and terpolymers, and with a preliminary pharmacological evaluation, on experimental animals, of pilocarpine inserts prepared with some of the polymers, alone or in admixture with PVA.

EXPERIMENTAL

a) <u>Preparation of the polymers</u> - The following commercial monomers, previously purified by standard procedures, were used: acrylic acid, AA (Carlo Erba); acrylamide, AAm (Fluka); hydroxyethyl acrylate, HEA (Merck); methyl acrylate, MA (Fluka), vinyl acetate, VAc (Merck); N-vinylpyrrolidone, NVP (Merck). Methoxytriethylene glycol acrylate, MTEGA, used as internal plasticizer, was prepared in 65% yield from acryloyl chloride and methoxytriethylene glycol in the presence of triethylamine. The product (b.p. 105-106 °C/1.0 mm Hg) was characterized by NMR analysis. Azobisisobutyronitrile, AIBN (Merck) was crystallized from methanol prior to use. Potassium persulfate and sodium thiosulfate (Carlo Erba) were used as received.

Polymers 1-6, 10 and 11 (Table I) were prepared in benzene, using AIBN as radical initiator. The polymerization reactions were carried out in sealed glass vials at 60 °C, under nitrogen; the reaction times ranged from 15 to 30 hr. Polymers 7-9, 12 and 13 (Table I) were prepared in deaerated water, using potassium persulfate and sodium thiosulfate as initiator and activator, respectively. The reaction mixtures were stirred under nitrogen, at 60 °C, for 15 to 65 hr.

All the polymeric materials were isolated from the reaction mixtures by precipitation with a non-solvent, and were purified by repeated precipitations with n-hexane from dioxane solutions, or with acetone from aqueous solutions. A final purification was carried out by dialysis against distilled water, using cellophane membranes with a cutoff MW = 8000 (Spectrapore 3787-F 35, A. Thomas Co.). The products were recovered from the water solutions by freeze-drying. The polymers were characterized by elemental analysis, ^{1}H-NMR, viscosity measurements and potentiometric titration.

b) <u>Preparation of inserts</u> - Pilocarpine nitrate, PiNO$_3$, m.p. 176-78 °C (Merck), poly(vinyl alcohol), PVA (Polyviol W/48-20, Wacker Chemie) and poly(acrylic acid), PAA (Carbomer 940, Goodrich Chemical Co) were used as received. Pilocarpine base, PiB, was obtained from the nitrate by alkalinization to pH 9.0 of the aqueous solution with ammonium hydroxide, and extraction with chloroform.

Polymers 10, 11, 12 and 13 were used for the preparation of inserts. Solutions (5.0% w/w) of the polymers in distilled water, containing the appropriate amounts of glycerol and of PiNO$_3$ (or PiB) were allowed to evaporate slowly (45°C) in Petri dishes

TABLE I.- Polymerization conditions and intrinsic viscosity values of copolymers of acrylic acid (AA) with various comonomers

Polymer	AA[a]	MA[a]	HEA[a]	VAc[a]	AAm[a]	NVP[a]	Solvent	Initiator[b]	Time (hours)	Yield %	$[\eta]^c$ (dl/g)
1	10			90			Benzene	AIBN	27	60	0.76
2	20			80			Benzene	AIBN	15	70	1.2
3	30			70			Benzene	AIBN	30	60	2.8
4	50			50			Benzene	AIBN	15	70	1.3
5	20	10	70				Benzene	AIBN	20	60	d
6	30	20	50				Benzene	AIBN	20	80	d
7	5				95		Water	$K_2S_2O_8/Na_2S_2O_3$	15	87	1.5
8	15				85		Water	$K_2S_2O_8/Na_2S_2O_3$	15	86	1.4
9	25				75		Water	$K_2S_2O_8/Na_2S_2O_3$	15	68	1.1
10	50[e]			50			Benzene	AIBN	15	74	1.9
11	50[e]			45	5		Benzene	AIBN	20	60	12
12	10				55	35	Water	$K_2S_2O_8/Na_2S_2O_3$	65	70	d
13	30				55	15	Water	$K_2S_2O_8/Na_2S_2O_3$	20	70	d

[a] Molar ratios of the monomers used in each polymerization run;   [b] AIBN = α,α'-Azobisisobutyronitrile;   [c] Intrinsic viscosity in DMSO at 30°C;   [d] Polymer insoluble in most common organic solvents;   [e] The reaction mixture contained also 5.0% w/w methoxy triethylene glycol acrylate (MTEGA).

(diameter, 70 mm). The resulting transparent, flexible films
(thickness, 0.4–0.5 mm) were cut in the form of small disks (diameter,
4.0 mm), each containing 1.0 ± 0.05 mg PiB (or the corresponding
amount of $PiNO_3$). Composite inserts containing polymer 13 and variable
amounts of PVA (5.0 to 40.0% w/w) besides all the other ingredients
were prepared by the same method.

A 13.5% w/w solution of $PiNO_3$ in pH 5.5 phosphate buffer
(denominated AS) and PVA inserts containing the PiB/PAA ionic complex
were used as reference standards. These inserts, prepared as described
in a previous paper (4), contained the same amount of PiB as those
under test. The composition of all inserts is reported in Table II.

TABLE II.– Composition of the inserts tested for miotic activity

| Insert[a] | Polymer | | PVA |
	type	amount(% w/w)	(% w/w)
1	10	95	0
2	11	95	0
3	11	45	50
4	12	95	0
5	13	95	0
6	13	95	0
7	13	90	5
8	13	85	10
9	13	75	20
10	13	65	30
11	13	55	40
12[b]	PAA	67	28

[a] All inserts contained 5.0% w/w glycerol and
1.0 mg PiB, to the exception of insert 6, which
contained an equivalent amount of $PiNO_3$.
[b] Cf. Ref.4.

The Pi content of all inserts was verified by HPLC analysis
(6) after thorough extraction with methanol. Water sorption tests
were carried out by suspending the inserts, contained in a small
stainless-steel wire basket rotating at 30 rpm, in distilled water
at 30°C. The inserts were withdrawn at appropriate intervals, dried
superficially and weighed, then were dried to constant weight at
40°C and weighed again. The percent water uptake, and possible losses
of polymeric material to the solution could thus be assessed.

c)   Biological tests – Miosis vs. time data on rabbits (male albino,
2.0 to 2.5 Kg) were obtained by placing 1 0 µl of solution AS, or
one insert into the lower conjunctival sac of one eye of the animals,
the other eye serving as control. The measurements were made at
intervals, under standard lighting conditions, by estimating to the

nearest 0.1 mm, with a micrometer always held by the same operator, the horizontal diameter of the pupil. Each type of preparation was tested at least on a set of 10 different animals. In no case eye irritancy or abnormal tissue reactions were observed.

RESULTS AND DISCUSSION

a)   Polymer synthesis - Copolymers of acrylic acid with various comonomers were prepared by free radical initiation at 60°C by either AIBN or $K_2S_2O_8$/$Na_2S_2O_3$. The polymerization data together with the intrinsic viscosity ($[\eta]$) values of the resulting materials are reported in Table I.

The characterization of the polymers was carried out by NMR and elemental analysis, and by potentiometric titration of the carboxyl groups. In all cases, the polymer composition corresponded (±10%) to the composition of the initial monomer mixture, which is not unexpected on consideration of the relatively high yields. No information, however, could be obtained on the distribution of the monomeric units along the polymer chains, even if a random distribution might be assumed on the basis of the reactivity ratios reported (7) for some of the present monomer pairs. Rather high molecular weights are indicated by the $[\eta]$ values, determined in DMSO at 30°C, that are in the range 0.76-12 dl/g.

All the polymeric materials prepared were submitted to a preliminary physical evaluation, by determining their capability of forming transparent, plastic films by casting from aqueous solution. Polymers 1-4 (AA-co-VAc), containing molar ratios of AA in the range 10-50, were poorly water-soluble, and gave opalescent, viscous solutions. The best solubility was displayed by 4, containing the lowest proportion of VAc. Acid hydrolysis of the acetate ester groups, carried out at 40°C in deaerated dioxane according to conventional methods (8) gave partially hydrolyzed (15-35%) derivatives, with slightly improved water solubility. These materials, however, proved to be unstable, and, unless thoroughly purified by dialysis, became insoluble upon prolonged exposure to air, and precipitated from their aqueous solutions. The AA-VAc copolymers were thus considered unsuitable for the preparation of inserts.

Polymers 5 and 6 (AA-ter-MA-ter-HEA), in which relatively large proportions of the more hydrophilic monomer HEA were introduced, represent an attempt to synthesize materials endowed with a greater solubility in water. These terpolymers, however, were discarded since they retained strongly the organic reaction solvent, giving amorphous, sticky solids, which became hard and poorly water-soluble after drying.

The AA-AAm copolymers 7-9 were fully water-soluble, and could be formed into clear, plastic films by casting. When submitted to a preliminary tolerance test on rabbits, however, these materials proved to be irritant and were also discarded.

Polymers 10 and 11, that can be considered as an evolution of polymer 4, contained 5.0% MTEGA as internal plasticizer. Polymer 11 contained also 5.0% AAm, besides AA and VAc. The two terpolymers 12 and 13 contained different proportions of AAm and NVP, besides AA. These last four materials, which proved the best of the whole series for solubility, film plasticity and tolerance, were selected for the preparation of inserts.

The compositions of the 11 different types of inserts prepared with these materials, and that of the reference insert (n.12),are indicated in Table II.

b)   <u>Biological tests</u> - Inserts made up with polymers 10, 11, 12 and 13 alone, containing 5.0% glycerol and 1.0 mg PiB (n.1,2,4 and 5, respectively), were submitted to miotic activity tests in comparison with an equivalent dose of $PiNO_3$ administered as solution (AS). Surprisingly, insert 1 showed an AUC (area under the miotic activity vs. time curve) value significantly lower than that of AS, while the other inserts produced miotic activity parameters not statistically different from those of the reference solution. The miotic activity of the four inserts, and of the reference solution, AS, is illustrated in Fig.1, while the main miotic activity parameters are summarized in Table III, together with the relevant statistical data.

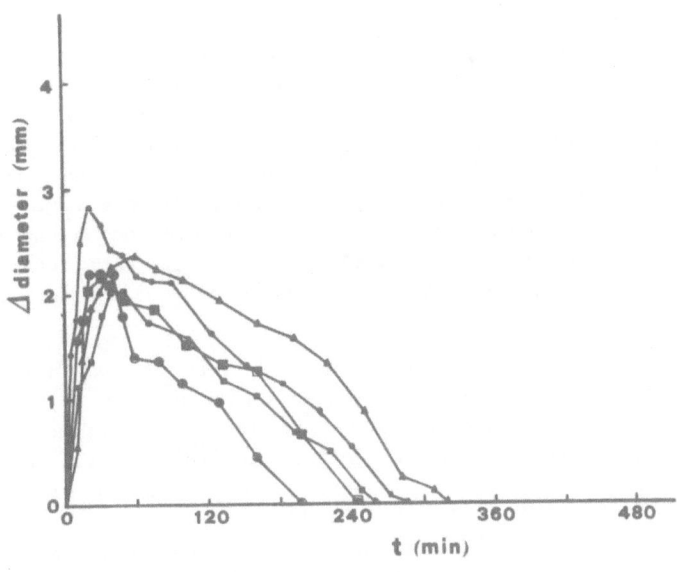

Fig. 1. Miotic effect in rabbits of AS (●) and of inserts 1 (◉), 2 (■), 4 (▲) and 5 (▣). The dose of PiB was in all cases 1.0 mg.

TABLE III.- Summary of the activity data in rabbits of pilocarpine (nitrate or polymer salt) administered in the reference aqueous solution (AS) and in the inserts

Sample		I_{max}[a]	TP[b]	$D_{1/2}$[c]	AUC[d]	Relative AUC
Solution AS		2.7 ± 0.5	20	144 ± 26	81.7 ± 19.7	1.00
Insert	1	3.08 ± 0.2	20	155 ± 21	43.5 ± 7.5	0.53
"	2	2.67 ± 0.4	40	198 ± 25	56.2 ± 13.2	0.69
"	3	3.25 ± 0.6	40	335 ± 32	100.2 ± 18.3	1.23
"	4	2.81 ± 0.4	60	232 ± 15	88.0 ± 21.9	1.08
"	5	3.41 ± 0.2	30	205 ± 36	59.2 ± 11.5	0.73
"	6	3.83 ± 0.3	20	172 ± 10	59.9 ± 10.7	0.74
"	7	3.75 ± 0.3	50	192 ± 38	73.8 ± 16.9	0.90
"	8	3.25 ± 0.7	40	201 ± 46	132.0 ± 32.7	1.61
"	9	3.33 ± 0.3	40	218 ± 27	146.0 ± 22.3	1.78
"	10	3.12 ± 0.9	50	269 ± 48	156.0 ± 39.3	1.91
"	11	3.50 ± 0.5	70	327 ± 28	208.0 ± 37.5	2.54
"	12	3.83 ± 0.3	30	250 ± 30	105.3 ± 12.8	1.29

[a]Peak height, mm ± 95% C.L.; [b]Time to peak, min; [c]Duration of miotic activity calculated at the point of half peak height, min ± 95% C.L.; [d]Area under the miosis vs. time curve, cm^2 ± 95% C.L.

It was speculated that the low degree of activity shown by the four inserts was possibly due to the high water solubility of the polymer matrices, causing the devices to be eliminated, together with the bound drug, before delivering their full drug content to the lacrimal fluid. As stated in the introduction, a prolonged retention of inserts in the eye. is an essential prerequisite for sustained activity. The above hypothesis was supported by the observation that a) insert 5 had practically the same activity as insert 6, which contained $PiNO_3$ instead of PiB, and b) an improvement of the activity parameters could be observed with insert 3, whose solubility with respect to 2 had been reduced by the addition of PVA (Cf. Table III).

Thus, for a subsequent series of experiments it was decided to prepare a series of inserts with polymer 13 and increasing amounts of PVA, in order to verify the effect of a progressively decreasing solubility of the matrix on the overall behaviour and biological activity of the inserts. When tested for miotic activity, these devices (n.7 to 11, PVA content 5.0 to 40.0%, Cf.Table III), showed indeed that the presence of PVA produced, relative to AS, a delayed peak time and a plateau of activity, typical of controlled release, the maximum overall bioavailability (2.54 times with respect to AS) being displayed by insert 11 (40.0% PVA): Cf. Fig.2 and Table III. The AUC values of all inserts tested (except insert 5, that

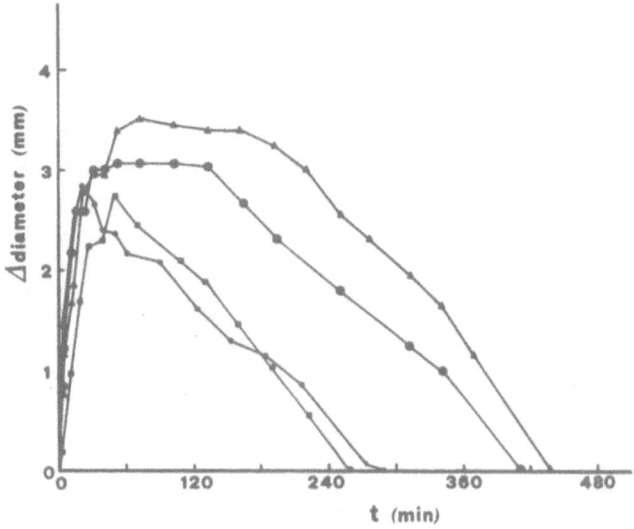

Fig. 2. Miotic effect in rabbits of AS (●) and of inserts 8 (■), 10 (◉)
and 11 (▲). The dose of PiB was in all cases 1.0 mg.

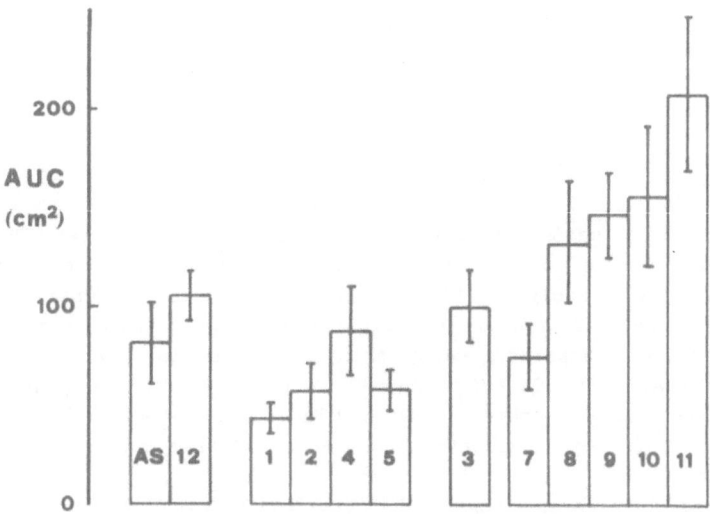

Fig. 3. Areas under the miotic activity vs. time curves (AUC) for the
different preparations described in the present study. Data for
insert 6 were practically identical with those of insert 5.
Vertical lines over bars indicate 95% confidence limits.

417

had practically the same value as 6) and of the reference solution AS, are shown in graphical form, together with the associated 95% confidence limits, in Fig.3.

Inserts 9, 10 and 11 showed also significantly greater AUC values than the reference insert 12 (PAA + 28.0% PVA), thus confirming the favourable characteristics of polymer 13 as matrix material for composite inserts. As shown in Fig.4, there appears

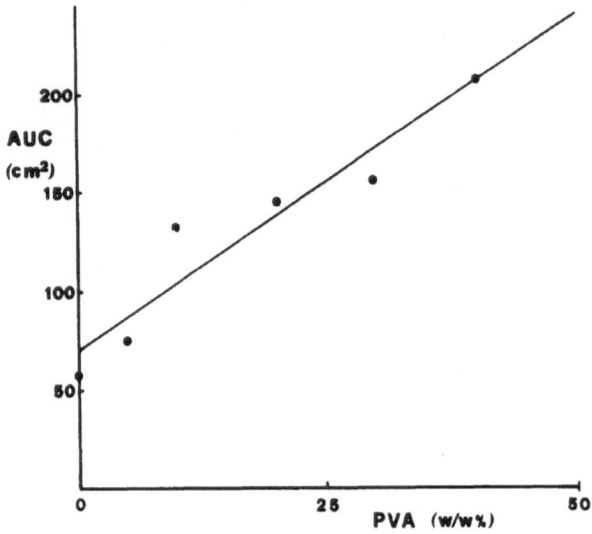

Fig. 4. Relationship between AUC values and percent w/w PVA content in composite inserts prepared with polymer 13 (inserts 5,7,8,9,10 and 11). The dose of PiB was in all cases 1.0 mg.

to exist a satisfactorily linear relationship between PVA content and AUC values, hence, pilocarpine bioavailability, of the inserts prepared with polymer 13 plus PVA. This is evidently due to a progressive reduction of the solubility of the matrices with increasing PVA content. Separate experiments proved indeed that inserts 5 and 7 (no PVA, and 5.0% PVA) underwent rapid transformation into a fluid, gel-like, soluble material when placed in water, while the others (8-11) underwent hydration and absorbed relatively large amounts of water, but maintained for several hours their integrity and shape. The absorbed water ranged from 540 to 720% of the original weight, with equilibration times ranging from 100 to 140 min. The considerable degree of swelling of the composite inserts 8-11 resulted in some cases in a largely increased diameter and thickness (up to 20 mm and 1.5 mm, respectively), and led to expulsion of the inserts from the eye.

In conclusion, the present results indicate two main, coexisting factors as responsible for a sustained biological effect in the inserts under investigation: a) the drug base should be present as an ionic complex, or salt with a polycarboxylic polymer; and b) the solubility in the lacrimal fluid of the insert matrix containing the ionic complex should be relatively low, and such as to allow release of the drug before elimination of the insert from the eye. In the present study, some inserts (1,2,4 and 5) evidently failed to induce a sustained miotic effect because the soluble polyanionic matrix was rapidly eliminated from the eye together with the strongly bound drug. Only when the drug/anionic polymer complex was incorporated in a less soluble PVA matrix, as in composite inserts 3 and 8-11, both factors became operative, and an activity pattern corresponding to sustained release of the drug base from the inserts was apparent.

Thus, it would appear that the control of the solubility of the hydrated polymeric matrix in the lacrimal fluid is not less important than the control of drug release from the matrix itself. The addition of PVA appeared as a satisfactory method to control solubility, but the method showed some drawbacks, such as an excessive swelling of the hydrated composite matrix, resulting in some cases in a premature expulsion of the insert, and the permanence in the eye of an empty (ghost) matrix at the end of the pharmacological activity. Alternative methods to prolong the time of residence in the eye of inserts containing polyanionic complexes of pilocarpine base are now under study.

ACKNOWLEDGMENT

This investigation was supported by CNR, Progetti Finalizzati Chimica Fine e Secondaria.

REFERENCES

1. A.Urtti, Deliverial and Pharmacokinetics Aspects of Ocular Pilocarpine Administration, Int. J. Pharm. in press.

2. Y.F.Maichuk, Ophthalmic Drug Inserts, Invest. Ophthalmol., 14: 87-90 (1975).

3. M.F.Saettone, B.Giannaccini, P.Chetoni, G.Galli and E.Chiellini, Polymeric Ophthalmic Drug Delivery Systems: Preparation and Evaluation of Pilocarpine-Containing Inserts, in:"Polymers in Medicine, Biomedical and Pharmacological Applications", E.Chiellini and P.Giusti, Eds., Plenum Press, New York, p.187-199 (1983).

4. M.F.Saettone, B.Giannaccini, P.Chetoni, G.Galli and E.Chiellini, Vehicle effects in ophthalmic bioavailability: an evaluation of polymeric inserts containing pilocarpine, J. Pharm. Pharmacol. 36: 229-234 (1984).

5. G.Odello, M.F.Saettone, B.Giannaccini, L.Mastrojeni and G.Meucci, Efficacia e durata dell'effetto di nuovi inserti congiuntivali contenenti pilocarpina, Communication, X° Conv. Soc. Oftalmol. Sicil., Mazara del Vallo, February 1985.

6. L.D.Dunn, B.S.Scott and E.D.Dorsey, Analysis of Pilocarpine and Isopilocarpine in Ophthalmic Solutions by Normal-Phase High-Performance Liquid Chromatography, J. Pharm. Sci. 70:446-449 (1981).

7. J.Brandrup and E.H.Immergut, Polymer Handbook, John Wiley & Sons, New York, 1975.

8. J.G.Pritchard, Poly(vinyl alcohol): Basic Properties and Uses, Gordon & Breach Science Publishers, London, p.1 (1970).

COMPARISON OF MICROENCAPSULATION PROCESSES

FOR CONTROLLED RELEASE OF DRUGS AND CHEMICALS

Robert E. Sparks

Department of Chemical Engineering
Washington University
St. Louis, Missouri

INTRODUCTION

Microencapsulation is but one of many methods of obtaining controlled release of drugs and chemicals. However, when microencapsulation processes are being considered for a particular use, it is difficult to gain enough perspective on them to decide which ones to study in detail. This is partly due to the fact that each technique has been developed by a different company and much of the processing information is proprietary know-how. In this paper, the major methods will be summarized and their advantages and disadvantages discussed in sufficient detail to help in choosing among them.

USES OF MICROENCAPSULATION

Many practical problems can be sometimes solved by microencapsulation. These include

>   Retarding degradation by moisture and oxygen
>   Prevention of interaction of components of mixtures
>   Decreasing rates of evaporation or sublimation
>   Controlling release profiles of drugs or chemicals
>   Masking unpleasant taste of drugs
>   Permitting easy handling of viscous liquids
>   Preventing agglomeration of hygroscopic solids
>   Stabilization of emulsions and suspensions
>   Obtaining a delay before activation of capsule contents

MAJOR MICROENCAPSULATION PROCESSES

Pan Coating

The oldest coating method which can be considered a microencapsulation process is that of pan coating, used extensively in the pharmaceutical industry. In this method, the particles to be coated are placed in a rotating chamber and rotated at such a speed that the particles tumble rapidly over each other, a coating material is then sprayed slowly on the tumbling particles so that they dry rapidly enough not to stick together. This usually requires that hot air be blown over or through the particles during the operation.

This process is particularly good for coating tablets and large irregular particles. A wide variety of coating solutions and melts can be used and water-based coatings can be applied to water-soluble core materials.

There are problems coating particles below 500 microns in diameter. Solutions or melts with viscosities above a few hundred centipoises cause difficulties and occasionally it is difficult to wet the core particles with the sprayed coating. The process must be run batchwise and usually requires a skilled operator.

Fluid-Bed Coating

For smaller particles which do not tumble well or dry rapidly enough in a pan coater, an upward moving stream of air can be used to fluidize the particles, with the coating being sprayed on the moving particles. Particularly if warm air is used for fluidization, the particles can be well coated without sticking together. A particularly useful form if this apparatus is employed at the Coating Place in Verona, Wisconsin, USA. In this apparatus, developed by Wurster (1), a central hollow cylinder is used, with most of the fluidization air directed through it. The solid particles flow from a packed bed of particles in the annular space outside the cylinder, through the gap separating the bottom of the cylinder from the perforated inlet air plate. As they accelerate in the rapidly moving air in the cylinder, the coating is sprayed on from the bottom. They dry as they move rapidly upward, disengage in a wider section above, then fall back into the annular packed bed. After circulating many times, they become uniformly coated.

Particles as small as 150 microns can be coated with this apparatus, but the coating rate is slow because the coating must be finely atomized and particles must be circulated slowly so the high surface area can be exposed properly to the spray. Large irregular particles are well coated in this apparatus and a wide variety of coating materials can be used. The process is relatively inexpensive for large particles.

It is difficult to coat particles smaller than 150 microns, and the coating of smaller particles is, in general, more expensive because of the time required. There are occasional problems with wetting of the core, and viscous wall materials do not handle well.

Complex Coacervation

This is considered the first "modern" microencapsulation technique, developed at National Cash Register Company in the mid-fifties to permit encapsulation of tiny droplets of solutions of crystal violet lactone, a dye precursor used in "carbonless" copy forms (2). The word "coacervation" comes from the field of colloid science and is used to refer to the formation of a second liquid phase from a colloidal dispersion or polymer solution.

In the classical microencapsulation by this method the core liquid is emulsified in a gelatin solution. The pH is adjusted to be below the isoelectric pH of the gelatin, which then becomes positively charged, and a solution of gum arabic (a polycarboxylic material) is added. The negative gum arabic reacts with the positive gelatin giving a high molecular weight polyionic complex which is no longer soluble and separates from solution as a gel-like coacervate. The droplets of coacervate wet the surface of the dispersed hydrophobic droplets, slowly forming a gelatinous layer of swollen polymer. The coating is hardened by cross-linking the gelatin portion of the polymer with glutaraldehyde and drying. Variations on this process, particularly by substitution of carboxymethyl cellulose or polyphosphate for the gum arabic are sill the basis of a sizable fraction of the microcapsules used for carbonless carbon paper, by far the largest industrial use for microencapsulation.

This highly developed process is excellent for the encapsulation of hydrophobic droplets, producing totally insoluble walls. The walls can be made very thin and have high integrity. Since the walls are still polar and have low free volume due to the cross-linking, they are highly effective for containing hydrophobic molecules, even such small volatile molecules as toluene and carbon tetrachloride.

A large volume of processing water is used in this process, making it unsuitable for encapsulation of core materials having even 1% water solubility. Wettability of the core by the coacervate is sometimes a problem.

The process is usually carried out batchwise and requires careful control. The process is relatively expensive per kilogram of finished microcapsules.

Organic Phase Separation

A similar technique was also developed by NCR for microencapsulation of materials with varying degrees of water solubility (3). This process begins with the core material being suspended in a solution of the wall-forming polymer. The typical system is ethylcellulose dissolved in cyclohexane. Because ethylcellulose has slight solubility in cyclohexane at room temperature, the suspension is heated to dissolve the polymer, then slowly cooled. Upon cooling, the polymer separates from solution as a coacervate. This process can be aided by adding a second polymer which has higher solubility in cyclohexane, such as low molecular-weight polyethylene. As in complex coacervation, the tiny coacervate droplets coalesce onto the surface of the suspended particles as a gelatinous layer which is subsequently hardened by addition of a nonsolvent.

This method of microencapsulation works well on a variety of solids, particularly when the polymer is ethylcellulose, which has the additional advantage of being approved by the Food and Drug Administration for use in foods and pharmaceuticals. The wall of the capsule formed by this process has high integrity and low water permeability, as evidenced by the fact that a 500-micron capsule of potassium chloride, containing 88% of the active ingredient, will require approximately two hours for the release of half the salt.

Wettability of the core particles by the coacervate is a frequent problem when attempting to apply this process to many materials. In this case, much of the coacervate will form separate particles rather than a coating on the core particles. Even more aggravating are problems with aggregation of the capsules during wall deposition or hardening. The process requires careful control of agitation, temperature and liquid addition rates and is usually run in batches of 200 gallons or less. Hence, it is an expensive process with costs of finished product running in the range of $8 to $15 per kilogram. This process does not work well for a variety of wall polymers.

Interfacial Polymerization

A more strictly chemical process is that of interfacial polymerization. The second major process used for producing the small liquid-core microcapsules needed for carbonless copy paper. In this process the wall is formed by carrying out a condensation polymerization directly at the surface of suspended droplets (4). If a polyamide wall is desired, the process begins with dissolving a diacyl chloride, e.g. sebacoyl chloride, in the hydrophobic core liquid. This solution would then be emulsified into water to obtain the desired droplet size. A diamine, e.g. hexamethylene diamine, is then added to the aqueous phase, along with a small amount of polyfunctional amine. The condensation polymerization begins instantaneously at the interface of the suspended droplets and and is allowed to continue for twenty to forty minutes as the reactants diffuse toward each other through the forming wall.

This process is highly developed for hydrophobic liquid core materials and forms totally insoluble walls of high integrity. Walls only a few hundred angstroms thick can be formed. The process can be used to encapsulate aqueous solutions by reversing the phases in the above description, but it is more difficult to control.

There is a strictly limited choice of wall polymers which can be formed in practice: polyamides, polyureas and mixed polyamide-polyureas. Other condensation polymers are too sticky or react too slowly. The process has not been successfully applied to solid cores, and it is not possible to produce thick protective walls. There is considerable sensitivity of the process to the nature of the core liquid and its surface properties. The raw chemicals are somewhat expensive. Of course, considerable expense can be spared if the material can be shipped and used as the aqueous suspension, as in the case of the microencapsulated pesticides produced by Pennwalt Corporation.

Microencapsulation Using an Annular Jet

A strictly physical method was developed by Southwest Research Institute for the encapsulation of liquid core materials (5). The core liquid is forced through an orifice as a jet, while the wall material is forced through a concentric orifice as annular jet at approximately the same velocity. The two-liquid jet breaks up by Rayleigh instability into droplets, with the material in the outer annular jet becoming the microcapsule wall surrounding core droplets of the inner jet. This method works well for droplets up to approximately 1.2 mm in diameter. For larger drops the technique is modified by adding a third liquid flow concentric to the jet to support the larger drops and aid in conveying them through a heat exchanger to cool the wall. This method can produce droplets over 2 mm in diameter.

For the formation of large liquid-core microcapsules, this is often the method of choice. There are seldom serious wettability problems and meltable materials such as waxes and wax-polymer blends handle well as wall materials. This is a simple, continuous physical process of low cost when the wall material is a melt. It can be used when solvent removal or chemical solidification of the wall is required, but it then becomes rather expensive.

It is difficult with this technique to form capsules much below 300 microns in diameter. There is difficulty in plugging of the jets and it is also difficult to use the process with viscous wall melts or solutions.

Capsules cannot usually be formed having payloads higher than 70%.

It is, of course, possible to form small particles in which an active substance is imbedded as a suspension, performing many of the functions associated with the simpler concept of a single particle surrounded by a shell. Such small particles are typically called "matrix particles", "microparticles" or "microspheres". One conceptually simple method for forming such particles is by emulsifying droplets of the suspension or solution in an immiscible liquid, followed by evaporation of solvent or cooling to solidify the droplets. Most active with this method are Fuji Photo Film Company in Japan and Southern Research Institute in the U.S.

An important advantage of this technique is the ability to use an active substance and a polymeric matrix which have similar solubilities. The methods based on phase separations cannot handle such systems. In solvent evaporation, the core material, e.g. a drug, is partially or totally dissolved in a solution of the matrix polymer, and then may crystallize as the solvent is removed. As opposed to many of the previous processes, a wide variety of core and matrix materials can be used with solvent evaporation.

Since the solvent is removed from the emulsified matrix droplets by a combination of temperature and vacuum, and must diffuse through an external liquid in which it has only small solubility, the solidification process may require 20 minutes to several hours. Due to the slowness of the process, crystallization of the dissolved drug may lead to large crystals and may be difficult to control. During the evaporation of the solvent, the matrix polymer usually passes through a concentration range in which it is sticky. Since the process is slow, a significant period of time may be required to pass through this concentration range, sometimes causing great difficulty with aggregation of the particles. Payloads above 35% are difficult to achieve and the batch process is likely to be rather expensive and difficult to scale up.

Starch-based Processes

Three processes based on starch chemistry have been developed by Shasha at the U.S. Department of Agriculture Laboratory in Peoria, Illinois. The processes are applicable primarily to hydrophobic liquids such as pesticides. In the first process the hydrophobic core liquid is emulsified into starch xanthate, formed by reaction of gelatinized starch with carbon disulfide. After the desired emulsion is obtained the starch xanthate is converted into the water-soluble xanthide by oxidation with a number of possible materials, such as sodium nitrite or hydrogen peroxide. The resulting material is moist but not sticky and can be ground to the desired final particle size. During the grinding operation the xanthide seals around the hydrophobic droplets allowing very little loss of the core material. The final particles, after drying, release the pesticide very slowly.

Since carbon disulfide is hazardous to handle, Shasha also developed a method based on the reaction of starch with boric acid, and a method based on the formation of the complex between calcium chloride and starch. These particles behave in a similar fashion but the process is simpler, less expensive, nonhazardous, and the calcium chloride process would give particles with walls suitable for ingestion.

The starch-based processes use inexpensive chemicals and give control of release of hydrophobic liquids. The materials would likely be approved by the Food and Drug Administration, and the process appears to be easy to scale-up.

The process has not yet been used to form particles less than 200 microns in diameter, nor is there experience with solid core particles. Payloads above 30-40% are unlikely to be successful.

Urea-formaldehyde Walls

Walls of urea-formaldehyde polymer have been formed by a number of companies, but 3M has been particularly successful. In this process, applicable primarily to hydrophobic liquids, the core liquid is emulsified in an aqueous solution of low molecular-weight urea-formaldehyde prepolymer. The pH is then reduced, causing the polymerization of the prepolymer, which deposits on the surface of the droplets.

This process works very well for liquids of low water solubility, forming walls which are completely insoluble and thermally stable. The process can be used to form the microcapsules needed for carbonless copy paper.

A small amount of water solubility of the core liquid renders the process impractical, and it requires control of pH and agitation.

Other Processes

There are also a number of processes not in widespread use, being limited primarily to the companies which developed them. Amonth these are

```
Hydroxypropyl cellulose walls - Mead Corporation
Liquid walls - Exxon Corporation
Reactive surfactants - Champion International Corporation
Clay-complex walls - Jack Ryan Group, Los Angeles
Parylene Process - Union Carbide Corporation
Encapcel Process - Damon Corporation
Direct Olefin Polymerization - National Lead Corporation
Gas Encapsulation - Materials Technology Corporation;
                        Pennsylvania Quartz Corporation
Nanoparticles - Speiser, Swiss Federal Institute;
                        Kreuter, Goethe University, Frankfurt
Cellulose acetate foams - Moleculon Corporation
```

Information on these processes can be found in reference 6.

Frequent Problems

There are a number of problems which often arise with the use of these processes. The most prevalent are:

```
Aggregation during wall formation and hardening
Limited choice of wall materials
Poor wetting of core particles or droplets
Tedious control of process steps
Difficulty in using meltable wall materials
Solvent handling and removal from product
Difficulty of scale-up
Cost: roughly $3 to $20 per kilogram
```

Research Needs

Several problems would benefit from careful attention. Examples are:

> Faster coating with coacervates
> FDA-acceptable condensation polymers
> Methods less sensitive to wettability
> Interfactial polymerization encapsulation of
>   solid particles
> Enteric coatings suitable for several encapsulation
>   methods
> Inexpensive methods for coating with melts.

References

1. U.S. Pat. 2,648,609 (Aug. 11, 1953); 2,799,241 (July 16, 1957), D.E. Wurster (to Wisconsin Alumni Research Foundation).
2. U.S. Pat. 2,800,456 (July 23, 1957), B.K. Green and L.S. Schleicher (to The National Cash Register Co.).
3. U.S. Pat. 3,155,590 (Nov. 3, 1964), R.E. Miller and J.L. Anderson (to The National Cash Register Co.).
4. E.E. Ivy, J. Econom. Etomol. 65,473 (1972).
5. J.T. Goodwin and G.R. Somerville, Chem. Tech. 4,623 (1974); J.T. Goodwin and G.R. Somerville in J.E. Vandegaer, ed., Microencapsulation: Process and Applications, Plenum Press, New York, 1974, pp. 155-163.
6. R.E. Sparks, "Microencapsulation" in Kirk-Othmer Encyclopedia of Chemical Technology, Vol. 15, 3rd Ed., John Wiley and Sons (1981).

CONTRIBUTORS

INDEX